Extracellular Matrix Protocols

METHODS IN MOLECULAR BIOLOGY™

John M. Walker, SERIES EDITOR

161. **Cytoskeleton Methods and Protocols**, edited by *Ray H. Gavon, 2001*
160. **Nuclease Methods and Protocols,** edited by *Catherine H. Schein, 2000*
159. **Amino Acid Analysis Protocols,** edited by *Catherine Cooper, Nicole Packer, and Keith Williams, 2000*
158. **Gene Knockoout Protocols,** edited by *Martin J. Tymms and Ismail Kola, 2000*
157. **Mycotoxin Protocols,** edited by *Mary W. Trucksess and Albert E. Pohland, 2000*
156. **Antigen Processing and Presentation Protocols,** edited by *Joyce C. Solheim, 2000*
155. **Adipose Tissue Protocols,** edited by *Gérard Ailhaud, 2000*
154. **Connexin Methods and Protocols,** edited by *Roberto Bruzzone and Christian Giaume, 2000*
153. **Neuropeptide Y Protocols**, edited by *Ambikaipakan Balasubramaniam, 2000*
152. **DNA Repair Protocols:** *Prokaryotic Systems*, edited by *Patrick Vaughan, 2000*
151. **Matrix Metalloproteinase Protocols,** edited by *Ian M. Clark, 2000*
150. **Complement Methods and Protocols,** edited by *B. Paul Morgan, 2000*
149. **The ELISA Guidebook**, edited by *John R. Crowther, 2000*
148. **DNA–Protein Interactions:** *Principles and Protocols* **(2nd ed.)**, edited by *Tom Moss, 2000*
147. **Affinity Chromatography:** *Methods and Protocols*, edited by *Pascal Bailon, George K. Ehrlich, Wen-Jian Fung, and Wolfgang Berthold, 2000*
146. **Protein and Peptide Analysis:** *New Mass Spectrometric Applications*, edited by *John R. Chapman, 2000*
145. **Bacterial Toxins:** ***Methods and Protocols***, edited by *Otto Holst, 2000*
144. **Calpain Methods and Protocols**, edited by *John S. Elce, 2000*
143. **Protein Structure Prediction:** *Methods and Protocols*, edited by *David Webster, 2000*
142. **Transforming Growth Factor-Beta Protocols**, edited by *Philip H. Howe, 2000*
141. **Plant Hormone Protocols**, edited by *Gregory A. Tucker and Jeremy A. Roberts, 2000*
140. **Chaperonin Protocols**, edited by *Christine Schneider, 2000*
139. **Extracellular Matrix Protocols**, edited by *Charles Streuli and Michael Grant, 2000*
138. **Chemokine Protocols**, edited by *Amanda E. I. Proudfoot, Timothy N. C. Wells, and Christine Power, 2000*
137. **Developmental Biology Protocols, Volume III**, edited by *Rocky S. Tuan and Cecilia W. Lo, 2000*
136. **Developmental Biology Protocols, Volume II,** edited by *Rocky S. Tuan and Cecilia W. Lo, 2000*
135. **Developmental Biology Protocols, Volume I,** edited by *Rocky S. Tuan and Cecilia W. Lo, 2000*
134. **T Cell Protocols:** *Development and Activation*, edited by *Kelly P. Kearse, 2000*
133. **Gene Targeting Protocols,** edited by *Eric B. Kmiec, 2000*
132. **Bioinformatics Methods and Protocols,** edited by *Stephen Misener and Stephen A. Krawetz, 2000*
131. **Flavoprotein Protocols,** edited by *S. K. Chapman and G. A. Reid, 1999*
130. **Transcription Factor Protocols,** edited by *Martin J. Tymms, 2000*
129. **Integrin Protocols,** edited by *Anthony Howlett, 1999*
128. **NMDA Protocols,** edited by *Min Li, 1999*
127. **Molecular Methods in Developmental Biology:** Xenopus *and Zebrafish*, edited by *Matthew Guille, 1999*
126. **Adrenergic Receptor Protocols,** edited by *Curtis A. Machida, 2000*
125. **Glycoprotein Methods and Protocols:** *The Mucins*, edited by *Anthony P. Corfield, 2000*
124. **Protein Kinase Protocols,** edited by *Alastair D. Reith, 2000*
123. ***In Situ* Hybridization Protocols (2nd ed.),** edited by *Ian A. Darby, 2000*
122. **Confocal Microscopy Methods and Protocols,** edited by *Stephen W. Paddock, 1999*
121. **Natural Killer Cell Protocols:** *Cellular and Molecular Methods*, edited by *Kerry S. Campbell and Marco Colonna, 2000*
120. **Eicosanoid Protocols,** edited by *Elias A. Lianos, 1999*
119. **Chromatin Protocols,** edited by *Peter B. Becker, 1999*
118. **RNA–Protein Interaction Protocols,** edited by *Susan R. Haynes, 1999*
117. **Electron Microscopy Methods and Protocols,** edited by *M. A. Nasser Hajibagheri, 1999*
116. **Protein Lipidation Protocols,** edited by *Michael H. Gelb, 1999*
115. **Immunocytochemical Methods and Protocols (2nd ed.),** edited by *Lorette C. Javois, 1999*
114. **Calcium Signaling Protocols,** edited by *David G. Lambert, 1999*
113. **DNA Repair Protocols:** *Eukaryotic Systems*, edited by *Daryl S. Henderson, 1999*
112. **2-D Proteome Analysis Protocols,** edited by *Andrew J. Link, 1999*
111. **Plant Cell Culture Protocols,** edited by *Robert D. Hall, 1999*
110. **Lipoprotein Protocols,** edited by *Jose M. Ordovas, 1998*
109. **Lipase and Phospholipase Protocols,** edited by *Mark H. Doolittle and Karen Reue, 1999*
108. **Free Radical and Antioxidant Protocols,** edited by *Donald Armstrong, 1998*
107. **Cytochrome P450 Protocols,** edited by *Ian R. Phillips and Elizabeth A. Shephard, 1998*
106. **Receptor Binding Techniques,** edited by *Mary Keen, 1999*
105. **Phospholipid Signaling Protocols,** edited by *Ian M. Bird, 1998*
104. **Mycoplasma Protocols,** edited by *Roger J. Miles and Robin A. J. Nicholas, 1998*
103. ***Pichia* Protocols,** edited by *David R. Higgins and James M. Cregg, 1998*
102. **Bioluminescence Methods and Protocols,** edited by *Robert A. LaRossa, 1998*
101. **Mycobacteria Protocols,** edited by *Tanya Parish and Neil G. Stoker, 1998*
100. **Nitric Oxide Protocols**, edited by *Michael A. Titheradge, 1998*
99. **Stress Response:** *Methods and Protocols*, edited by *Stephen M. Keyse, 2000*
98. **Forensic DNA Profiling Protocols**, edited by *Patrick J. Lincoln and James M. Thomson, 1998*
97. **Molecular Embryology:** *Methods and Protocols*, edited by *Paul T. Sharpe and Ivor Mason, 1999*

METHODS IN MOLECULAR BIOLOGY™

Extracellular Matrix Protocols

Edited by

Charles H. Streuli

and

Michael E. Grant

The Wellcome Trust Centre for Cell-Matrix Research, School of Biological Sciences, University of Manchester, Manchester, UK

Humana Press ✷ **Totowa, New Jersey**

999 Riverview Drive, Suite 208
Totowa, New Jersey 07512

The content and opinions expressed in this book are the sole work of the authors and editors, who have warranted due diligence in the creation and issuance of their work. The publisher, editors, and authors are not responsible for errors or omissions or for any consequences arising from the information or opinions presented in this book and make no warranty, express or implied, with respect to its contents.

Cover design by Patricia Cleary

For additional copies, pricing for bulk purchases, and/or information about other Humana titles, contact Humana at the above address or at any of the following numbers: Tel: 973-256-1699; Fax: 973-256-8341; E-mail: humana@humanapr.com or visit our Website at www.humanapress.com

Photocopy Authorization Policy:
Authorization to photocopy items for internal or personal use, or the internal or personal use of specific clients, is granted by Humana Press Inc., provided that the base fee of US $10.00 per copy, plus US $00.25 per page, is paid directly to the Copyright Clearance Center at 222 Rosewood Drive, Danvers, MA 01923. For those organizations that have been granted a photocopy license from the CCC, a separate system of payment has been arranged and is acceptable to Humana Press Inc. The fee code for users of the Transactional Reporting Service is: [0-89603-624-3/00 $10.00 + $00.25].

Printed in the United States of America. 10 9 8 7 6 5 4 3 2 1

Library of Congress Cataloging in Publication Data

Extracellular matrix protocols/edited by Charles H. Streuli and Michael E. Grant.
p.cm.—(Methods in molecular biology; v. 139)
Includes bibliographical references and index.
ISBN 0-89603-634-3
1. Extracellular matrix—Laboratory manuals. 2. Extracellular matrix proteins—Laboratory manuals. I. Streuli, Charles H. II. Grant, Michael E. III. Series.
QP88.23 .E955 2000
571.6—dc21

00-026153
CIP

Preface

It is now widely accepted that much of the dynamic function of cells and tissues is regulated from outside the cell by the extracellular matrix. In addition to its conventional role in providing a scaffold for building tissues, the extracellular matrix acts as a directional highway for cellular movement and provides instructional information for promoting survival, proliferation, and differentiation. Indeed, the extracellular matrix is beginning to take a starring role in the choreography of cell and tissue function.

The diverse roles of the extracellular matrix are reflected in its highly complicated structure, consisting of an ever increasing number of components. Yet the mechanisms of extracellular matrix assembly and how they influences cell behavior are only just beginning to be understood. In order to solve these problems new methodologies are, of necessity, being developed. Many of these technologies are highly sophisticated and are currently available only in a handful of laboratories. However, we believe that they can readily be transported and established by other researchers. Thus, the purpose of *Extracellular Matrix Protocols* is to present some of these complicated techniques in a style that is relatively easy to reproduce.

The approach we have taken is to divide *Extracellular Matrix Protocols* into four sections dealing broadly with biochemical, biophysical, molecular biological, and cell biological methods. Each chapter deals with one specific method developed and/or used by the contributing laboratory, but all are readily adaptable. Moreover, they are nearly all written by postdoctoral scientists who are using these methods on a day-to-day basis, and are therefore tried and tested methods that work. We have made a concerted effort not to reproduce methods that appear in an earlier ECM protocol book published by IRL Press, and indeed our volume is designed to complement it.

A brief glance at the contents list is indicative of the breadth of techniques now available to study extracellular matrix structure and function. We have focused on areas that are particularly appropriate for studying extracellular matrix and areas that are likely to have an important impact over the next few years. Many of the methods will become standard protocols in all ECM labs, and we regard this book as an essential benchtop item. We greatly hope that our book will open novel horizons for many in the field. Enjoy!

Charles H. Streuli
Michael E. Grant

Contents

Contributors

VITALI ALEXEEV • *Department of Dermatology and Cutaneous Biology, Jefferson Medical College, Philadelphia, PA*

FIORELLA ALTRUDA • *Universita di Torino, Torino, Italy*

NICHOLAS C. AVERY • *Division of Molecular and Cellular Biology, University of Bristol, Langford, Bristol, UK*

ALLEN J. BAILEY • *Division of Molecular and Cellular Biology, University of Bristol, Langford, Bristol, UK*

MARA BRANCACCIO • *Universita di Torino, Torino, Italy*

MICHAEL D. BRIGGS • *School of Biological Sciences, University of Manchester, Manchester, UK*

JEREMY R. BRIGHT • *Department of Biochemistry, University of Oxford, Oxford, UK*

NEIL J. BULLEID • *School of Biological Sciences, University of Manchester, Manchester, UK*

IAIN D. CAMPBELL • *Department of Biochemistry, University of Oxford, Oxford, UK*

CHRISTOPHER S. CHEN • *Childrens Hospital, Harvard Medical School, Boston, MA*

DAVID R. CLEMMONS • *Department of Medicine, University of North Carolina, Chapel Hill, NC*

SARAH L. DALLAS • *Department of Medicine, University of Texas Health Science Center at San Antonio, San Antonio, TX*

DANIEL E. EMERLING • *Department of Molecular and Cell Biology, University of California, Berkeley, CA*

JÜRGEN ENGEL • *Biocenter of the University of Basel, Basel, Switzerland*

EVA-MARIA ERB • *Biocenter of the University of Basel, Basel, Switzerland*

JOSÉ MARÍA FRADE • *Department of Neurobiochemistry, Max Planck Institute of Neurobiology, Martinsried, Germany*

CHARLES FFRENCH-CONSTANT • *Wellcome Trust Cancer Research Campaign Institute, University of Cambridge, Cambridge, UK*

EMMA E. FROST • *Wellcome Trust Cancer Research Campaign Institute, University of Cambridge, Cambridge, UK*

EMILIANA GIACOMELLO • *Division for Experimental Oncology, The Reference Center for Oncology, Aviano, Italy*
CHRIS J. GILPIN • *School of Biological Sciences, University of Manchester, Manchester, UK*
MICHAEL E. GRANT • *School of Biological Sciences, University of Manchester, Manchester, UK*
HELEN K. GRAHAM • *School of Biological Sciences, University of Manchester, Manchester, UK*
PHILIP GRIBBON • *School of Biological Sciences, University of Manchester, Manchester, UK*
TIMOTHY HARDINGHAM • *School of Biological Sciences, University of Manchester, Manchester, UK*
EMILIO HIRSCH • *Universita di Torino, Torino, Italy*
DAVID F. HOLMES • *School of Biological Sciences, University of Manchester, Manchester, UK*
MARTIN J. HUMPHRIES • *School of Biological Sciences, University of Manchester, Manchester, UK*
DONALD E. INGBER • *Childrens Hospital, Harvard Medical School, Boston, MA*
KARL E. KADLER • *School of Biological Sciences, University of Manchester, Manchester, UK*
MICHÈLE KEDINGER • *INSERM, Strasbourg, France*
PAUL D. KEMP • *Stockport, UK*
CAY M. KIELTY • *School of Biological Sciences, University of Manchester, Manchester, UK*
TERESA C. M. KLINOWSKA • *School of Biological Sciences, University of Manchester, Manchester, UK*
PAUL KREBSBACH • *National Institute of Health, Bethesda, MD*
ARTHUR D. LANDER • *Department of Molecular and Cell Biology, University of California, Berkley, CA*
FRIEDRICH LAUB • *Brookdale Center for Developmental and Molecular Biology, Mt. Sinai Hospital, New York, NY*
YING LIU • *National Institute of Health, Bethesda, MD*
TODD L. MATHUS • *Robert Wood Johnson Medical School, Piscataway, NJ*
ROGER S. MEADOWS • *School of Biological Sciences, University of Manchester, Manchester, UK*
BARBARA MERKL • *Institute for Biochemistry, University of Cologne, Cologne, Germany*
RICHARD MILNER • *Wellcome Trust Cancer Research Campaign Institute, University of Cambridge, Cambridge, UK*

GRADIMIR N. MISEVIC • *Universite des Sciences et Technologies de Lille, Villeneuve D'Ascq, France*
PAUL A. MOULD • *School of Biological Sciences, University of Manchester, Manchester, UK*
JOHANNA MYLLYHARJU • *Collagen Research Unit, University of Oulu, Oulu, Finland*
UWE ODENTHAL • *Institute for Biochemistry, University of Oxford, Oxford, UK*
EMANUELE OSTUNI • *Childrens Hospital, Harvard Medical School, Boston, MA*
MATS PAULSSON • *Institute for Biochemistry, University of Cologne, Cologne, Germany*
ROBERTO PERRIS • *Division for Experimental Oncology, The Reference Center for Oncology, Aviano, Italy*
ANDREW R. PICKFORD • *Department of Biochemistry, University of Oxford, Oxford, UK*
JENNIFER R. POTTS • *Department of Biochemistry, University of Oxford, Oxford, UK*
FRANCESCO RAMIREZ • *Brookdale Center for Developmental and Molecular Biology, Mount Sinai Hospital, New York, NY*
ANTHONY RATCLIFFE • *Advanced Tissue Sciences, La Jolla, CA*
RONDA E. SCHREIBER • *Advanced Tissue Sciences, La Jolla, CA*
MICHAEL J. SHERRATT • *School of Biological Sciences, University of Manchester, Mancherster, UK*
PATRICIA SIMON-ASSMANN • *INSERM, Strasbourg, France*
TREVOR J. SIMS • *Division of Molecular and Cellular Biology, University of Bristol, Langford, Bristol, UK*
NEIL SMYTH • *Institute for Biochemistry, University of Cologne, Cologne, Germany*
PAOLA SPESSOTTO • *Division for Experimental Oncology, The Reference Center for Oncology, Aviano, Italy*
CHARLES H. STREULI • *School of Biological Sciences, University of Manchester, Manchester, UK*
HIDEAKI SUMIYOSHI • *Brookdale Center for Developmental and Molecular Biology, Mount Sinai Hospital, New York, NY*
ALFREDO RODRIGUEZ TÉBAR • *Department of Neurobiochemistry, Max Planck Institute of Neurobiology, Martinsried, Germany*
NORIYUKI TSUMAKI • *National Institute of Health, Bethesda, MD*
GEORGE M. WHITESIDES • *Childrens Hospital, Harvard Medical School, Boston, MA*
RICHARD R. WILSON • *School of Biological Sciences, University of Manchester, Manchester, UK*

YOSHIHIKO YAMADA • *National Institute of Health, Bethesda, MD*
KYONGGEUN YOON • *Department of Dermatology and Cutaneous Biology, Jefferson Medical College, Philadelphia, PA*
PETER D. YURCHENCO • *Robert Wood Johnson Medical School, Piscataway, NJ*
BO ZHENG • *Department of Medicine, University of California, Berkeley, CA*

I

Biochemistry of Extracellular Matrix

1

Semipermeabilized Cells to Study Procollagen Assembly

Richard R. Wilson and Neil J. Bulleid

1. Introduction

This chapter will describe the preparation and use of a semipermeabilized (SP) cell system that reconstitutes the initial stages in the assembly and modification of proteins entering the secretory pathway *(1)*. The procedure involves treating cells grown in culture with the detergent digitonin and isolating the cells free from their cytosolic component *(2)*. The expression of proteins in an SP cell system allows protein assembly to be studied in an environment that more closely resembles that of the intact cell. As this is an in vitro system, the individual components can be manipulated easily, providing a means by which cellular processes can be studied under a variety of conditions. In addition, membrane-permeable chemical crosslinking reagents can be added in order to facilitate the study of interaction between proteins within the endoplasmic reticulum (ER) lumen *(1,3)*. Furthermore, as the ER remains morphologically intact, the spatial localization of folding and transport processes within the reticular network may also be examined.

The basic protocol involves translation of an mRNA transcript encoding the protein of interest in a rabbit reticulocyte lysate supplemented with the SP HT1080 cells prepared as outlined below (**Subheadings 3.1.** to **3.3.**). This particular cell line was initially chosen because it can carry out the complex co- and post-translational modifications required for the assembly of procollagen molecules into thermally stable triple helices *(4)*. Other cell lines have been used to study the initial stages in the biosynthesis of a wide range of proteins demonstrating the flexibility of this approach. The mRNA transcript coding for the protein of interest is translated in the presence of a radio-labeled amino acid (^{35}S-methionine) such that the protein synthesized can be visual-

From: *Methods in Molecular Biology, vol. 139: Extracellular Matrix Protocols*
Edited by: C. Streuli and M. Grant © Humana Press Inc., Totowa, NJ

ized by autoradiography. As the RNA can be synthesized in vitro from cloned cDNA, the effect of manipulating the primary amino acid sequence on folding and assembly can be evaluated rapidly. Here we outline the procedures for preparing SP cells, transcribing cloned cDNAs, and translation of the RNA transcripts generated in translation systems optimized for folding reactions. We also describe some procedures for characterizing the product of translation, both in terms of its incorporation into the ER of the SP cells, and folding status.

To illustrate this approach we will describe the synthesis, translocation, and assembly of procollagen *(5,6)*, however, it should be stressed that these techniques are applicable to all proteins entering the secretory pathway. The following data demonstrate that when added to the SP cell translation system, procollagen RNA can be translated into procollagen chains that are translocated into the lumen of the ER and fold and assemble to form interchain disulfide-bonded trimers. Thus, when an exogenous protease is added to the translation reaction after translation, all the untranslocated procollagen chains are digested leaving only chains that have been translocated into the ER lumen (**Fig. 1A**, lanes 2 and 3). The material left after protease treatment is protected from proteolysis by being segregated within the ER as can be demonstrated by the complete digestion of this material following disruption of the ER membrane by the addition of detergent (**Fig. 1A**, lane 4). To demonstrate correct folding of the synthesized procollagen chains, a time-course of translation was carried out and the products of translation separated by SDS-PAGE under reducing or nonreducing conditions. Proteins containing intrachain disulfide bonds migrate with a faster relative mobility than the fully reduced protein and proteins forming interchain disulfide bonds have a correspondingly slower electrophoretic mobility *(7)*. At early time-points, the newly synthesized procollagen chains form intrachain disulfides indicating correct folding of the monomeric chains (**Fig. 1B**, compare lane 1 with lane 7, and lane 2 with lane 8). These monomeric chains quickly associate to form interchain disulfide-bonded trimers (**Fig. 1B**, lanes 9–12). We have also shown, by protease resistance, that these molecules have formed a correctly aligned triple helix *(5,8)*. Thus, the SP cell system faithfully reproduces the initial stages in the folding and assembly of procollagen.

2. Materials

2.1. Preparation of SP Cells

1. HT1080 cells (75 cm^2 flask of subconfluent cells).
2. Phosphate-buffered saline (Gibco-BRL).
3. 1X trypsin-EDTA solution (Gibco-BRL).
4. KHM buffer: 110 m*M* KOAc, 2 m*M* MgOAc, 20 m*M* HEPES, pH 7.2.

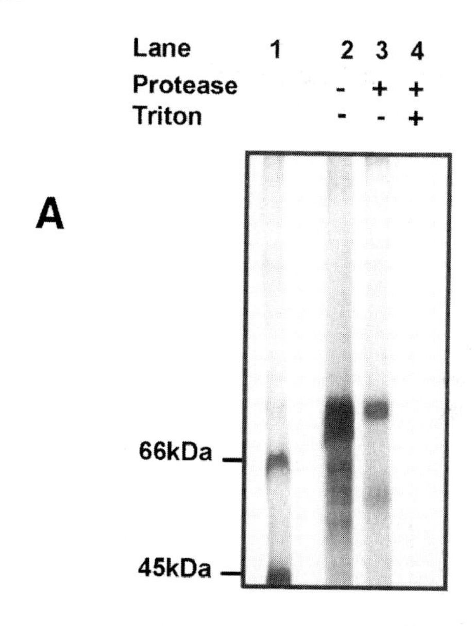

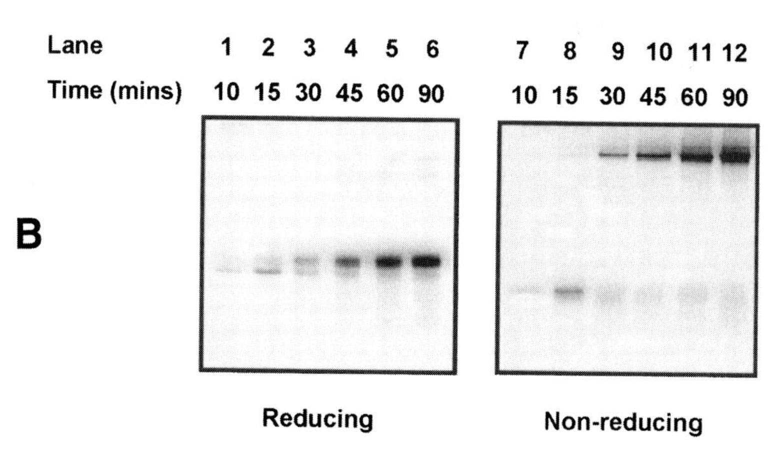

Fig. 1. Procollagen folding occurs within semipermeablized cells: RNA coding for procollagen was translated in the presence of semipermeabilized cells (*see* text for details).

5. HEPES buffer: 50 m*M* KOAc, 50 m*M* HEPES, pH 7.2.
6. 50 mg/mL soybean trypsin inhibitor, in sterile water stored at –20°C (Sigma, St. Louis, MO).
7. 40 mg/mL digitonin in DMSO, stored at –20°C (Calbiochem, La Jolla, CA).
8. 0.4% Trypan blue solution.
9. 0.1 *M* $CaCl_2$ (stored at –20°C).

10. 1 mg/mL micrococcal nuclease in sterile water, stored at –20°C.
11. 0.4 *M* EGTA (stored at –20°C).

2.2. Transcription In Vitro

1. 10 µg linearized plasmid DNA, containing gene of interest downstream form a viral polymerase promoter, in RNase-free water.
2. 5X transcription buffer (400 m*M* HEPES buffer, pH 7.4, 60 m*M* $MgCl_2$, 10 m*M* Spermidine).
3. Nucleotide triphosphates (ATP, UTP, CTP, and GTP) (25 m*M* each) (Boehringer Mannheim, Mannheim, Germany).
4. 100 m*M* DTT (Sigma).
5. T3/T7 RNA polymerase (50 U/µL) (Promega, Madison, WI).
6. RNase inhibitor (Promega).

2.3. Translation In Vitro

1. Flexi™ rabbit reticulocyte lysate.
2. Amino acid mix (minus methionine).
3. 2.5 *M* KCl.
4. EasyTag™ ^{35}S-methionine.

All reagents were supplied by Promega and stored at –70°C, except the ^{35}S-methionine that is supplied by NEN Dupont and stored at 4°C.

2.4. Proteinase K Treatment

1. 2.5 mg/mL proteinase K in sterile H_2O.
2. 0.1 *M* $CaCl_2$.
3. 10% Triton X-100 stored at 4°C.
4. 0.1 *M* PMSF in isopropanol.
5. All reagents stored at –20°C.

3. Methods

3.1. Preparation of SP Cells

This procedure uses a modified protocol based on that of Plutner et al. *(2)*, which has been adapted for the cell-free expression of proteins *(1)*. Treatment of mammalian cells with a low concentration of the detergent digitonin renders the plasma membrane permeable to the components of the cell-free translation system whereas retaining the ER membrane in a functionally intact state. This selective permeabilization of the plasma membrane is a consequence of the cholesterol-binding properties of digitonin. As cholesterol is only a minor constituent of the internal membrane system of the cell, the ER and Golgi networks remain intact, although some swelling of the ER is observed (*see* **Note 1**).

1. Rinse HT1080 cells in flask with 2 × 10 mL PBS to remove medium that could inhibit trypsin. Drain and add 2 mL of trypsin solution (prewarmed to room temperature) and incubate at room temperature for 3 min. Cells should now be detached and can be disrupted by gently tapping the flask. Add 8 mL of KHM buffer and 20 µL soybean trypsin inhibitor (final concentration 100 µg/mL) to the tissue-culture flask. Transfer cell suspension to a 15-mL Falcon tube on ice.
2. Pellet cells by centrifugation at 350 × *g* for 3 min at 4°C. Aspirate the supernatant from the cell pellet.
3. Resuspend cells in 6 mL of ice-cold KHM. Add 6 µL digitonin (from 40 mg/mL stock, i.e., final concentration 40 µg/mL) and mix immediately by inversion and incubate on ice for 5 min (*see* **Note 2**).
4. Adjust the volume to 14 mL with ice-cold KHM and pellet cells by centrifugation as in **step 2**.
5. Aspirate the supernatant and resuspend cells in 14 mL ice-cold HEPES buffer. Incubate on ice for 10 min and pellet cells by centrifugation as in **step 2**.
6. Aspirate the supernatant and resuspend cells carefully in 1 mL ice-cold KHM (use a 1-mL Gilson pipet and pipet gently up and down). Place on ice.
7. Transfer a 10-µL aliquot to a separate 1.5-mL microcentrifuge tube and add 10 µL of trypan blue.
8. Count cells in a hemocytometer and check for permeabilization, i.e., whether the trypan blue permeates into the cell.
9. Transfer cells to a 1.5-mL microcentrifuge tube and spin for 30 s at 15,000 × *g*. Aspirate supernatant and resuspend the cells in 100 µL KHM using a pipet.
10. Treat the cells with a calcium-dependent nuclease to remove the endogenous mRNA. Add 1 µL of 0.1 *M* $CaCl_2$ and 1 µL of monococcal nuclease and incubate at room temperature for 12 min.
11. Add 1 µL of 0.4 *M* EGTA to chelate the calcium and inactivate the nuclease. Isolate the cells by centrifuging for 30 s in a microcentrifuge and resuspend in 100 µL of KHM.
12. Use approximately 10^5 cells per 25-µL translation reaction (approx 4 µL of the 100 µL obtained).

3.2. Transcription In Vitro

The cDNA encoding the protein of interest is ligated into a mammalian expression vector, such as pBluescript, upstream of a suitable promotor containing an RNA polymerase binding site from which transcription is initiated. Prior to transcription, the cDNA clone must be linearized by restriction endonuclease digestion to generate a template for mRNA synthesis. This method is a modification of a method described previously ***(9)***.

1. Prepare a 100-µL reaction mixture containing 44 µL H_2O, 10 µL linearized DNA (5–10 µg), 20 µL transcription buffer (5X), 10 µL 100 m*M* DTT, 1 µL RNasin (20 U), 3 µL of each nucleotide.

2. Add 3 μL of the appropriate RNA polymerase (150 U) and incubate at 37°C for 2 h (*see* **Note 3**).
3. The RNA can be extracted with phenol/chloroform 1:1, then twice with chloroform and precipitate by adding NaOAc, pH 5.2 to a final concentration of 300 m*M* and 3 vol of ethanol. The RNA pellet is resuspended in 100 μL RNase-free H_2O containing 1 m*M* DTT and 1 μL RNasin.
4. To assess the yield of RNA, a 1-μL aliquot should be removed and analyzed on a 1% agarose gel (*see* **Note 4**).

3.3. Translation In Vitro

The translation of proteins in vitro can be performed using either wheat germ extracts or rabbit reticulocyte lysates that contain ribosomes, tRNAs, and a creatine phosphate-based energy regeneration system.

1. Prepare a 25-μL reaction mixture containing 17.5 μL Flexi™ lysate, 0.5 μL amino acids, 0.5 μL KCl, 1.5 μL EasyTag ^{35}S-methionine, 1 μL mRNA, and 4 μL SP cells (*see* **Notes 5** and **6**). Incubate the translation sample at 30°C for 60 min and then place on ice.
2. Prepare the translation sample for SDS-PAGE by adding 2 μL of the product to 15 μL SDS-PAGE sample buffer (0.0625 *M* Tris/HCl pH 6.8, SDS (2% w/v), glycerol (10% v/v), and bromophenol blue) plus 2 μL DTT (1 *M*) and boiling the sample for 5 min.
3. The samples should be separated through a SDS-PAGE gel appropriate for the expected molecular weight for the protein of interest. After running, the gel should be dried and exposed to autoradiography film (*see* **Note 7**).

3.4. Proteinase K Protection

A "protease protection" assay is a method used to determine whether the nascent chains are targeted to the ER membrane and translocated into the ER lumen of the SP cells. The translation samples are treated posttranslationally with proteinase K, which rapidly digests any nontranslocated translation products, whereas fully translocated proteins are protected by the lipid bilayer of the ER membrane. A control sample is incubated in the presence of proteinase K and Triton X-100, which solubilizes the ER membrane, and, therefore renders the translocated protein susceptible to proteinase K digestion (*see* also **Note 8**).

1. Prepare a 25-μL translation reaction including freshly prepared SP cells as described in **Subheading 3.3.** After translation, place the sample on ice and gently disperse the SP cells using a pipet tip.
2. Divide the translation mixture into three microcentrifuge tubes containing 3 × 8-μL aliquots. One sample is used as a nontreated control. To the other two tubes, add 1 μL $CaCl_2$ (100 m*M*) and 1 μL Proteinase K (2.5 mg/mL). To one of these samples, also add Triton X-100 to a concentration of 1% (v/v).

3. Incubate the three samples on ice for 20 min, followed by a further incubation of the samples on ice for 5 min with 1 m*M* PMSF to inhibit the proteinase K.
4. Prepare the samples for electrophoresis by adding 5 μL of each reaction to 15 μL of SDS-PAGE buffer containing 2 μL DTT (1 *M*).
5. The samples should be separated through an SDS-PAGE gel appropriate for the expected molecular weight for the protein of interest. After running, the gel should be dried and exposed to autoradiography film (*see* **Note 8**).

3.5. Analysis of Disulfide Bond Formation

The native conformation of proteins entering the secretory pathway is often stabilized by disulfide bonds. The formation of intrachain disulfide bonds in a particular domain of a nascent polypeptide may represent a key event in the folding pathway. In the case of fibrillar procollagen chains, formation of the correct intrachain disulfide bonds in the carboxy-terminal domain (the C-propeptide) is necessary for the folding of these domains and is a prerequisite for trimer formation *(10,11)*. The trimers in turn are stabilized by interchain disulfides. The formation of disulfide bonds during folding can be monitored over time by trapping folding intermediates using the alkylating reagent N-ethyl maleimide (NEM) *(12)*. The formation of intrachain disulfide bonds generally increases the electrophoretic mobility of proteins during SDS-PAGE analysis, provided the sample is analyzed under nonreducing conditions. In contrast, multisubunit proteins that are stabilized by interchain disulfide bonds have a faster migration when the protein is treated with reducing agents that cause dissociation into constituent monomers.

1. Prepare a 100-μL translation mix and divide this into four aliquots of 25 μL in separate microcentrifuge tubes
2. At intervals of 15 min, remove one of the tubes and add NEM to a final concentration of 20 m*M* and place on ice for the remainder of the time-course.
3. Isolate and "wash" the SP cells as described in **steps 1** and **2** (**Subheading 3.5.**).
4. Solubilize each of the washed cell pellets in 50 μL SDS-PAGE buffer and then transfer 25 μL of each sample into fresh tubes containing 2 μL DTT. Boil the samples for 5 min prior to electrophoresis (*see* **Note 9**).

4. Notes

1. The procedure takes approximately 1 h and should be carried out immediately prior to using the SP cells for translation in vitro, SP cells do not efficiently reconstitute the translocation of proteins after storage. It is also advisable to use a minimum of one 75 cm^2 flask of cells (75–90% confluent) as it is difficult to work with a smaller quantity of cells. The size of the cell pellet will usually decrease during the procedure because of loss of the cell cytosol that is accompanied by a decrease in cell volume.

2. The digitonin concentration has been optimized for permeabilization of HT1080 cells (i.e., the lowest concentration of digitonin that results in 100% permeabilization). If a different cell line is used, the concentration of digitonin required for permeabilization should be assessed by titration. It is not essential to trypan blue stain each batch of SP cells, although this is recommended if the procedure is not used routinely.
3. The yield of mRNA can be increased by a further addition of RNA polymerase (1 µL) after 1 h.
4. To minimize degradation of the mRNA, the use of sterile pipet tips, microcentrifuge tubes is recommended. If the yield is low or the RNA is partially degraded it is possible that apparatus or solutions have been contaminated with RNases.
5. The authors recommend that the translation protocol is optimized for each different mRNA transcript as the optimal salt concentration (KCl and MgOAc) may vary.
6. To test the translation efficiency of a new RNA preparation, a single 25-µL reaction including 4 µL of sterile water instead of SP cells can be prepared.
7. If there are no protein bands then the RNA may need to be heated to 60°C for 10 min prior to translation in order to denature any secondary structure. Additional products with molecular weights smaller than the major translation product may be observed because of ribosome binding to "false" start sites downstream of the initiation codon.
8. The translocated and nontranslocated forms of the protein usually migrate differently on a reducing SDS-PAGE gel owing to modification of the nascent chain in the ER lumen. The translation products not treated with proteinase K will contain a mixture of these forms, the ratio of which is dependent upon the efficiency of translocation. As the nontranslocated polypeptides are selectively degraded by addition of proteinase K, it is possible to assess which translation product corresponds to each form. The nontranslocated form will also comigrate with the polypeptide synthesized in the absence of SP cells.

   In the case of transmembrane proteins, treatment with proteinase K results in an increase in electrophoretic mobility corresponding to loss of the cytoplasmic domain that is accessible to the enzyme. The translocated polypeptide may migrate faster than the nontranslocated form resulting from signal peptide cleavage that occurs when the nascent chain enters the ER lumen. However, this may only be detected if no other covalent modification of the polypeptide occurs. Usually, the translocated polypeptides will exhibit decreased electrophoretic mobility because of glycosylation. In the case of procollagens, hydroxylation of proline and lysine residues also results in decreased electrophoretic mobility.
9. The reduced and nonreduced samples should be separated by a gap of two lanes in order to prevent reduction of the nonreduced samples by DTT that may diffuse across the gel matrix during electrophoresis.

References

1. Wilson, R., Oliver, J., Brookman, J. L., High, S., and Bulleid, N. J. (1995) Development of a semi-permeabilised cell system to study the translocation, folding, assembly and transport of secretory proteins. *Biochem. J.* **307,** 679–687.

2. Plutner, H., Davidson, H. W., Saraste, J., and Balch, W. E. (1992) Morphological analysis of protein transport from the ER to Golgi membranes in digitonin-permeabilized cells: role of the p58 containing compartment. *J. Cell Biol.* **119,** 1097–1116.
3. Elliott, J. G., Oliver, J. D., and High, S. (1997) The thiol-dependent reductase ERp57 interacts specifically with N-glycosylated integral membrane proteins. *J. Biol. Chem.* **272,** 13,849–13,855.
4. Pihlajaniemi, T., Myllyla, R., Alitalo, K., Vaheri, A., and Kivirikko, K. I. (1981) Posttranslational modifications in the biosynthesis of type IV collagen by a human tumor cell line. *Biochemistry* **20,** 7409–7415.
5. Bulleid, N. J., Wilson, R., and Lees, J. F. (1996) Type III procollagen assembly in semi-intact cells: chain association, nucleation and triple helix folding do not require formation of inter-chain disulfide bonds but triple helix nucleation does require hydroxylation. *Biochem. J.* **317,** 195–202.
6. Bulleid, N. J. (1996) Novel approach to study the initial events in the folding and assembly of procollagen. *Seminars Cell Dev. Biol.* **7,** 667–672.
7. Goldenberg, D. P. and Creighton, T. E. (1984) Gel electrophoresis in studies of protein conformation and folding. *Anal. Biochem.* **138,** 1–18.
8. Bruckner, P. and Prockop, D. J. (1981) Proteolytic enzymes as probes for the triple-helical conformation of procollagen. *Anal. Biochem.* **110,** 360–368.
9. Gurevich, V. V., Pokrovskaya, I. D., Obukhova, T. A., and Zozulya, S. A. (1991) Preparative in vitro RNA synthesis using SPG and T7 RNA polymerases. *Anal. Biochem.* **195,** 207–213.
10. Lees, J. F. and Bulleid, N. J. (1994) The role of cysteine residues in the folding and association of the COOH-terminal propeptide of types I and III procollagen. *J. Biol. Chem.* **269,** 24,354–24,360.
11. Schofield, D. J., Uitto, J., and Prockop, D. J. (1974) Formation of interchain disulfide bonds and helical structure during biosynthesis of procollagen by embryonic tendon cells. *Biochemistry* **13,** 1801–1806.
12. Creighton, T. E., Hillson, D. A., and Freedman, R. B. (1980) Catalysis by protein disulphide isomerase of the unfolding and refolding of proteins with disulphide bonds. *J. Mol. Biol.* **142,** 43–62.

2

Quantitative Determination of Collagen Crosslinks

Trevor J. Sims, Nicholas C. Avery, and Allen J. Bailey

1. Introduction

The primary functional role of collagen is as a supporting tissue and it is now well established that the aggregated forms of the collagen monomers are stabilized to provide mechanical strength by a series of intermolecular crosslinks. These links are formed by oxidative deamination of the ε-amino group of the single lysine in the amino and carboxy-telopeptides by lysyl oxidase. The aldehyde thus formed reacts with an ε-amino group of a lysine at a specific point in the triple helix because of the quarter-staggered end-overlap alignment of the molecules in the fibers. The chemistry of these crosslinks is dependent on both the nature and age of the collagenous tissue *(1,2)*. Differences in the crosslinks are because of the degree of hydroxylation of both the telopeptide and the specific lysine in the triple helix. Thus, the amounts of intermediate crosslinks present in immature tissue, dehydro-hydroxylysinonorleucine (Δ-HLNL), and hydroxylysino-keto-norleucine (HLKNL) may vary considerably between tissues, e.g., rat tail tendon and skin contain Δ-HLNL whereas cartilage and bone contain predominantly HLKNL.

These divalent crosslinks are only intermediates and are subsequently converted into stable trivalent crosslinks that accumulate in the tissue as collagen turnover decreases during maturation *(2)*. The Schiff base aldimine Δ-HLNL is stabilized by reaction with histidine to form the trivalent crosslink, histidino-hydroxlysinonorleucine (HHL). The keto-imine HLKNL, on the other hand, reacts with a second hydroxylysyl aldehyde to form the pyridine derivatives, hydroxylysyl-pyridinoline (Hyl-Pyr), and lysyl-pyridinoline (Lys-Pyr), **Fig. 1**. The proportion of these three known mature crosslinks, again varies with age and with the type of tissue. For example, HHL is the major, known, mature crosslink of human and bovine skin *(3)*, whereas the pyridinolines are the

From: *Methods in Molecular Biology, vol. 139: Extracellular Matrix Protocols*
Edited by: C. Streuli and M. Grant © Humana Press Inc., Totowa, NJ

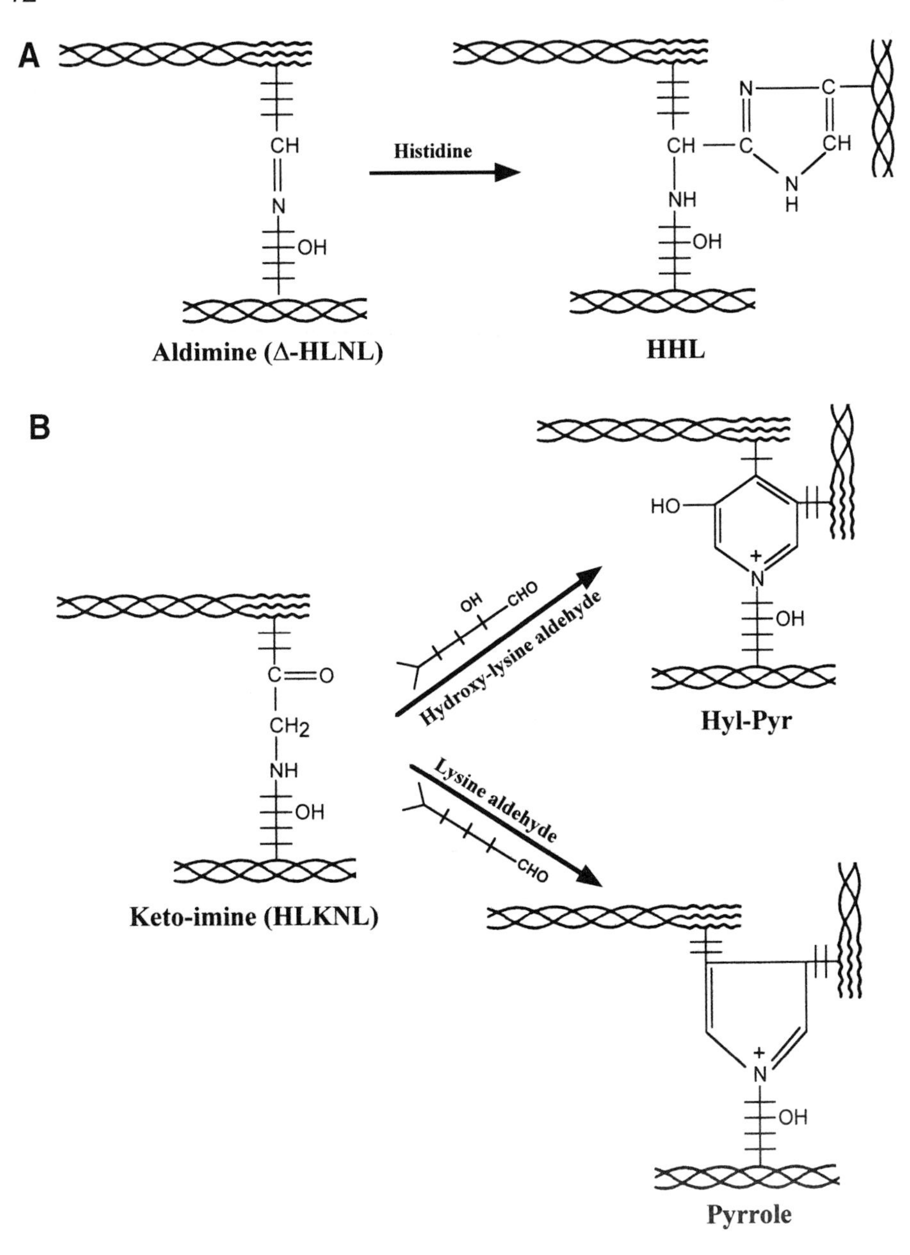

Fig. 1. Schematic representation of the formation of collagen crosslinks. **(A)** The divalent Schiff base (aldimine) dehydro-hydroxylysinonorleucine (Δ-HLNL) which subsequently reacts with histidine to give the trivalent crosslink histidinohydroxylysinonorleucine. **(B)** The intermediate keto-imine crosslink (hydroxy-lysinoketonorleucine; HLKNL) can react with hydroxylysine aldehyde to give hydroxylysyl-pyridinoline (Hyl-Pyr) or with lysine aldehyde to give the hydroxylysyl-pyrrole.

major, mature crosslinks of bone and cartilage *(4)*; tendon, however, contains a mixture of both mature crosslinks. An additional mature crosslink may be formed if the keto-imine reacts with a lysine-aldehyde in which case a pyrrole structure is favored (*see* **Fig. 1**). Although the structure of the so-called "pyrrole" crosslink has not been confirmed there is mounting evidence for its presence in bone and tendon collagen *(2)*. An accurate determination of the ratio of the immature to mature crosslinks provides a valuable indication of the degree of turnover of a collagenous tissue. It is, therefore, important in any study of changes to the collagen in a pathological tissue to understand the nature of the normal, age-related changes that occur in the particular tissue under investigation.

The intermediate crosslinks may be radio-labeled by reduction of the tissue with tritiated sodium borohydride, thus facilitating their location and identification during subsequent chromatography *(5)*. However, their quantification requires either ninhydrin, or a similar post-column derivatization technique, following their separation from the acid hydrolysate of the tissue by ion-exchange chromatography. Precolumn derivatization of these polyvalent crosslinks for subsequent analysis by reversed-phase high performance liquid chromatography (HPLC) can produce multiple derivatives that elute as separate peaks throughout the subsequent analysis and is therefore not recommended. The mature crosslinks HHL, Hyl-Pyr, and Lys-Pyr can be simultaneously quantified using the same ion-exchange column *(6)*. Hyl-Pyr and Lys-Pyr can also be determined, with greater sensitivity, by HPLC utilizing their natural fluorescence to facilitate their detection and quantification *(7)*. It has not yet been possible to analyze the pyrrole crosslink by ion exchange or HPLC chromatography and until this is possible a rather nonspecific colorimetric method is employed *(8)*.

The other major connective tissue protein, elastin, is also stabilized by lysine-derived crosslinks based on the same enzymic mechanism, but yields two tetravalent pyridine compounds, desmosine (DES) and iso-desmosine (I-DES) *(10)*. Both of these compounds can be detected by ninhydrin after elution, under the same conditions, from the same ion-exchange column.

A second crosslinking mechanism occurs when the turnover of collagenous tissues decreases following maturation and involves the reaction of glucose with the ε-amino group of lysine and subsequent oxidation reactions *(9)*. Generally known as glycation, the addition of glucose is nonenzymic, adventitious, and possibly random. Crosslinks formed by this mechanism, such as pentosidine, could provide good biomarkers of low metabolism and possible damage to the functional properties of collagen during aging and in diabetic subjects. However, to date, none of the glycation crosslinks has been related to changes in the functional properties of collagen, hence, we have only considered pentosidine.

2. Materials

Unless stated otherwise, all reagents should be of Analar grade.

1. Sodium borohydride should be dissolved in 0.01 *M* sodium hydroxide solution at 5°C immediately prior to use. The dry solid is deliquescent producing an explosive gas (hydrogen) when wet, consequently, care with storage and handling is essential.
2. The hydrochloric acid used for protein hydrolysis is a constant boiling mixture. This can be purchased commercially (BDH, Poole, UK) or prepared in the laboratory by distillation of a 50% mixture of hydrochloric acid with distilled water and collecting the distillate that separates at 110°C.
3. Fibrous cellulose, CF-1, is a commercially available product from Whatman (Maidstone, Kent, UK).
4. Filters, for sample preparation, and for both HPLC and amino acid analyzer buffer filtration, are commercially available (HPLC Technology, Macclesfield, UK). 4 mm or 13 mm PVDF syringe filters are used for sample filtration.
5. A steel 'mortar and pestle'. The 'mortar' consists of a cylindrical block of stainless steel (40 mm × 40 mm) with a flat-bottomed 10-mm diameter hole drilled into it to a depth of 30 mm. The "pestle" is also made from stainless steel and measures 9.5 mm in diameter and 100 mm in length. These dimensions provide a close sliding fit into the mortar.
6. A source of liquid nitrogen.
7. An amino acid analyzer equipped with a post-column ninhydrin detection system.
8. A high performance liquid chromatography (HPLC) system linked to a fluoresence detector.
9. Ideally, both of the above should be equipped with computer-based chromatography data handling software or a computing integrator.
10. The reagents and buffers for use with the amino acid analyzer are best purchased from the equipment supplier. Any alteration to the concentration or pH of such buffers should be done with great care and any buffers modified in this way should be passed through a 0.2 µm filter prior to use to remove particulate matter. The buffers should incorporate 0.01% phenol to prevent bacterial spoilage and should be stored at 15–20°C.
11. All HPLC reagents need to be HPLC grade and 0.2 µm filtered prior to use.
12. A microtiter plate reader fitted with a 570 nm filter and a number of flat-bottomed 96-well microtiter plates are required for the pyrrole assay.

    The following reagents are also required for measuring the pyrrole crosslink, and can all be obtained from Sigma-Aldrich (Poole, UK).
13. TAPSO Buffer: 2.81 g of 3-[N-tris(hydroxymethyl)methylamino]-2-hydroxypropanesulphonic acid (TAPSO) is dissolved in 80 mL of distilled water, adjusted to pH 8.2 with 1 *M* sodium hydroxide and then made up to 100 mL.
14. TPCK/TAPSO enzyme inactivator: 1.6 mg N-tosyl-L-phenylalanine chloromethyl ketone (TPCK) is dissolved in 100 µL of ethanol and then taken up to 5 mL in TAPSO buffer. This solution may be turbid, but the turbidity will be removed later in the assay by filtration.

15. TPCK-Trypsin reagent: TPCK is added to trypsin to inactivate any residual chymotrypsin activity, which would otherwise destroy the pyrrole. This reagent should be freshly prepared on the day of use. Trypsin is dissolved in the TPCK inactivator solution (1000 U/200 μL) and left at room temperature for 25 min to inactivate any chymotrypsin.
16. DAB reagent: 500 mg of 4-dimethylaminobenzaldehyde (DAB) is dissolved in 4.4 mL 60% perchloric acid and made up to 10 mL with water. Reagent blank is prepared as above minus the DAB.
17. Pyrrole standards: A standard curve is prepared using 1-methyl pyrrole. This is obtainable as a liquid from Aldrich (Cat no. M7, 880-1). 11.1 μL is made up to 2.5 mL in ethanol from which 20 μL is diluted to 100 mL in TAPSO/TPCK reagent to give a final concentration of 10 μ*M* pyrrole. This stock solution is used to produce a series of solutions in the concentration range 1–5 μ*M* by dilution according to the following table:

10 μ*M* Stock pyrrole, μL	20	40	60	80	100
TAPSO/TPCK buffer, μL	180	160	140	120	100
Pyrrole conc., μmol/L	1	2	3	4	5

3. Methods (*see* Note 1)

3.1. Borohydride Reduction of Sample

1. The weighed sample is finely comminuted (*see* **Note 2**) and evenly dispersed in a volume of phosphate buffered saline (0.15 *M* sodium chloride, 0.05 *M* sodium phosphate pH 7.4) equal to between 5 and 10 times the volume of the sample (*see* **Note 3**).
2. A weight of sodium borohydride equal to 1% of the sample wet weight is dissolved in 0.001 *M* sodium hydroxide at 4°C and this is added to the sample (*see* **Note 4**).
3. The temperature of the reduction mixture is raised to approximately 20°C and reduction allowed to proceed for 1 h in a fume hood with occasional stirring. After this period the mixture is acidified to approximately pH 3.0 by addition of glacial acetic acid (pH paper accuracy is sufficient). (*See* **Note 5.**)
4. The acidified reducing reagents are now discarded either by filtration or after centrifugation, however, it can often be simply achieved by carefully decanting the reagents. The sample is then washed three times with distilled water in order to remove both the acetic acid and salts from the reduction mixture prior to freeze-drying (*see* **Note 6**).

3.2. Hydrolysis of the Sample

1. The weighed, dry sample is placed in a suitable vessel and hydrolysed in a volume of constant boiling hydrochloric acid to give a concentration of approx 5 mg of sample / mL of acid (*see* **Notes 7** and **8**).
2. Seal the hydrolysis vessel and heat to 110°C for 24 h (*see* **Note 9**).

3. After hydrolysis the sample is allowed to cool, the seal is broken and the sample then brought to between –20°C and –80°C prior to lyophilization to remove all residual 6 *N* hydrochloric acid (*see* **Note 10**).
4. After drying, the sample can be rehydrated in water (usually 0.5 mL) and divided according to the requirements of the subsequent analyses (*see* **Note 11**).

3.3. Measurement of Hydroxyproline (see Note 12)

1. A portion of the rehydrated hydrolysate, which is estimated to contain about 16 nmol of hydroxyproline (representing 15 μg of collagen), can be analyzed on the same ion-exchange column as that used for the analysis of the collagen crosslinks (*see* **Subheading 3.6.**), but using a different set of elution buffers.
2. The column is equilibrated in 0.2 *M* sodium citrate buffer pH 2.65 and held at 50°C throughout the run. After the sample has been applied, the column is eluted with 0.2 *M* sodium citrate buffer pH 3.20.
3. Hydroxyproline elutes early from the column (before aspartic acid) and reacts with ninhydrin to produce a yellow color that can be detected at 440 nm. The column and detection system are calibrated using an external standard consisting of a solution of pure hydroxyproline of known concentration.
4. Collagen is generally considered to contain 14% hydroxyproline by weight, so the collagen content of the sample can now be calculated from the measured hydroxyproline value.

3.4. Preparation of a CF-1 Prefractionation Column (see Note 13)

1. The CF1 cellulose powder (50 g) is first thoroughly wetted with 400 mL distilled water in a 2-L measuring cylinder, to which is subsequently added 400 mL glacial acetic acid, and finally 1600 mL of butan-1-ol. The resulting 2.5 L of thin slurry is shaken carefully to ensure complete mixing and suspension of the cellulose and then left to settle (about 20 min) until the bulk of the cellulose is below the 500 mL mark. The supernatant containing suspended cellulose fines is then poured to waste leaving approximately 600–800 mL of the 4:1:1 organic mixture of butan-1-ol:acetic acid:water containing the bulk of the original 50 g of CF-1. The slurry is now topped up to 1 L with fresh 4:1:1 organic mixture to yield an approximately 5% CF-1 slurry, which is decanted into a screw-cap container and stored at room temperature until required (*see* **Note 14**).
2. The prefractionation procedure requires the production of a minicolumn of CF-1. The top of a 3 mL plastic, Pasteur pipet bulb is cut off and the flow from the tip reduced, but not blocked, with glass wool or nonabsorbent cotton wool. The CF1 slurry is poured into the pipet through the cut bulb and the cellulose is allowed to settle, adding more slurry as necessary to produce a settled bed height of 8 cm. The 3-mL graduation mark on the pipet is a useful guide. Care should be taken to avoid fluid-filled cavities in the column bed, as this will adversely affect the chromatographic properties of the column. The newly prepared column should then be conditioned by passing 2 × 3 mL of fresh 4:1:1 eluant through the column. CF-1 columns made in this fashion do not readily dry out, although this should be guarded against.

3.5. CF-1 Prefractionation

1. The dried sample is rehydrated in 0.5 mL water followed by 0.5 mL glacial acetic, and finally 2 mL butan-1-ol. It is necessary to ensure thorough mixing of the sample by using a vortex mixer after the addition of each component of the solvent (*see* **Note 15**).
2. The 3 mL of sample is applied to the CF-1 column and its containment vessel washed with 2 x 1 mL of fresh 4:1:1 eluant, which is also loaded onto the column after the initial sample load has run to waste.
3. 6 × 3 mL of 4:1:1 eluant are now run through the column resulting in the elution of the bulk of the standard amino acids while the collagen crosslinks remain adsorbed to the cellulose. This portion of the eluate can therefore be discarded.
4. After the 4:1:1 eluant has passed through the column, a collection vessel is placed under the Pasteur pipette and 3 × 3 mL of water passed through the column to desorb the crosslinks from the cellulose. This aqueous eluate should now be taken to dryness (*see* **Note 16**).

3.6. Ion-Exchange Chromatography with Ninhydrin Detection

1. The freeze-dried aqueous phase eluate from the CF-1 column is reconstituted in 120 µL of 0.01 *M* hydrochloric acid by thorough vortex mixing of the tube to ensure complete dispersion of the solution around the walls of the vessel.
2. The tube is then centrifuged briefly (30 s) to bring the solution to the bottom of the tube and thus ensure maximum recovery.
3. The sample should be passed through a 4 or 13 mm 0.2-µm PVDF syringe filter to remove particulate matter. The sample is now ready for analysis on the amino acid analyzer (*see* **Note 17**).
4. The analytical column used for the analysis is the high resolution column supplied by Pharmacia (Uppsala, Sweden) measuring 270 × 4 mm and filled with their UltroPac 8 resin in the sodium form. The column should be maintained at 90°C throughout the analysis.
5. Prior to application of the sample, the column should be equilibrated in 0.2 *M* sodium citrate buffer pH 4.25.
6. After the sample has been applied, the column should be eluted with 0.4 *M* sodium citrate buffer pH 5.25 (*see* **Note 18**) for 46 min during which time data are collected (*see* **Note 19**). The column is then washed for 6 min in 0.4 *M* sodium hydroxide and regenerated for 23 min in 0.2 *M* sodium citrate buffer pH 4.25 when it is ready for running the next sample.
7. At the completion of the run, the area of each peak is computed from the collected data and the concentration of each crosslink is determined by comparison with the peak area of a leucine external standard of known concentration (*see* **Note 20** for a detailed explanation of the calculations).
8. A typical elution profile using authentic collagen crosslinking amino acids is shown in **Fig. 2**.

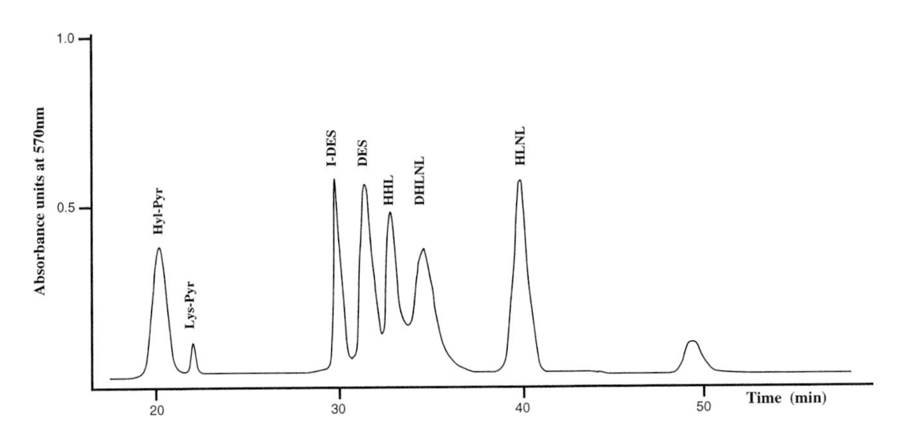

Fig. 2. Relative elution positions of hydroxylysyl-pyridinoline, (Hyl-Pyr); lysyl-pyridinoline (Lys-Pyr); isodesmosine (I-DES); desmosine (DES); histidino-hydroxylysinonorleucine (HHL); dihydroxylysinonorleucine (DHLNL); and hydroxylysinonorleucine (HLNL) on an ion-exchange amino acid analyzer using modified buffers. (DHLNL and HLNL are the sodium borohydride reduction products of hydroxylysino-ketonorleucine, HLKNL and dehydro-hydroxylysinonorleucine, Δ-HLNL, respectively).

3.7. HPLC Techniques

1. After hydrolysis and lyophilization each sample is rehydrated in an acidic, aqueous solution at a concentration of approximately 1 μg of collagen per μL. The HPLC system used for analysis dictates the solvent used for rehydration.
2. This laboratory originally used a 250 × 4.6 mm, 5 μm, octadecyl silane (ODS) column eluted with a 5–35% acetonitrile (MeCN) gradient in water at 1 mL/min over 70 min (0.5%/mL/min) (a modification of ***(4)***). Both aqueous and organic solvents contained 0.05 *M* heptafluorobutyric acid (HFBA), ion pairing agent. Samples destined for this system were hydrated in 5% or 10% HFBA (*see* **Note 21**).
3. Currently, we use a Shandon (Runcorn, UK) Hypercarb S, 100 × 4.6 mm, graphitic carbon column eluted with a 0 – 12% tetrahydrofuran (THF) gradient in water at 1 mL/minute. Both aqueous and organic solvents contain 0.5% trifluoroacetic acid (TFA). Samples destined for analysis by this system are hydrated in 1% TFA (*see* **Note 22**).
4. Following rehydration the samples are 0.2 μm filtered into tapered glass sample vials (Chromacol, Welwyn Garden City, UK), sealed and stored at 4°C. An aliquot of up to 90 μL is loaded onto the analytical column via an autosampler or a larger volume can be manually loaded via a Rheodyne valve with a 500 μL sample loop (Anachem, Luton, UK).
5. Prior to use the HPLC buffers are degassed either under vacuum (for 10 min) or by helium sparging (1–2 min).

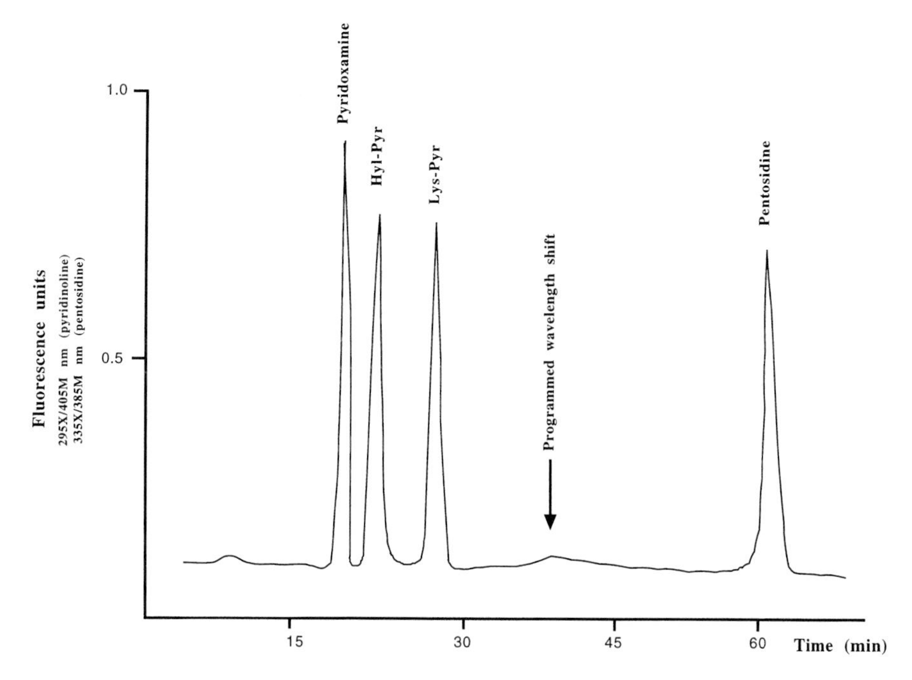

Fig. 3. Relative elution positions of the standards, pyridoxamine, hydroxylysyl-pyridinoline (Hyl-Pyr); lysylpyridinoline (Hyl-Pyr) and the glycation crosslink pentosidine on a Hypercarb S reversed phase HPLC column using fluorescence detection.

6. After an 8 min isocratic period in water, a 0–12% THF linear gradient in water is applied over 60 min at a flow rate of 1 mL/min (0.2%/mL/min). Hydroxylysyl and lysyl-pyridinoline elute at approx 26 and 29 min, respectively, and the glycation crosslink pentosidine elutes at 62 min.
7. The pyridinium crosslinks are detected by means of their natural fluorescence at 405 nm emission after excitation at 295 nm. Pentosidine is also naturally fluorescent but at 385 nm emission after excitation at 335 nm. We program a wavelength shift into our Perkin-Elmer LS-5 fluorimeter (Bucks, UK) to take place after the pyridinolines have eluted. A typical HPLC elution profile of pyridinoline and pentosidine standards is shown in **Fig. 3**.
8. In this laboratory, data are collected during the analytical run using a computing integrator and stored to disk at the end of the analysis.
9. The area under each peak of interest is calculated as a proportion of that derived from known concentrations of standards prepared within this laboratory or purchased commercially. Where possible, the concentration of the standards should be confirmed by amino acid analysis.
10. Column integrity and fluorimeter efficiency is confirmed by regularly running a standard mixture (every 8–10 samples) and calculating the fluorescence yield per pmol of each fluorophore.

11. In addition, pyridoxamine, a commercially available fluorophore not naturally present in protein hydrolysates, can be added to all samples. Fluorescing at the same wavelengths as pyridinoline, but migrating earlier, it acts as an internal fluorescent standard to provide a constant monitor of both column and fluorimeter efficiency.

3.8. Pyrrole Crosslink (see Note 23)

1. Powdered bone or tendon (*see* **Note 2**) is placed in a glass universal and repeatedly extracted to remove lipids by gentle agitation in acetone at room temperature until the acetone no longer forms an opaque emulsion when mixed with water.
2. 40–50mg of the bone powder defatted in this way is then suspended in 5 mL of 0.5 *M* acetic acid, sealed in a tube and left to decalcify for about 2 wk at 4°C. The acid is then decanted and the sample washed with 4 mL distilled water and then combined with 400 µL of TAPSO buffer pH 8.2.
3. The container should be resealed and the sample denatured by heating in an oven or dry-block at 110°C for 35 min, following which it is placed in a shaking water bath and left to equilibrate at 37°C.
4. 200 µL of the trypsin-TPCK solution is added to the sample tube and this digest mixture is shaken gently for 18 h at 37°C.
5. The digest is then centrifuged for 10 min at 10,000*g* to remove particulate material.
6. 40 mL of DAB reagent is added to 200 µL of the digest sample and to each of the calibration standards. Similarly, 50 µL of reagent blank is added to further 200 µL of digest, to provide a sample blank. After 10 min the samples and blanks are passed through a 13 mm 0.2 µm filter and 180 µL of each sample, blank, and calibration standard pipeted into the wells of a 96-well microtiter plate.
7. The plate is scanned using a microtiter plate reader set at 570 nm and sample blank values subtracted from the test sample values. A standard curve is prepared from the calibration standard readings by plotting absorbance against concentration of 1-methyl pyrrole and corrected values for the test samples are read off this curve.

4. Notes

1. Before undertaking the collagen crosslink analysis of an extracellular matrix, consideration should be given to the amount of tissue available for the multiple analytical procedures and the likely collagen content of that tissue.

   There exists an approximately 10-fold difference in sensitivity between HPLC using fluorescence detection and amino acid analysis using ninhydrin detection in favor of HPLC and, in addition, the HPLC procedure is nondestructive allowing complete recovery of sample for further analysis. However, the intermediate crosslinks cannot be readily quantified by HPLC without prior derivatization, with the previously discussed associated problems of multiple peaks for each component. The advanced glycation end-product pentosidine cannot be quantified by ion-exchange chromatography as it is retained on the column. Ideally both analytical procedures should be adopted, but occasionally samples are so small, e.g., at biopsy, that the limitations of the HPLC procedure alone have to be accepted. In such a case, reduction of the sample with borohydride and CF1 pretreatment can be ignored but it will not then be possible to quantify the intermediate crosslinks, deH-HLNL and HLKNL, as these are destroyed by acid hydrolysis. However, the mature crosslinks hydroxylysyl and lysyl pyridinoline, and the advanced glycation end-product pentosidine can all be quantified after hydrolysis without prior reduction.

In summary, amino acid analysis by ion-exchange chromatography using ninhydrin detection can be used to assay the following crosslinks:

Hyl-Pyr, Lys-Pyr, HHL, DES, and I-DES.

HLNL and HLKNL can only be assayed after reduction of the sample with borohydride prior to acid hydrolysis.

Reversed-phase HPLC using fluorescence detection can be used to assay Hyl-Pyr, Lys-Pyr, and pentosidine. (None of these crosslinks require prior reduction with borohydride, nor is it essential to use the CF1 clean-up procedure, though its use will remove other fluorescent compounds resulting in a chromatogram, which is easier to interpret.)

2. Various methods of homogenization are available and the one chosen should be appropriate to the tissue under analysis:

   Skin or *Hide*. These should first be cleaned of subcutaneous fat and adhering muscle, and any hair removed with a scalpel or razor blade. The cleaned skin can then be chopped very finely with a blade or alternatively, homogenized in phosphate-buffered saline (PBS) (*see* **Subheading 3.1.**). We use a Polytron (Kinematica AG, Lucerne, Switzerland) for this purpose, which works well for most soft tissues, except *tendon*, which has a tendency to accumulate at the end of the homogenizer probe. Tendon is best treated by being chopped finely with a blade.

   Bone and *cartilage*. These are probably best comminuted in a steel "mortar and pestle" (*see* **Subheading 3.6.**) at the temperature of liquid nitrogen. The sample, in a cryothermic container, is frozen in a bath of liquid nitrogen as is the steel mortar and pestle. This usually takes about 5–10 min. After removal from the nitrogen bath the sample is placed in the mortar and the pestle, hammered onto the sample, causing it to shatter. With a suitably sized mortar and pestle, very small samples such as biopsies can be handled in this way with good recovery of the powdered sample.

   Fatty tissues. Lipids are readily removed from tissues with a very high fat content by overnight extraction in 3:1 chloroform:methanol at 4°C. The chloroform mixture can then be decanted and the tissue rehydrated by several extractions in PBS.

   Muscle. The collagen content of muscle is very low (1–5%), therefore it is necessary to remove the bulk of the myofibrillar proteins prior to analysis. This can be achieved by brief ultrasonic homogenization in Hasselbach-Schneider buffer consisting of 0.6 *M* potassium chloride, 0.1 *M* disodium hydrogen phosphate, 0.01 *M* sodium pyrophosphate, 0.001 *M* magnesium chloride, and 0.005 *M* dithiothreitol according to the method of Avery and Bailey (***11***). The insoluble collagenous network that remains is recovered by filtration through a 380-μ*M* copper sieve.

3. This volume is not critical, although too much could result in under reduction because of over dilution, and too little could result in the sample being carried out of the reducing reagent by gaseous hydrogen.
4. The volume of hydroxide should be as small as possible (μL) to avoid altering the pH of the sample buffer. The weight of borohydride used for reduction assumes 30% dry matter in the sample and 30% collagen. As a consequence, the proportion of sodium borohydride to collagen will be 1 part to 10 parts of collagen. For convenience, with multiple reductions, the requisite amount of borohydride for all samples can be dissolved in a volume of the ice-cold sodium hydroxide and then appropriate volumes immediately pipeted into each sample as required.
5. At this point, excess sodium borohydride will rapidly release gaseous hydrogen with a potential risk of sample loss. This should be kept in mind when selecting the vessel for the reduction procedure.
6. If the sample is very small, it is better to carry out all the above procedures in a vessel suitable for subsequent acid hydrolysis to avoid loss of sample.
7. The hydrolysis vessel is classically of borosilicate glass and can be reused after cleaning with chromic acid.
8. The ratio of sample to volume of acid is not critical provided the sample concentration does not exceed 10 mg/mL when certain resistant peptide bonds may not be cleaved.
9. It is customary to perform hydrolysis under a barrier of nitrogen gas (elimination of oxygen), however, the presence of oxygen is not known to influence collagen crosslink assays, though certain other amino acids are effected.
10. Even at –80°C the hydrolysate will not be frozen, however, without prior chilling there is a considerable risk of sample loss owing to boiling in the reduced pressure of the dryer. The freeze-drying apparatus must be rigorously defended from attack by hydrochloric acid vapor that will destroy seals, welds, and pump valves very rapidly. We use a glass vapor trap at –110°C to protect the vacuum pump, which itself is of a specialist design to resist corrosion. We also maintain a strict monthly vacuum pump oil-change regime.
11. 50 μg of collagen is sufficient for measurement of crosslinks by HPLC, but for ion-exchange 1 mg of collagen or more is required with a CF-1 prefractionation step. It is important that the CF-1 column is not overloaded, so it is recommended that no more than 30 mg dry weight of sample be run on a CF1 column.
12. The amino acid hydroxyproline is almost unique to collagen; present in mammalian collagen at about 95 residues per 1000, it is used to determine the total collagen content. This determination is crucial to subsequent procedures because the final crosslink quantification is expressed as moles of crosslink per mole of collagen. The most accurate method to use is the ion-exchange column used for crosslink analysis, but using the standard buffer gradient for amino acid analysis as described in the text. However, other analytical techniques are available, for

example, an automated flow analyzer (Burkard Scientific, Uxbridge, UK) based on the method of Grant ***(12)*** or a microtiter plate method ***(13)***, employing the same chemistry, can also be used for rapid determination of multiple samples.
13. The collagen crosslinks represent about 1 mol per mole of collagen, consequently locating these novel amino acids among the excess of normal amino acids has historically proved difficult. Prefractionation is therefore carried out to enhance the relative proportion of the crosslinking amino acids. The preferred method used in this laboratory is fibrous cellulose although several other methods have been reported with varying success, Sell and Monnier, ***(14)***; Dyer et al. ***(15)***; Takahashi et al. ***(16)***; Avery ***(17)***.
14. It is important to record the date of the slurry production as prolonged storage, e.g., longer than 2 mo, gives rise to an aggregated product that is unusable and must be discarded. On long-term storage, the 4:1:1 eluant also tends to separate into two layers and must not be used once this has occurred as the two layers will not remix into a single phase.
15. Rehydration in this fashion prevents the formation of an "oily" residue that occasionally occurs if the samples are hydrated with the 4:1:1 eluant directly.
16. This is best done in a centrifugal evaporator as the sample is then maintained as a small volume at the bottom of the tube. However, if an evaporator is not available, then the sample can be freeze-dried in the following manner. The vessel containing the column effluent is capped and a small hole pierced in the lid. The vessel should then be frozen at an angle of 45° in a –80°C freezer. This minimizes the chance of sample loss resulting from its rising up the container during the freeze-drying process.
17. The technique is based on the use of an automatic amino acid analyzer; we use an AlphaPlus II (Pharmacia) as previously described ***(20)***, but the technique can be applied to other amino acid analyzers. The supplier of such equipment obviously provides instructions on its use, so detailed explanations will not be provided here except for information specific to the analysis of crosslinks.
18. This can be prepared by dilution of Pharmacia's 1.2 *M* sodium citrate buffer pH 6.45 to a molarity of 0.4 *M* with water containing 0.1% phenol followed by adjustment to pH 5.25 with concentrated hydrochloric acid.
19. Our laboratory uses the AI-450 data handling software from Dionex UK Ltd. (Camberley, Surrey, UK) (their current version is called 'PeakNet'), to collect and manipulate the data generated by the amino acid analyzer, although any chromatography data-handling software would do. A simple strip-chart recorder linked to the analyzer would suffice, however, integration of the peak areas would then have to be performed by manual measurement of the peaks or by cutting out and weighing the peaks.
20. The crosslink peaks should be identified by comparison with authenticated crosslink standards and expressed as moles of crosslink per mole of collagen or as the reciprocal of this value, i.e., 1 crosslink molecule every "×" molecules of collagen. The elastin crosslinks Desmosine and iso-Desmosine should be determined as nmols of crosslink per mg of tissue. The following equation is used

to calculate the amount of each collagen crosslink as moles of crosslink/mol of collagen.

$$(A \times RF_{(Leu)} \times V_{(HCl)}) / (V_{(anal)} \times W_{(coll)} \times L \times 3.3) \quad (1)$$

A = Area under the crosslink peak (this value is obtained from the data-handling software or can be determined from a chart recorder connected to the analyzer by measuring the dimensions of the peak and calculating the sum of the peak height and the peak width at half the height).

$RF_{(Leu)}$ = Response Factor for leucine (obtained from the calibration of the analyzer using an external leucine standard of known concentration) and is calculated as follows:

$$\text{nmols of leucine run on the analyzer/measured area under the leucine peak} \quad (2)$$

$V_{(HCl)}$ = Volume (in µL) of 0.01 *N* hydrochloric acid used to dissolve the sample after CF-1 chromatography.

$V_{(anal)}$ = Volume (in mL) of this sample solution run on the amino acid analyzer.

$W_{(coll)}$ = Weight of collagen (in µg) contained in the sample applied to the CF-1 column (this is calculated from the measured hydroxyproline content of the hydrolyzed sample prior to CF-1).

L = Ninhydrin Leucine Equivalence value for each crosslink; these are:

HLNL 1.8; HLKNL 1.8; Lys-Pyr 1.7; Hyl-Pyr 1.7; HHL 1.97; I-DES 3.4; DES 3.4.

Worked example:

Let us assume that a sample of hydrolyzed bone was run on a CF-1 column and that this sample contained 11.4 mg of collagen ($W_{(coll)}$) obtained from measurement of its hydroxyproline content. After CF-1 chromatography, the aqueous eluate was dried and redissolved in 120 µL ($V_{(HCl)}$) of 0.01 *M* hydrochloric acid, of which 60 µL ($V_{(anal)}$) was run on the amino acid analyzer. A peak was obtained on the analyzer for hydroxylysyl-pyridinoline (Hyl-Pyr) with an area of 681,731 arbitrary units (A), as obtained from the data-handling software connected to the analyzer. A previous calibration run on the analyzer with a standard solution of leucine showed that 174,848 units of area were equivalent to 1 nmol of leucine. Therefore, using **Eq. 2**, $RF_{(Leu)} = 1/174{,}848 = 5.719 \times 10^{-6}$. The leucine equivalence value (L) for Hyl-Pyr is 1.7.

Therefore, using **Eq. 1**, we have:

$$\frac{681{,}731 \times 5.719 \times 10^{-6} \times 120}{60 \text{ x } 11.4 \text{ x } 1.7 \text{ x } 3.3} = 0.122 \text{ mols Hyl-Pyr / mol of collagen}$$

21. Silica-based columns are degraded by prolonged exposure to low pH solvents. The working life of such columns is very short (months) although they can be regenerated once or twice by a repacking procedure Avery and Light ***(18)***.

22. TFA and HFBA are strong (fuming) organic acids, additionally, HFBA has a pungent smell, MeCN and THF are both flammable, consequently, buffer preparation should be in a fume cupboard and care during handling is important.
23. The structure of the so-called pyrrole crosslink has yet to be determined and, at the present time, is quantified by a modification of the procedure for Ehrlich chromogens ***(9)***. Tryptic digests of the collagen are reacted with 4-dimethylaminobenzaldehyde (DAB) to give a pink/purple reaction product indicating the presence of pyrroles. The method is not specific, for pyrroles colored products also being formed with imidazoles, polyhydroxyphenols, and indoles ***(19)***.

References

1. Bailey, A. J., Light, N. D., and Atkins, E. D. T. (1980) Chemical cross-linking restrictions on models for the molecular organization of the collagen fibre. *Nature* **288,** 408–410.
2. Knott, L. and Bailey, A. J. (1998) Collagen cross-links in mineralising tissues: A review of their chemistry, function and clinical relevance. *Bone* **22,** 181–187.
3. Yamauchi, M., London, R. E., Guemat, C., Hashimoto, F., and Mechanic, G. L. (1987) Structure and function of a stable histidine-based tri-functional cross-link in skin collagen. *J. Biol. Chem.* **262,** 11,428–11,434.
4. Eyre, D. R. and Oguchi, H. (1980) The hydroxypyridinium cross-link of skeletal collagen. *Biochem. Biophys. Res. Commun.* **92,** 403–410.
5. Light, N. D. and Bailey, A. J. (1982) Covalent cross-links in collagen, in *Methods in Enzymology*, vol. 82A, pp. 360–372.
6. Sims, T. J. and Bailey, A. J. (1992) Quantitative analysis of collagen and elastin cross-links using a single-column system. *J. Chromatog.* **582,** 49–55.
7. Robins, S. P. (1982) Analysis of the cross-linking components in collagen and elastin. *Methods Biochem. Anal.* **28,** 329–379.
8. Scott, J. E., Hughes, E. W., and Shuttleworth, A. (1981) A collagen associated Ehrlich chromogen: A pyrrollic cross-link? *Biosci. Rep.* **209,** 263–264.
9. Paul, R. G. and Bailey, A. J. (1996) Glycation of collagen. The basis of its central role in the late complications of ageing and diabetes. *Interntl. J. Biochem. Cell Biol.* **28,** 1297–1310.
10. Partridge, S. M. (1970) Isolation and characterisation of elastin, in *Chemistry and Molecular Biology of the Intercellular Matrix,* vol. 1, Academic, New York, pp. 593–616
11. Avery, N. C. and Bailey, A. J. (1995) An efficient method for the isolation of intramuscular collagen. *Meat Sci.* **41,** 97–100.
12. Grant, R. A. (1964) Application of the auto-analyser to connective tissue analysis. *J. Clin. Pathol.* **17,** 685–691.
13. Riley, G., Harrall, R. L., Constant, C. R. Chard, M. D., Cawston, T. E., and Hazleman, B. L. (1994) Tendon degeneration and chronic shoulder pain: changes in the collagen composition of the human rotator cuff tendons in rotator cuff tendinitis. *Ann. Rheum. Dis.* **53,** 359–366.

14. Sell, D. R. and Monnier, V. M. (1989) Structure elucidation of a senescence cross-link from human extracellular matrix-Implication of pentoses in the ageing process. *J. Biol. Chem.* **264,** 21,597–21,602.
15. Dyer, D. G., Blackledge, A., Thorpe, S. R., and Baynes, J. W. (1991) Formation of pentosidine during non-enzymatic browning of proteins by glucose. *J. Biol. Chem.* **268,** 11,654–11,660.
16. Takahashi, M., Ohishi, T., Aoshima, H., Kushida, K, Inoue T., and Horiuchi, K. (1993) Prefractionation with cation exchanger for determination of intermolecular cross-links, pyridinoline and pentosidine, in hydrolysates. *J. Liq. Chrom.* **16,** 1355–1370.
17. Avery, N. C. (1996) The use of solid phase cartridges as a pre-fractionation step in the quantitation of intermolecular collagen cross-links and advanced glycation end-products. *J. Liq. Chrom.* **19,** 1831–1848.
18. Avery, N. C. and Light, N. D. (1985) Re-packing reversed-phase high performance liquid chromatography columns as a means of regenerating column efficiency and prolonging packing life. *J. Chrom.* **328,** 347–352.
19. Dawson, R. M. C., Elliott, D. C., Elliott, W. H., and Jones, K. M. (1986) *Data for Biochemical Research.* 3rd Ed. Clarendon, Oxford.
20. Bailey, A. J., Sims, T. J., Avery, N. C., and Halligan, E. P. (1995) Non-enzymic glycation of fibrous collagen; reaction products of glucose and ribose. *Biochem. J.* **305,** 385–390.

3

Analysis of Laminin Structure and Function with Recombinant Glycoprotein Expressed in Insect Cells

Todd L. Mathus and Peter D. Yurchenco

1. Introduction

Recent developments in the application of eukaryotic recombinant protein techniques have provided new tools with which to dissect and map functional activities in basement membrane glycoproteins. This has been particularly valuable in the case of laminins where the relationship between structure and function has been difficult to establish. Several characteristics of the laminins lie at the heart of the problem. First, the molecules are very large, each assembled from three multidomain subunits joined in a long-coiled coil (**Fig. 1**). Second, laminins bear substantial disulfide and carbohydrate modifications that are crucial for proper conformation. Many, perhaps most, laminin activities present in native laminin are lost upon heat- or chaotropic-denaturation. Native activities are often not retained in short peptides or even in recombinant fragments generated in prokaryotic cells. Furthermore, there is evidence to suggest that synthetic laminin peptides can exhibit "cryptic" activities not found in native protein.

Traditional ways of evaluating proteolytic fragments by biochemical and ultrastructural techniques have provided useful information in the form of a low-resolution functional map (**Fig. 1**). However, these methods are limited by the relatively large size of active fragments (50 – 450 kDa) with a failure of proteases to cleave at most interdomain junctions under conditions that preserve native structure. Whereas prolonged proteolysis of the standard fragments introduces additional cleavages in the peptide backbone, noncovalent interactions frequently keep the peptide moieties together as a single unit. Denaturation then is required to induce dissociation, a method that can kill native

From: *Methods in Molecular Biology, vol. 139: Extracellular Matrix Protocols*
Edited by: C. Streuli and M. Grant

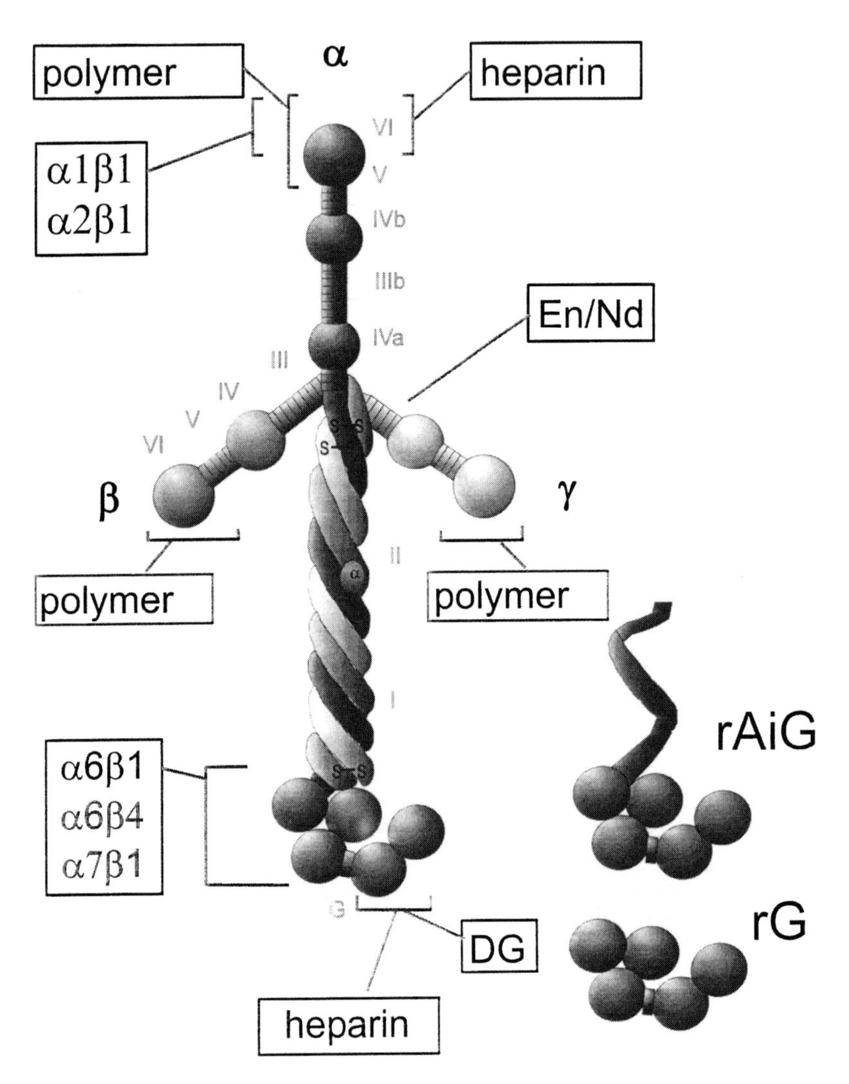

Fig. 1. Structure and functions of laminin-1 and it's proteolytic derivatives. Laminins are heterotrimeric glycoproteins composed of α, β, and γ chains joined in a coiled-coil to form a cruciform-shaped molecule. The carboxyl-terminal G (globular) domain of the α chain extends from the coiled-coil. Roman numerals indicate domains that are found within each subunit chain. Laminin-1 can support integrin-mediated adhesion of cells (α1β1, α2β1, α6β1, α6β4, α7β1), undergo self-polymerization (polymer), and bind entactin/nidogen (En/Nd), α-dystroglycan (αDG), and heparin. Note that several domains have overlapping activities. Recombinant G domain (rG) and recombinant G domain with proximal α chain coiled-coil (rAiG) have been purified and shown to have activities similar to those found in native laminin-1 purified from tissue ***(5,6,8)***.

activity. One example of this is found in fragment E1', a large N-terminal fragment containing the complete α1 and γ1 short arms and a proximal portion of the β1 short arm. The α1-chain domain VI, an N-terminal globule, bears an elastase cleavage site near the domain VI/V junction but remains noncovalently bound to the rest of the fragment. This fragment (E35) can be dissociated in 6 *M* urea; however, the resulting fragment can no longer bind to the α1β1 integrin *(2)*.

We have recently found that eukaryotic recombinant approaches, applying DNA technology to protein chemistry, are effective means to address issues of tertiary structure preservation. In recent years, we have produced functionally active recombinant laminin and laminin fragments using insect and mammalian expression systems *(2,4–8)*. In addition, we have found that laminin fragments generated in prokaryotes, while generally not functionally active, can sometimes be useful for the preparation of specific antigens for antibody generation or characterization (*see* **Note 1**). Here we present selected protocols that we have employed for the generation of laminin G domain recombinants using the baculovirus expression system, as well as highlighting advantages and disadvantages of using this and other systems for the generation of recombinants. We have also attempted to guide the reader through lessons learned through our trials and tribulations.

1.1. Choosing a Recombinant Expression System

Bacterial Expression: Prokaryotic recombinant expression systems have long been used to generate proteins and protein fragments for analysis. However, these systems have more limited useful application in the case of extracellular matrix molecules whose carbohydrate and disulfide-bonding may be critical for structure and function. Our experience with recombinant laminin fragments generated in *Escherichia coli* is that they tend to be associated with inclusion bodies and are insoluble after bacterial lysis. Whereas the recombinant fragment can often be solubilized with urea or guanidine-HCl, the resulting solution tends to be one of inactive and, apparently, incorrectly folded protein. Furthermore, there is a tendency for the recombinant protein to reprecipitate when the protein is dialyzed into urea- or guanidine-free buffer.

Baculovirus Expression: We have had considerable success using the baculovirus expression system in insect cells to generate recombinant C-terminal laminin fragments *(4–6,8)*. However, baculovirus generation of N-terminal fragments, however, has been problematic for us for unclear reasons. The C-terminal fragments have been relatively straightforward to purify with generation of up to several milligrams per liter of conditioned medium. These proteins have been useful to generate function-blocking antibodies that react with native protein with the recombinant protein capable of exhibiting normal biological activity. Post-translational modifications of glycosylation and disulfide isomerization can be found in those proteins that are secreted. However, it should be realized that the N-linked carbohydrate adducts are of a simple mannose-rich type rather than the complex-type found in mammalian glycoproteins *(8)*. Methodological details are described in detail (*see* **Subheading 3.1.**).

Mammalian Expression: The growing number of commercially available vectors with high-efficiency promoters has made it possible to express reasonable amounts of recombinant laminins and other basement membrane glycoproteins (typically ~1 µg/mL). Vectors driven by the CMV (cytomegalovirus) promoter have been used to express nidogen and laminin, both intact and fragments ***(1–3,7)***. Laminin and laminin fragments expressed in this manner contain complex N-linked oligosaccharide and have been found to have native function. In essence, this progress has opened the door for the future of the study of laminin structural-functional relationships by demonstrating the potential for purifying to homogeneity recombinant laminins with point-specific mutations, alterations, or truncations that will perhaps be attributable to disease.

1.2. Strategy for Designing and Generating Laminin Recombinants

It is thought that the domains within a multidomain protein fold largely independently of each other. Because of concern that partial domains or motifs may fail to fold properly and may be more susceptible to degradation, we encourage a consideration of the known or predicted domain structure to guide construct design. Additionally, the choice of expression system used to produce a recombinant should be determined by the purpose for creating the recombinant; that is, why should this recombinant be made and/or what will it be used for? Consider the following when choosing an expression system to make any recombinant: level of expression and yield, time requirements, ease of manipulating the system, costs, and end use of the recombinant. Bear in mind that this is not an all-inclusive list of factors that should be considered. We outline below, detailed protocols for the production of laminin G domain recombinants in insect cells using the baculovirus expression system. This system has been very useful in our laboratory and satisfies the majority of the above criteria.

2. Materials

2.1. Baculovirus Expression System: Generation and Purification of Recombinant Glycoproteins

1. Spodoptera frugiperda cells (Invitrogen, San Diego, CA; B825-01) (*see* **Note 2**).
2. pVL1393 baculovirus expression vector (Invitrogen; V1392-20) (*see* **Note 3**).
3. BaculoGold Transfection Kit (Pharmingen, 21100K) (*see* **Note 4**).

4. Sf900II-serum free medium for insect cell culture (Gibco-BRL, Gaithersburg, MD, 10902-088) (*see* **Note 5**).
5. Fetal bovine serum (Gemini Bio-Products; 100 – 106), heat-inactivated sterile-filtered, stored at –20°C (*see* **Note 6**).
6. Gentamycin sulfate (Gibco-BRL; 15710-015).
7. Amphotericin B (Fungizone, Gibco-BRL; 15295-017).
8. Tissue culture plastic: 25cm^2-, 75cm^2-vented caps, 96-well plates (Corning, NY).
9. Magnetic spinner flasks: 50 mL, 250 mL, 500 mL, and 1 L (Bellco Glass, Inc).
10. Multi-channel magnetic stir plate (Bellco; Multi-Stir 9 model) (*see* **Note 7**).
11. Incubator stringently maintained at 27°C.
12. Protease inhibitor stock solutions: 1 *M* EDTA, pH 8.6 in dd water; 150 m*M* PMSF in isopropanol.
13. Dialysis tubing (Spectrum; 132703).
14. Elution gradient mixer (>200 mL per chamber) with magnetic stir bar apparatus.
15. Conventional heparin-sepharose CL-6B column (Pharmacia, Uppsala, Sweden), bed volume >100 mL, equipped with UV monitor and fraction collector.
16. Conventional DEAE-sepharose CL-6B column (Sigma, St. Louis, MO, DCL-6B-100 CL-6B), bed volume >25 mL, equipped with UV monitor.
17. Aquacide (Calbiochem; 17851) (*see* **Note 8**).
18. Programmable HPLC apparatus: TSK-Gel Heparin-5PW (Toso-Haas; 14444) column, dimensions (7.5 cm × 8 mm inner diameter).
19. Buffers:
    a. 50 m*M* NH_4HCO_3 pH 8.0, 5 m*M* EDTA, 1 m*M* PMSF.
    b. with 1 *M* NaCl.
    c. 150 m*M* Tris HCl pH 7.4, 1 m*M* EDTA, 0.1 m*M* PMSF.
    d. (c) with 1 *M* NaCl.
    e. Tris Buffered Saline (TBS) pH 7.4, 0.02% NaN_3.

3. Methods

3.1. Cloning and Transfection

1. Clone the cDNA encoding G domain downstream of the promoter and appropriate signal sequence using standard molecular cloning protocols (*see* **Table 1**).
2. Seed a 25-cm^2 flask with 1 × 10^6 Sf9 cells in log phase and allow the cells to attach for at least 30 min (*see* **Note 9**).
3. Aspirate SF900II medium and replace with 1.5 mL BaculoGold Transfection Buffer A (supplied in the BaculoGold Transfection Kit) making sure the entire dish surface is covered.
4. In a sterile 5 mL-polycarbonate tube, combine 5 μg recombinant plasmid with 0.5 μg BaculoGold baculovirus DNA. Incubate 15 min at room temperature.
5. Add 1.5 mL BaculoGold Transfection Buffer B to the DNA mixture and mix by inversion. *Add dropwise with agitation* to Sf9 cells in Buffer A (*see* **Note 10**). Allow the transfection to continue for 4 h at 27°C.
6. Aspirate transfection medium and replace with 5 mL fresh SF900II medium.

Table 1
Time Requirements for Baculovirus Expression

Week 1	Design and construction of recombinant in baculoviral vector
Week 2	Sf9 cell culture, transfection, and isolation of P_0
Week 3–4	Recombinant baculovirus cloning and expansion
Week 5	Preparation of clonal recombinant baculovirus stocks
Week 6	Infection for large-scale protein production
Week 7	Protein purification is complete

7. After 5 d collect conditioned medium, and remove cell debris by centrifugation of (2000*g*, 10 min, room temperature). The medium should be labeled P_0 and stored at 4°C (*see* **Note 11**).

3.2. Recombinant Baculovirus Expansion and Clonal Selection

8. Add 100 µL P_0 to a 75-cm^2-flask previously seeded with 1×10^7 adherent cells in 10 mL medium. Incubate as in **step 7** and label the cleared medium P_1 (*see* **Note 12**).
9. Seed several 96-well plates with Sf9 cells at half maximal density (5×10^3 cells/well). The volume of each well should not be <100 mL. Assuming a titer of 10^8 viral particles/mL for P_1, infect the cells with 10, 5, 1, 0.5, and 0.1 viral particles/well on separate plates by dilution of P_1 in medium and adding directly to the plates (*see* **Note 13**). Incubate at 27°C for 3 d.
10. Score the wells on each plate for signs of infection (*see* **Note 12**). Screen for recombinant protein by Western dot blot using half the volume of each well (<50 µL).
11. To obtain a pure virus stocks, choose a clone from a single well on a plate that yielded the fewest number of positive wells scored by both means (*see* **Note 14**).
12. Amplify the recombinant baculovirus clone of choice by repeating **step 8**, using the remainder of the medium in the well as inoculum. Be careful to note increasing passage numbers throughout (*see* **Note 15**). Monitoring of the recombinant protein by Western blotting is recommended after each passage.
13. Prepare a viral stock that will be used for future manipulations and determine the actual viral titer by end point dilution. Typically, viral stocks of 500 mL are sufficient for most purposes, and should be stored at 4°C. Shelf-life of viral stocks should be considerable, with titers being recalculated upon prolonged storage.

3.3. Generation and Purification of Recombinant G Domain

14. Expand Sf9 cells in 3 1 L spinner flasks to a cell density of $1 - 2 \times 10^6$ cells/mL. The culture volume should not exceed 600 mL/spinner flask (*see* **Note 7**).
15. Add the viral stock to each of the flasks such that the ratio of recombinant baculovirus:Sf9 cells is 5:1. Allow the virus to adsorb 1 h, undisturbed, at room temperature.
16. Incubate 72 h at 27°C while stirring at approx 100 rpm.
17. Pellet the cells by centrifugation (875*g*, 15 min, 4°C) (*see* **Note 16**). Transfer the medium to a prechilled flask and keep on ice.

18. Add EDTA and PMSF to a final concentration of 5 m*M* and 3 m*M*, respectively. *The following steps should be carried out in the cold (4°C) using prechilled reagents and equipment.*
19. Immediately dialyze against 20 L of 50 m*M* NH_4HCO_3 pH 8.0, 5 m*M* EDTA, 1 m*M* PMSF with at least two changes of buffer (*see* **Note 17**).
20. Centrifuge the dialyzed medium at 16,500*g* for 30 min (4°C) to remove any cell debris or precipitates that may remain. Load onto a 100-mL conventional Heparin-Sepharose column. The unbound can either be discarded or passed through the column again for optimal recovery of recombinant, so long as the maximum column binding capacity has not been reached. Wash column with an equal volume of the same buffer (*see* **Note 18**).
21. While following A_{280}, elute bound material from the column with a 200-mL linear 0–1 *M* NaCl gradient, collecting 150 drop fractions. If you are not equipped with a conductivity measurement device to follow salt concentration, determine the conductivity of every other fraction of eluant. Pool fractions from 18 mS to the end of the A_{280} peak. Alternatively, fractions can be analyzed by SDS-PAGE and protein staining and pooled based upon the position of the recombinant. For a typical elution profile, *see* **Fig. 2A**.
22. Dialyze pooled fractions versus 150 m*M* Tris HCl pH 7.4, 1 m*M* EDTA, 0.5 m*M* PMSF (*see* **Note 19**).
23. Load onto a conventional 25 mL DEAE-sepharose column equilibrated in the same buffer, collecting the unbound protein (**Fig. 2B**).
24. Centrifuge *unbound* protein at 230,000*g* for 20 min (4°C).
25. Inject onto a Heparin 5PW FPLC column with a flow rate of 1 mL/min and elute bound material with a programmed linear gradient of 0 – 1 *M* NaCl in the same buffer (*see* **Note 20**). For a typical elution profile *see* **Fig. 2C** (*see* **Note 21**). The rG peak elutes at conductivity of approx 32 mS. Pool fractions containing the recombinant, concentrate with Aquacide and dialyze against TBS_{50}, 0.02% NaN_3. Determine recombinant protein concentration, homogeneity, aliquot, snap freeze in liquid nitrogen, and store at –120°C.

4. Notes

1. Antibodies were most useful for immunoblotting rather than other immunological techniques.
2. The lepidopteran cell line used in our laboratory is Sf9, derived from the parental cell line Sf21ovarian cells of *S. frugiperda*. We have compared baculovirus expression in Sf9 and High Five cells (*Trichoplusia ni; Invitrogen*) and found no significant differences for recombinant laminin expression. Sf9 cells grow at 27°C with ambient CO_2 and can be weaned from serum-dependence for growth. Additionally, they can be reversibly maintained in culture as adherent monolayers or suspension cultures. These growth conditions and characteristics are optimal for recombinant protein expression and purification: the absence of serum in growth medium reduces overall protein levels permitting easier purification of the recombinant while still providing carrier protein; the ability to grow in

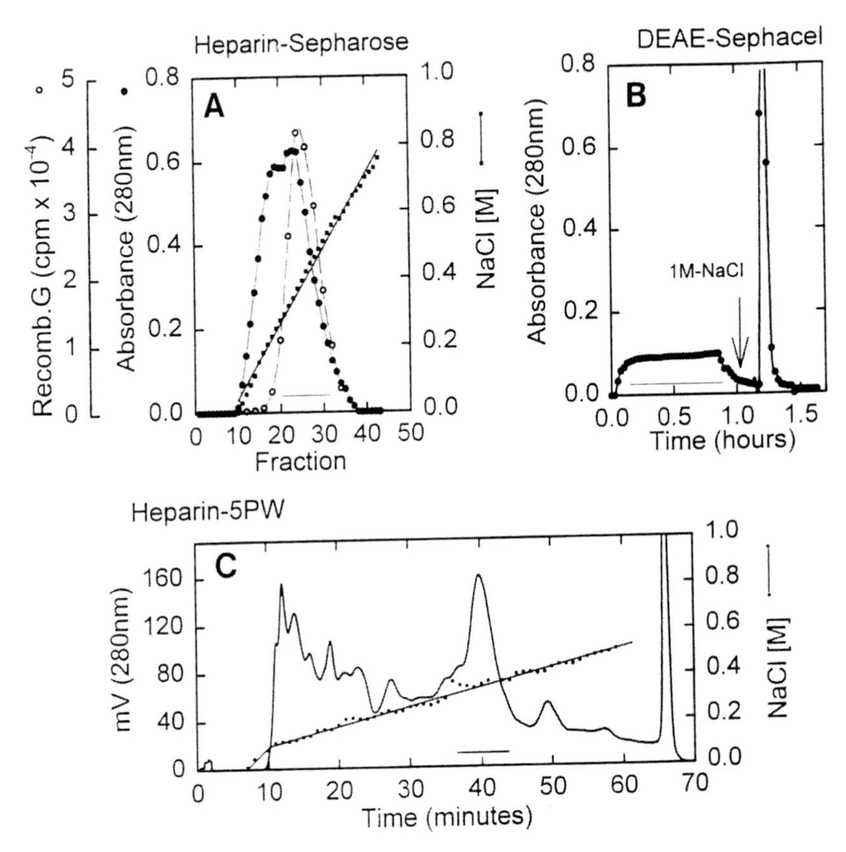

Fig. 2. Purification steps for the purification of recombinant laminin-1 α1 G domain. **(A)** Recombinant G domain (○) elutes within or as a shoulder off the main protein peak (●) of a heparin-sepharose column. Molar concentrations of the NaCl eluant are indicated (■). **(B)** Although recombinant G domain remains in the unbound fraction of the DEAE-Sephacel column, it removes other unwanted proteins. **(C)** Elution profile for recombinant G domain off of a heparin-5PW HPLC column. Pooled fractions 36–44 contain recombinant G domain whose purity is greater than 90%. Bars in each panel represent fractions pooled at each stage of purification.

suspension, as opposed to monolayers, greatly increases cell numbers that will reflect greater levels of recombinant produced. However, growth at 27°C is also optimal for yeast and fungi, so careful aseptic techniques should be used at all times.

3. The pVL1392/pVL1393 series of baculovirus expression vectors has been excellent for production of recombinant laminin domains in Sf9 cells in our laboratory, provided an appropriate signal sequence has been cloned in the vector. The fibronectin signal sequence has successfully used to obtain secretion of a number

of laminin fragments. In recent years, several new baculovirus expression systems have been developed and proven to be equal to those described herein.

4. We use the BaculoGold transfection kit for simplicity and ease of isolation of recombinant clones. We have substituted unsupplemented Grace's medium, pH 6.1 (Gibco-BRL; 11590-056) for transfection Buffer A and sterile-filtered 25 m*M* HEPES, 140 m*M* NaCl, 125 m*M* $CaCl_2$, pH 7.1 for transfection Buffer B and purchased BaculoGold DNA rather than the complete transfection kit. Traditional methods of calcium phosphate precipitation with wild-type AcMNPV virus have also proven useful, but require longer periods of time to isolate recombinant virus that exclude AcMNPV that has not undergone homologous recombination. Transfections using cationic lipids such as Cellfectin (Gibco-BRL; 10362-010) have also been successful for isolation of recombinant baculovirus from infected Sf9 cells.
5. Grace's insect cell culture medium (Gibco-BRL; 11605-094) is an alternative medium for cell culture.
6. Serum components have been known to vary with lot number. As with all cells grown in culture, it is advised to screen several lots of serum for optimal cell growth, transfection efficiency, and recombinant protein production and stability.
7. The major problem encountered with growth of Sf9 cells in suspension is aeration. Insufficient aeration can give rise to decreased cell viability and subsequent release of cytosolic proteinases upon cell lysis. Spinner flasks should be maintained at or near 100 rpm and never be filled above two-thirds of flask volume. Circulation greater than 100 rpm may result in additional cell death by "beating the cells to death". Should cell viability fall below 95% given these culture conditions, decrease the flask volume to 1/2 and/or add surfactants such as Pluronic F-68 (BASF Corp.; 588280).
8. We prefer to use this method for the concentration of proteins in dilute solution to preserve native functions; it is the gentlest means of protein concentration. Traditional methods of such as lyophilization, salting out, or acid/acetone precipitation may irreversibly denature some proteins, especially a concern if conformational stability is required for function. Caution should be exercised so as to not over-concentrate proteins in solution because of the risk of precipitation, and should never be allowed to dry in dialysis tubing.
9. Sf9 cells can be weaned of serum-requirements for growth by twofold reduction of serum content with passaging the cells. A minor lag phase in cell division is usually evident after the first three serial dilutions, but later is overcome through adaptation to reduced serum content in the growth medium. We have found that Sf9 cells grown in suspension are more easily weaned of serum-dependence than those grown as monolayers.
10. DNA precipitation is usually evidenced by turbidity.
11. It had been suggested that recombinant baculovirus is light-sensitive; hence, we typically store viral stocks in the dark or wrapped in aluminum foil to reduce exposure to light. Long-term storage of recombinant baculovirus at 4°C is not recommended because viral titer decreases with prolonged storage. Aliquots of

important viruses should be frozen at –120°C in 20% FCS, 10% DMSO, 70% medium. Viral stocks that become contaminated can be sterile-filtered (0.22 µ*M* pore) and retitered.

12. Characteristic changes in cell morphology are noticeable upon baculovirus infection. When viewed by phase microscopy, the nucleus tends to be larger and more prominent in infected cells, and at later stages of infection occupies most of the intracellular space. It is recommended to compare cell and nuclear morphology to an uninfected Sf9 control to notice sometimes subtle differences. Cell division also ceases after infection, and a comparison of cell number with that of uninfected controls can also be used to as a sign of infection.
13. Alternatively, recombinant baculovirus can be mixed with cells in suspension and plated directly.
14. To insure that your recombinant baculovirus is indeed clonal, repeat dilutional cloning again (preferably twice) using the single positive well.
15. As is observed for the growth of primary explants or transformed cell lines in culture, viruses appear to diverge from the original clone from which they were derived with increasing passage/expansion in culture. To prevent problems that may arise upon prolonged usage, we strongly recommend documenting each passage number. If and when problems begin to become apparent, return to the earliest possible passage to re-establish baculoviral stocks.
16. Care should be taken when decanting the supernatant. A low G-force spin pellets cells without damaging membranes and prevents the release of intracellular proteinases. Therefore, the pellet will not be a compact aggregate and be relatively loose.
17. Equilibration rates are directly proportional to tubing diameter; large-diameter dialysis tubing requires more time than small-diameter tubing to reach equilibrium. In the initial purification steps we use NH_4HCO_3 so that later fractions can be rapidly lyophilized and recombinant followed by SDS-PAGE and Western blotting. Substitution of 50 m*M* Tris pH 8.0, which provides better overall buffering capacity, can be substituted once several large-scale purifications have been successful.
18. Be sure to regenerate any column prior to use with buffer containing 1 *M* NaCl, or as recommended by the manufacturer. This removes any residual protein that may have remained on the column after previous runs and allows optimal binding capacity for recombinant.
19. For Tris buffered reagents, pH is *inversely* proportional to temperature. As a general rule, for each °C the pH changes approx 0.03 pH units. For example, for preparation of a Tris buffer at room temperature (25°C) that will ultimately be used at in the cold (4°C), the difference in temperature is 21°C: 21 × 0.03 pH units = 0.63 pH units that the buffer will increase upon decreasing temperature. So, if pH 7.4 is required at 4°C, at room temperature the pH should be adjusted to 6.77.
20. Buffers used for HPLC should always be filtered and de-gassed to prevent column clogging and the formation of air bubbles in the column matrix, leading to an increase in the lifetime of the column.

21. The elution profile shown is for a specific recombinant and may not represent elution positions of other laminin G domain recombinants. For recombinants whose elution position is unknown, collected fractions should be analyzed by SDS-PAGE to determine relative heparin affinities and fractions then pooled.

References

1. Colognato, H., MacCarrick, M., O'Rear, J. J., and Yurchenco, P. D. (1997) The laminin alpha2-chain short arm mediates cell adhesion through both the alpha1beta1 and alpha2beta1 *Integrins. J. Biol. Chem.* **272,** 29,330–29,336.
2. Colognato-Pyke, H., O'Rear, J. J., Yamada, Y., Carbonetto, S., Cheng, Y. S., and Yurchenco. P. D. (1995) Mapping of network-forming, heparin-binding, and alpha 1 beta 1 integrin-recognition sites within the alpha-chain short arm of laminin- 1. *J. Biol. Chem.* **270,** 9398–9406.
3. Fox, J. W., Mayer, U., Nischt, R., Aumailley, M., Reinhardt, D., Wiedemann, H. et al. (1991) Recombinant nidogen consists of three globular domains and mediates binding of laminin to collagen type IV. *EMBO. J.* **10,** 3137–3146.
4. Rambukkana, A., Salzer, J. L., Yurchenco, P. D., and Tuomanen, E. I. (1997) Neural targeting of Mycobacterium leprae mediated by the G domain of the laminin-alpha2 chain. *Cell* **88,** 811–821.
5. Sung, U., O'Rear, J. J., and Yurchenco, P. D. (1993) Cell and heparin binding in the distal long arm of laminin: identification of active and cryptic sites with recombinant and hybrid glycoprotein [published erratum appears in *J. Cell Biol.* (1993) **Dec;123**(6 Pt 1):1623]. *J. Cell. Biol.* **123,** 1255–1268.
6. Sung, U., O'Rear, J. J., and Yurchenco, P. D., (1997) Localization of heparin binding activity in recombinant laminin G domain. *Eur. J. Biochem.* **250,** 138–43.
7. Yurchenco, P. D., Quan, Y., Colognato, H., Mathus, T., Harrison, D., Yamada, Y., and O'Rear, J. J. (1997) The alpha chain of laminin-1 is independently secreted and drives secretion of its beta- and gamma-chain partners. *Proc. Natl. Acad. Sci. USA* **94,** 10189–10194.
8. Yurchenco, P. D., Sung, U., Ward, M. D., Yamada, Y., and O'Rear, J. J. (1993) Recombinant laminin G domain mediates myoblast adhesion and heparin binding. *J. Biol. Chem.* **268,** 8356–8365.

4

Recombinant Collagen Trimers from Insect Cells and Yeast

Johanna Myllyharju

1. Introduction

At least 19 proteins have now been defined as collagens *(1,2)*, but many of those recently discovered are present in tissues in such small amounts that their isolation for characterization at the protein level has so far been impossible. Some of the fibril-forming collagens are now in medical use, in applications ranging from the use of type I collagen as a biomaterial and a delivery system for several drugs to trials for the potential of type II collagen as an oral tolerance-inducing agent for the treatment of rheumatoid arthritis *(3,4)*. An efficient recombinant expression system for collagens can thus be expected to have numerous scientific and medical applications. The systems commonly used for expressing other proteins in lower organisms are not suitable as such for the production of recombinant collagens, however, as bacteria and yeast have no prolyl 4-hydroxylase activity and insect cells have insufficient levels of it. Prolyl 4-hydroxylase, an $\alpha_2\beta_2$ tetramer in vertebrates, plays a central role in the synthesis of all collagens, as 4-hydroxyproline-deficient collagen polypeptide chains cannot form triple helices that are stable at 37°C *(5,6)*. All attempts to assemble an active prolyl 4-hydroxylase tetramer from its subunits in vitro have been unsuccessful, but active recombinant human prolyl 4-hydroxylase has been produced in insect cells and yeast by coexpression of its α- and β-subunits *(7,8)*. We have recently shown that recombinant human type III collagen with a stable triple helix can be efficiently produced in insect and yeast cells by simultaneous coexpression with the recombinant human prolyl 4-hydroxylase *(8,9)*. This chapter describes detailed procedures for the production of stable recombinant human type III collagen in insect cells and in the yeast *Pichia*

From: *Methods in Molecular Biology, vol. 139: Extracellular Matrix Protocols*
Edited by: C. Streuli and M. Grant © Humana Press Inc., Totowa, NJ

pastoris. These methods can be applied to the expression of other recombinant collagen types, both homotrimeric and heterotrimeric ***(10,11)***.

2. Materials

2.1. Materials for Expression of Recombinant Human Type III Procollagen in Insect Cells

1. pVL1392 and p2Bac baculovirus expression vectors (Invitrogen, San Diego, CA).
2. Full-length cDNAs for the α- ***(12)*** and β- ***(13)*** subunits of human prolyl 4-hydroxylase and the proα1 chain of human type III procollagen [proα1(III)] ***(14)***.
3. Wizard Plus Maxiprep DNA purification system (Promega, Madison, WI).
4. Sf9 and High Five (H5) insect cells (Invitrogen).
5. TNM-FH medium (Sigma, St. Louis, MO) supplemented with 10% fetal bovine serum (FBS).
6. Six-well plates and 60- and 100-mm tissue-culture Petri dishes.
7. BaculoGold DNA and Transfection Buffers A: Grace's Medium with 10% FBS and B: 25 m*M* HEPES, pH 7.1, 125 m*M* $CaCl_2$, 140 m*M* NaCl (Pharmingen, San Diego, CA).
8. 3% seaplaque low-melting-point agarose (FMC, Rockland, ME) in H_2O (autoclaved).
9. L-ascorbic acid phosphate (Wako, Osaka, Japan).
10. PBS: 0.15 *M* NaCl, 20 m*M* phosphate, pH 7.4.
11. Homogenization buffer (300 m*M* NaCl, 0.2% Triton X–100, and 70 m*M* Tris, pH 7.4).
12. SDS-polyacrylamide gel electrophoresis (SDS-PAGE) apparatus.
13. PIIINP radioimmunoassay (Farmos Diagnostica, Turku, Finland).
14. Pepsin (Boehringer Mannheim, Mannheim, Germany), trypsin type XIII (Sigma), chymotrypsin type VII (Sigma), trypsin inhibitor type II-S (Sigma).

2.2. Materials for Expression of Recombinant Human Type III Procollagen in the Yeast Pichia pastoris

1. pAO815, pPIC9, and pPICZ B *Pichia pastoris* expression vectors (Invitrogen). pYM25 containing the *Saccharomyces cerevisiae ARG4* selection marker (obtained from Dr. James Cregg, Keck Graduate Institute of Applied Life Sciences, Claremont, CA.
2. *his4, arg4 Pichia pastoris* host strain (obtained from Dr. James Cregg).
3. Yeast extract peptone dextrose medium (YPD: 1% yeast extract, 2% peptone, 2% dextrose), minimal dextrose medium (MD: 1.34% yeast nitrogen base with ammonium sulfate and without amino acids [YNB], 4×10^{-5}% biotin, 1% dextrose), buffered glycerol-complex medium (BMGY: 0.1% yeast extract, 0.2% peptone, 100 m*M* potassium phosphate, pH 6.0, 1.34% YNB, 4×10^{-5}% biotin, 1% glycerol), buffered minimal methanol medium (BMM: 100 m*M* potassium phosphate, pH 6.0, 1.34% YNB, 4×10^{-5}% biotin, 0.5% methanol). Recipes for

various *Pichia* media can be found in Version 3.0 of the Invitrogen *Pichia* Expression Kit Manual ***(15)***.

4. Zeocin (Invitrogen).
5. Triple-baffled Erlenmeyer flasks and silicone sponge closures (Sigma).
6. Electroporation device and cuvets.
7. Acid-washed glass beads (Sigma).
8. 1 *M* sorbitol.
9. 5% glycerol, 50 m*M* sodium phosphate buffer, pH 7.4.

3. Methods

3.1. Expression of Recombinant Human Type III Procollagen in Insect Cells

3.1.1. Generation of Recombinant Baculovirus Expression Vectors

1. Generate a double promoter baculovirus expression vector for recombinant human prolyl 4-hydroxylase (p2Bac4PHαβ) by cloning the cDNAs for its α- and β-subunits into the *Not*I site downstream of the p10 promoter and the *Bam*HI site downstream of the polyhedrin promoter of p2Bac, respectively. Generate a *Not*I site in the α-subunit cDNA 46 bp upstream of the translation initiation (ATG) codon by PCR (*see* **Note 1**).
2. Create a *Bgl*II site 16 bp upstream of the translation initiation codon to the full-length cDNA for the proα1 chain of human type III procollagen (*see* **Note 1**). Digest the cDNA with *Bgl*II and *Xba*I and ligate the insert to *Bgl*II-*Xba*I-digested pVL1392, generating pVLrhproCIII.
3. Purify the recombinant expression vectors p2Bac4PHαβ and pVLrhproCIII using the Wizard Plus Maxiprep DNA purification system and sterilize them with a 0.22-μm syringe filter.

3.1.2. Generation of Recombinant Baculoviruses

1. Generate the recombinant baculoviruses 4PHαβ and rhproCIII by cotransfection with BaculoGold baculovirus DNA (*see* **Note 2**). Seed 2×10^6 Sf9 cells on 60- mm tissue-culture Petri dishes. Allow cells to attach for at least 30 min. Remove the culture medium from the plates and add 1 mL of Transfection Buffer A. Mix BaculoGold DNA, 0.5 μg, with 5 μg of the baculovirus expression vector DNA and incubate for 5 min at room temperature. Mix 1 mL of Transfection Buffer B well with the DNA mixture, and add dropwise to the 60-mm insect cell plate. Gently rock the plate after the addition of every 3–5 drops of the transfection solution.
2. Incubate the plates at 27°C for 4 h, remove the transfection solutions, and add 4 mL of TNM-FH medium supplemented with 10% FBS.
3. Incubate the plates at 27°C for 4 d. Collect the medium containing the recombinant virus and amplify once by infecting 6×10^6 Sf9 cells in a 100-mm Petri dish with 500 μL of the collected transfection medium. Incubate the plates at 27°C for 3 d before harvesting the amplification medium.

4. Plaque-purify the amplified recombinant viruses (*see* **Note 3**). Seed 2×10^6 Sf9 cells on 60-mm Petri dishes and allow to attach for at least 30 min. Prepare 3 mL serial dilutions (10^{-4}, 10^{-5}, 10^{-6}, 10^{-7}) of the recombinant viruses in TNM-FH supplemented with 10% FBS. Remove the medium from the 60-mm plates and add 1 mL of the virus dilutions on duplicate plates. Incubate the plates at 27°C for 1 h. Melt 3% SeaPlaque agarose in a microwave oven, cool to 37°C, and add two volumes of the culture medium preheated to 37°C to obtain a 1% agarose overlay. Remove the virus dilutions and add 4 mL of the agarose overlay. Incubate the plates in a humid environment at 27°C for 6 d. Remove agar plugs containing single clear plaques with a sterile Pasteur pipet (*see* **Note 4**) into 1 mL of the culture medium and elute the virus particles by incubation at 4°C overnight.
5. Amplify the plaque-purified viruses three times for 3 d at 27°C (amplifications I, II, and III) to obtain high-titer virus stocks (*see* **Note 5**). After the first amplification, perform test infections to screen the viruses for the production of the recombinant proteins of interest (*see* **Subheading 3.1.3.**). AI (amplification I): seed 1×10^6 Sf9 cells on the wells of a 6-well plate and use 200 µL of the eluted plaque-purified virus for infection, AII: seed 6×10^6 Sf9 cells on a 100-mm Petri dish and infect with 100 µL of the AI virus, and AIII: seed 6×10^6 Sf9 cells on a 100-mm Petri dish and infect with 5–10 µL of the AII virus. The titer of the AIII virus should be about 10^8 PFU (plaque forming units/mL).

3.1.3. Expression of Recombinant Human Type III Procollagen in Insect Cells

1. Seed 6×10^6 H5 cells (*see* **Note 6**) on a 100-mm Petri dish and infect with the rhproCIII and 4PHαβ viruses. Use the former in a 5 to 10-fold excess over the latter. Incubate plates at 27°C and add L-ascorbic acid phosphate (80 µg/mL) to the culture medium daily.
2. Harvest cells (*see* **Note 7**) 72 h after infection, wash with PBS and homogenize in the homogenization buffer (500 µL/6×10^6 cells). Centrifuge the cell homogenates at 10,000*g* for 20 min and collect the soluble fractions. Analyze samples with SDS-PAGE under reducing conditions, followed by staining with Coomassie brilliant blue or Western blotting using polyclonal antibodies to the α- and β-subunits of human prolyl 4-hydroxylase ***(16)***, the N-propeptide of human type III procollagen (Farmos Diagnostica, Turku, Finland) or a monoclonal antibody 95D1A recognizing the collagenous region of various collagen chains (A. Snellman and T. Pihlajaniemi, unpublished observations). Estimate the amount of recombinant human type III procollagen by means of a PIIINP radioimmunoassay for the trimeric N-propeptide of human type III procollagen, and assay the amount of prolyl 4-hydroxylase activity by a method based on the hydroxylation-coupled decarboxylation of 2-oxo[1-^{14}C]glutarate ***(17)***.
3. The triple-helical conformation of the recombinant human type III procollagen can be examined by pepsin digestion ***(18)***. Lower the pH of the samples to 2.5, add pepsin (1.5 mg/mL in 10 m*M* acetic acid) to a final concentration of 0.2 mg/mL, and digest the samples for 1–4 h at 22°C. Inactive pepsin by adjusting the pH

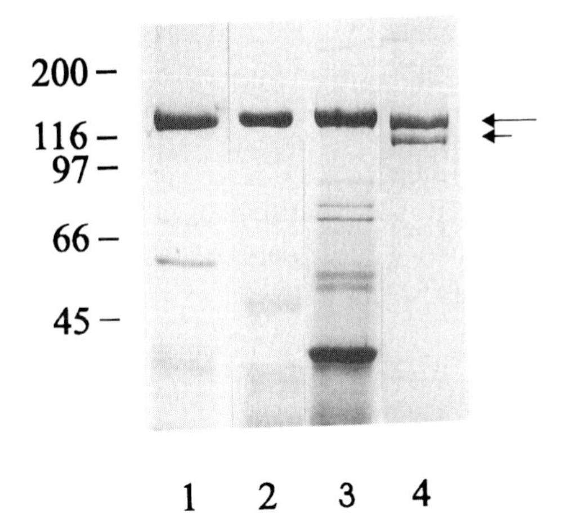

Fig. 1. SDS-PAGE analysis of pepsin digested recombinant human collagens expressed in H5 insect cells. Long arrow indicates the α1 chains of type III (lane 1) and type II collagens (lane 2), the type I collagen homotrimer (lane 3), and the type I collagen heterotrimer (lane 4). Short arrow indicates the α2 chain of type I collagen heterotrimer.

back to 7.4 and analyze the samples by SDS-PAGE (**Fig. 1**). The thermal stability of the pepsin digested recombinant human type III collagen can be studied by trypsin-chymotrypsin digestion ***(19)***. Preheat samples at the selected temperatures for 5 min and digest for 2 min with a mixture of 100 μg/mL trypsin and 250 μg/mL chymotrypsin. Terminate the digestion by adding trypsin inhibitor to a final concentration of 0.5 mg/mL and analyze the samples by SDS-PAGE.

4. Expression of recombinant human type III procollagen in H5 cells can be scaled up by using suspension cultures in shaker flasks or bioreactors. The largest amount of recombinant type III procollagen we have obtained from H5 cells cultured in shaker flasks has been about 60 mg/L (*see* **Note 8**). The recombinant type III collagen produced in H5 cells was found to be very similar in its 4-hydroxyproline content and T_m to type III collagen extracted from various tissues, whereas its hydroxylysine content was found to be about 60% of that of the nonrecombinant type III collagen ***(9)***.

3.2. Expression of Recombinant Human Type III Procollagen in Pichia pastoris

3.2.1. Generation of Recombinant Pichia Expression Vectors

1. Generate a modified *Pichia* expression vector pARG815 by replacing the *HIS4* selection marker in pAO815 with the *S. cerevisiae ARG4* selection marker. Digest pYM25 containing the *ARG4* gene with *Hpa*I, and ligate the *ARG4 Hpa*I fragment into *Eco*RV-digested pAO815.

2. Synthesize the 5' end of the human prolyl 4-hydroxylase α-subunit cDNA, extending from the translation initiation codon to an internal *Hind*III site, with *Hind*III and *Sma*I sites flanking the initiation codon, and the 3' end, extending from an internal *Pst*I site to the translation stop codon, with *Sma*I and *Bam*HI sites following the stop codon, by PCR. Use these PCR fragments to replace the 5' and 3' ends of the original α-subunit cDNA to generate a cDNA without any 5' and 3' untranslated regions. Ligate the *Sma*I-*Sma*I α-subunit insert into the *Eco*RI site of pARG815, generating pARG815α.
3. Replace the signal sequence of the human prolyl 4-hydroxylase β-subunit with the *S. cerevisiae* α mating factor (αMF) pre-pro sequence (*see* **Note 9**). Synthesize human prolyl 4-hydroxylase β-subunit cDNA, extending from the codon for the first amino acid after the signal peptide cleavage site to the stop codon and flanked by *Eco*RI sites, by PCR and ligate into the *Eco*RI site following the (αMF) pre-pro sequence of pPIC9, generating pPIC9β.
4. Replace the 3' end of the proα1(III) cDNA used to generate recombinant baculovirus pVLrhproCIII (*see* **step 2** in **Subheading 3.1.1.**) by a PCR fragment extending from an internal *Eco*RI site to the translation stop codon, with a *Xba*I site following the stop codon. Ligate the *Bgl*II-*Xba*I proα1(III) cDNA into the *Eco*RI-*Xba*I site of pPICZ B, generating pPICZ Bproα1(III).

3.2.2. Generation of the Recombinant Pichia Strain and Expression of the Recombinant Human Type III Procollagen

1. To obtain a recombinant *P. pastoris* strain expressing human prolyl 4-hydroxylase, linearize pARG815α and pPIC9β with *Dra*III and *Stu*I, respectively, and cotransform into the *his4, arg4 P. pastoris* host strain by the electroporation method ***(15)***. Culture 5 mL of the *his4, arg4 P. pastoris* host strain in YPD in a 50-mL conical tube at 30°C overnight. Inoculate 500 mL of YPD with 0.1–0.5 mL of the overnight culture and grow to an OD_{600} of 1.3–1.5 (*see* **Note 10**). Centrifuge the cells at 4°C at 1 500*g* for 5 min and wash twice, with 500 mL and 250 mL of ice-cold sterile water. Resuspend in 20 mL of ice-cold 1 *M* sorbitol, centrifuge, and resuspend in ice-cold 1 *M* sorbitol to obtain a final volume of 1.5 mL. Mix 3 μL (approx 0.6 μg) of the linearized pARG815α and pPIC9β with 40 μL of the *P. pastoris* cells, transfer to an ice-cold electroporation cuvet and incubate on ice for 5 min. Pulse the cells using the parameters 1500 kV, 25 μF, and 400 Ω (*see* **Note 11**), and add 1 mL of ice-cold 1 *M* sorbitol immediately afterwards. Spread aliquots of 50–200 μL of the cells on MD plates and incubate at 30°C until colonies appear. Repeat selection for His^+, Arg^+ colonies twice by streaking single colonies on MD plates.
2. Culture the recombinant His^+, Arg^+ colonies obtained above at 30°C in 25 mL of BMGY in 250 mL shaker flasks to an OD_{600} of 5–10. Centrifuge the cells at 1 500*g* for 5 min and resuspend to an OD_{600} of 1–3 in BMM to induce expression. Add methanol every 24 h to a final concentration of 0.5%. Harvest cells after a 60-h methanol induction, wash once and resuspend to an OD_{600} of 250–350 in cold 5% glycerol in 50 m*M* sodium phosphate buffer, pH 7.4. Add an equal

volume of glass beads and vortex the samples 8 times for 30 s at 4°C with 30 s intervals. Centrifuge at 15,000*g* for 30 min at 4°C, collect the supernatants and screen for expression of the prolyl 4-hydroxylase α- and β-subunits and for the amount of prolyl 4-hydroxylase activity as in **step 2** of **Subheading 3.1.3.** Select a recombinant prolyl 4-hydroxylase expressing strain and term it α/βα-MF.

3. Generate a recombinant strain coexpressing human prolyl 4-hydroxylase and proα1(III) chains by transforming *Pme*I-linearized pPICZ Bproα1(III) into the α/βα-MF strain by electroporation as in **step 1**. Select the transformants in YPD (+ 100 µg/mL zeocin) (*see* **Note 12**). Culture the cells, induce and harvest as in **step 2**, and assay expression of human prolyl 4-hydroxylase and type III procollagen in the soluble fraction of cell lysates as in **step 2** of **Subheading 3.1.3.** (*see* **Note 13**). Select a recombinnt human type III procollagen producing strain based on these assays and term it α/βα-MF/proα1(III).
4. For the production of recombinant human type III procollagen, culture the α/βα-MF/proα1(III) *Pichia* strain in shaker flasks and induce with methanol. Harvest cells 60 h after methanol induction and brake with glass beads as in **step 2**. The purified recombinant type III collagen produced in shaker flasks in our experiments was essentially identical in its amino acid composition to nonrecombinant human type III collagen, except that the degree of 4-hydroxylation of the proline residues was 44.2% whereas the corresponding value for nonrecombinant type III collagen is 51.6% *(8)*. However, fully hydroxylated recombinant human type III procollagen can be produced in the α/βα-MF/proα1(III) *Pichia* strain in bioreactors under optimal oxygenation conditions (unpublished results). The *Pichia*-derived recombinant human type III collagen contains no hydroxylysine, whereas nonrecombinant type III collagen has five hydroxylysine residues per 1000 amino acids *(8)*.

4. Notes

1. In the p2Bac and pVL1392 vectors, the ATG translation initiation codons of polyhedrin and p10 have been altered to ATT, which means that the inserts must provide their own ATG initiation codons. The 5' cloning site in the inserts should be as close to the translation initiation codon as possible (less than 100 bp) *(20)*. Because translation starts from the first ATG codon downstream of the polyhedrin and p10 promoters, no additional ATG codons should exist upstream of the initiation codon of the gene of interest. For example, the polyhedrin multicloning site in the p2Bac vector has a *Nco*I recognition sequence (CCATGG), and, therefore, none of the downstream cloning sites can be used. The length or sequence of the 3' untranslated region of an insert usually has no effect on the expression levels. The p2Bac and pVL1392 vectors provide the polyhedrin or SV40 polyadenylation signals, and therefore the genes of interest do not have to include any polyadenylation signals.
2. BaculoGold DNA is a modified baculovirus DNA which contains a lethal deletion *(20)*, and viable virus particles are obtained only by cotransfection with a complementing baculovirus expression vector. When BaculoGold DNA is used,

the expression vectors must therefore include at least 1.7 kbp of baculovirus DNA downstream of the polyhedrin stop codon to counteract the lethal deletion. The recombination frequency with BaculoGold DNA is at least 99%, which is a major improvement over the 0.1% recombination frequency obtained with the wild-type baculovirus DNA.

3. In spite of the high recombination frequency with the BaculoGold DNA and the fact that wild-type baculovirus DNA expressing polyhedrin cannot be formed during recombination, it is still advisable to plaque-purify the recombinant virus stocks, as aberrant crossing over during transfection can lead to total or partial deletions of the recombinant gene of interest. The resultant virus stock after plaque-purification is derived from a single virus clone.
4. If plaques are not clearly visible, the plaque assay plates can be stained with Neutral Red or MTT (thiazolyl blue), for example ***(21)***.
5. The MOI (multiplicity of infection = PFU/cell number) should be below 1 during all amplifications, as values greater than 1 can lead to deletions in the recombinant virus particles ***(20)***. The PFU/mL of a virus stock can be obtained from a plaque assay by the formula: PFU/mL = 1/virus dilution (e.g., 10^{-5}) × number of plaques × 1/mL virus dilution added to the plate.
6. H5 cells consistently give 3- to 10-fold higher expression levels than Sf9 cells for recombinant human type III procollagen ***(9)***.
7. Most of the recombinant fibril-forming human type I, II, or III procollagens produced (70–90%) is retained within the insect cells ***(9–11)***.
8. The highest expression levels obtained for type I and II collagens in suspension culture have been 20–40 mg/L and 50 mg/L, respectively ***(10,11)***.
9. The signal sequence of the prolyl 4-hydroxylase β subunit was found to play a major role in enzyme assembly in *Pichia* ***(8)***. The authentic human signal sequence was ineffective for transport of the β-subunit into the lumen of the endoplasmic reticulum, as only trace amounts of an active $\alpha_2\beta_2$ enzyme tetramer were produced. The highest tetramer assembly level was obtained with the *S. cerevisiae* α mating factor pre-pro sequence, whereas that obtained with the *P. pastoris* acid phosphatase 1 signal sequence was about 40% of this value.
10. The doubling time of the *his4, arg4 P. pastoris* strain is approximately 2 h in YPD.
11. *P. pastoris* cells can be pulsed using the parameters suggested for *S. cerevisiae* by the manufacturer of the electroporation device.
12. Zeocin is active only at a low salt concentration (<90 m*M*) and at pH 7.5. Therefore, screening for His^+, Arg^+ phenotype and zeocin resistance must take place in parallel plates, as the pH of MD plates is not suitable for zeocin selection.
13. As in the case of insect cells (*see* **Note 7**), the vast majority of the recombinant human type III procollagen in the *Pichia* expression system, too, is retained within the cells and can thus be obtained from the soluble fraction of the cell lysates.

References

1. Prockop, D. J. and Kivirikko, K. I. (1995) Collagens: Molecular biology, diseases, and potentials for therapy. *Annu. Rev. Biochem.* **64,** 403–434.

2. Kielty, C. M., Hopkinson, I., and Grant, M. E. (1993) The collagen family: structure, assembly and organization in the extracellular matrix, in *Connective Tissue and its Heritable Disorders. Molecular, Genetic and Medical Aspects* (Royce, P. M. and Steinmann, B., eds.), Wiley-Liss, New York, pp. 103–147.
3. Ramshaw, J. A. M., Werkmeister, J. A., and Glattauer, V. (1996) Collagen-based biomaterials. *Biotechnol. Genet. Eng. Rev.* **13,** 335–382.
4. Barnett, M. L., Kremer, J. M., St. Clair, E. W., Clegg, D. O., Furst, D., Weisman, M., et al. (1998) Treatment of rheumatoid arthritis with oral type II collagen. Results of a multicenter, double-blind, placebo-controlled trial. *Arthr. Rheum.* **41,** 290–297.
5. Kivirikko, K. I. and Pihlajaniemi, T. (1998) Collagen hydroxylases and the protein disulfide isomerase subunit of prolyl 4-hydroxylases. *Adv. Enzymol. Related Areas Mol. Biol.* **72,** 325–398.
6. Kivirikko, K. I. and Myllyharju, J. (1998) Prolyl 4-hydroxylases and their protein disulfide isomerase subunit. *Matrix Biol.* **16,** 357–368.
7. Vuori, K., Pihlajaniemi, T., Marttila, M., and Kivirikko, K. I. (1992) Characterization of the human prolyl 4-hydroxylase tetramer and its multifunctional protein disulfide-isomerase subunit synthesized in a baculovirus expression system. *Proc. Natl. Acad. Sci. USA*, **89,** 7467–7470.
8. Vuorela, A., Myllyharju, J., Nissi, R., Pihlajaniemi, T., and Kivirikko, K. I. (1997) Assembly of human prolyl 4-hydroxylase and type III collagen in the yeast *Pichia pastoris*: formation of a stable enzyme tetramer requires coexpression with collagen and assembly of a stable collagen requires coexpression with prolyl 4-hydroxylase. *EMBO J.* **16,** 6702–6712.
9. Lamberg, A., Helaakoski, T., Myllyharju, J., Peltonen, S., Notbohm, H., Pihlajaniemi, T., and Kivirikko, K. I. (1996) Characterization of human type III collagen expressed in a baculovirus system. Production of a protein with a stable triple helix requires coexpression with the two types of recombinant prolyl 4-hydroxylase subunit. *J. Biol. Chem.* **271,** 11988–11995.
10. Myllyharju, J., Lamberg, A., Notbohm, H., Fietzek, P. P., Pihlajaniemi, T., and Kivirikko, K. I. (1997) Expression of wild-type and modified proα chains of human type I procollagen in insect cells leads to the formation of stable $[\alpha 1(I)]_2\alpha 2(I)$ collagen heterotrimers and $[\alpha 1(I)]_3$ homotrimers but not $[\alpha 2(I)]_3$ homotrimers. *J. Biol. Chem.* **272,** 21824–21830.
11. Nokelainen, M., Helaakoski, T., Myllyharju, J., Notbohm, H., Pihlajaniemi, T., Fietzek, P. P., and Kivirikko, K. I. (1998) Expression and characterization of recombinant human type II collagens with low and high contents of hydroxylysine and its glycosylated forms. *Matrix Biol.* **16,** 329–338.
12. Helaakoski, T., Vuori, K., Myllylä, R., Kivirikko, K. I., and Pihlajaniemi, T. (1989) Molecular cloning of the α-subunit of human prolyl 4-hydroxylase: The complete cDNA-derived amino acid sequence and evidence for alternative splicing of RNA transcripts. *Proc. Natl. Acad. Sci. USA* **86,** 4392–4396.
13. Pihlajaniemi, T., Helaakoski, T., Tasanen, K., Myllylä, R., Huhtala, M.-L., Koivu, J., and Kivirikko, K. I. (1987) Molecular cloning of the β-subunit of human prolyl 4-hydroxylase. This subunit and protein disulphide isomerase are products of the same gene. *EMBO J.* **6,** 643–649.

14. Tromp, G., Kuivaniemi, H., Shikata, H., and Prockop, D. J. (1989) A single base mutation that substitutes serine for glycine 790 of the α1(III) chains of type III procollagen exposes an arginine and causes Ehlers-Danlos syndrome IV. *J. Biol. Chem.* **264,** 1349–1352.
15. *A Manual of Methods for Expression of Recombinant Proteins in Pichia pastoris. Version 3. 0.* Invitrogen, San Diego, CA.
16. Veijola, J., Pihlajaniemi, T., and Kivirikko, K. I. (1996) Co-expression of the α subunit of human prolyl 4-hydroxylase with BiP polypeptide in insect cells leads to the formation of soluble and insoluble complexes. *Biochem. J.* **317,** 613–618.
17. Kivirikko, K. I. and Myllylä, R. (1982) Posttranslational enzymes in the biosynthesis of collagens: Intracellular enzymes. *Meth. Enzymol.* **82,** 245–304.
18. Bruckner, P. and Prockop, D. J. (1981) Proteolytic enzymes as probes for the triple-helical conformation of procollagen. *Anal. Biochem.* **110,** 360–368.
19. Sieron, A. L., Fertala, A., Ala-Kokko, L., and Prockop, D. J. (1993) Deletion of a large domain in recombinant human procollagen II does not alter the thermal stability of the triple helix. *J. Biol. Chem.* **268,** 21,232–21,237.
20. Crossen, R. and Gruenwald, S. (1993) *Baculovirus Expression Vector System 8 Manual.* Pharmingen, San Diego, CA.
21. *Guide to Baculovirus Expression Vector Systems (BEVS) and Insect Cell Culture Techniques. Instruction Manual.* Gibco BRL, Life Technologies.

5

Eukaryotic Expression and Purification of Recombinant Extracellular Matrix Proteins Carrying the Strep II Tag

Neil Smyth, Uwe Odenthal, Barbara Merkl, and Mats Paulsson

1. Introduction

For recombinant expression of extracellular matrix (ECM) proteins or their individual domains, the use of transformed mammalian cells offers two major advantages. First, eukaryotic expression can be expected under optimum conditions to produce a large proportion of correctly folded molecules. ECM proteins are made from a group of 25 structurally known *(1)* and about 200 cDNA derived domains many of which regularly reappear in the different proteins. These have often a complex secondary structure, maintained by multiple disulfide bonds. Whereas by denaturing and then carefully renaturing, an approximation to the native structure may be obtained using prokaryotic expression systems, the best that may be expected is that a small percentage of the protein folds into such a conformation. Second, most ECM proteins are at least to some extent glycosylated and often heavily so, and the use of the mammalian system offers the best approximation to the sugar structures present in the native form of the molecule.

While the use of the mammalian systems has the above advantages, it also has major drawbacks. They are relatively slow to set up as opposed to those using yeast, bacteria, or baculovirus/insect cells. It takes up 2 wk to obtain 1 L of culture medium from mammalian cells, as opposed to 8 to 12 h from bacteria. Tissue culture over such periods is labor intensive and expensive. Finally, the level of the secreted protein in the media is variable (50 μg–3 mg per L of medium) and where no antibody exists to the protein or domain, this can lead to problems in its identification and in following its progress during subsequent purification steps. A tag fused either on the N' or C' terminus can act as

From: *Methods in Molecular Biology, vol. 139: Extracellular Matrix Protocols*
Edited by: C. Streuli and M. Grant © Humana Press Inc., Totowa, NJ

an aid in the immunological identification of the expressed protein. It may also be used in a variety of affinity chromotography methods leading to its purification.

We use a modified version of the system described previously for expression of a variety of matrix proteins in human embryonic kidney 293 cells *(2–4)*. The human cytomegalovirus immediate early gene enhancer-promoter is used to drive the expression of the recombinant protein. This gives potentially high expression of the product in primate cells, however, levels in rodent lines may be minimal. The BM40 (SPARC/osteonectin) signal peptide is placed downstream of the promoter and fused in frame to the protein with the tag. The plasmid contains the cassette for the puromycin N-acetyltransferase gene allowing selection in transfected eukaryotic cells and also an ampicillin resistance gene for use in selection and plasmid manipulation in *Escherichia coli*. Further, these plasmids also contain the Epstein Barr virus (EBV) origin of replication. This is sensitive to the action of the EBV nuclear antigen, which is expressed in 293-EBNA cells leading to the maintenance of the plasmid extrachromosomally, with it replicating independently of the host cells division. Hence, such plasmids can be used either to give stably integrated clones or to produce cells with possibly higher protein production by expression of the protein coded on the episomal plasmid. By maintaining the selection pressure from the puromycin, these episomal plasmids are retained for prolonged periods in such cells. There is, however, the possibility that with time the plasmid may be lost leading to a progressive reduction in the yield of recombinant protein, though we have failed to see this even over 30 cell passages.

Many fusion tags have been successfully used to aid in the purification of recombinant proteins. The tag peptide is a small sequence, usually 6–10 amino acids long, which acts either as an antibody epitope or as an affinity ligand or both. The first problem encountered with the use of a tag is that its activity depends either upon it being immunogenic or participating in some other form of molecular interaction. As generally the production of recombinant proteins is to derive antisera or monoclonal antibodies or to study interactions, this may be a handicap. This problem may be overcome by the addition of a protease cleavage site between the tag and the protein to allow the removal of the tag (*see* **Note 1**). Second, the position of the tag may affect the activity of the expressed protein by altering folding or other post-translational modifications. Finally, the tag itself may be masked by the tertiary structure of the protein. Hence, care should be taken in deciding whether the tag should be placed to the N- or C-terminus. We have successfully used the Strep II tag to give a single step purification of recombinant proteins from the media of 293-EBNA cells. This tag was initially found by screening a peptide library with streptavidin and is highly specific, not binding to the closely related protein avidin ***(5)***.

Crystallization of the peptide with its ligand showed that it bound to the same pocket as biotin, though in a more superficial position. The original peptide WRHPQFGG only bound strongly when placed C-terminally, as the final Gly-Gly-COO^- was used to form a salt bridge with the streptavidin molecule. By mutating the penultimate glycine to a glutamic acid, which can donate a COO^- to form such a bridge, a new form of the peptide was produced that is active both C- and N-terminally *(6)*. Further mutations led to the formation of an optimized binding sequence of WSHPQFEK (Strep II) *(6)*. Random mutation of the gene for streptavidin has since produced a form of this protein "StrepTactin," which binds to the Strep II tag also highly specifically but with a greater affinity *(7)*. Binding can be competed with biotin or its derivatives leading to the elution of the Strep-tagged protein under gentle conditions giving a native protein. Whereas the biotin–StrepTactin interaction is for practical purposes irreversible, desthiobiotin, a homolog of biotin, which can also displace the Strep tag from StrepTactin, binds relatively weakly to StrepTactin. It may then be eluted simply allowing the reuse of the StrepTactin matrix.

We have produced two vectors CMVNstrep and CMVCstrep with the same frame at the 5' end (**Fig. 1**) for N- or C-terminal tagging. cDNA is amplified by PCR with primers containing modified ends. The 5' end of the sense primer carries a *Spe*I, *Nhe*I, or *Xba*1 site placed to bring the codon alignment of the amplified fragment in frame with the signal peptide (+/– tag). On the antisense primer a stop codon is followed by a *Not*1 site in the case of CMVNstrep and a *Not*1 site, bringing the linking alanines of the tag into frame in the case of the CMVCstrep. The choice of the primers should be checked carefully, first, to keep frame and second, as the selection of the wrong domain borders will probably lead to misfolding of the expressed protein and possibly lead to its intracellular degradation. Where the borders are unknown, we suggest making a series of constructs of increasing size integrating successive likely border cysteines and leaving a small linker region between the cysteine and the tag or signal peptide. The PCR is carried out on preexisting cDNA using a proofreading enzyme to limit polymerase-induced errors and, after cloning into the relevant vector, it is sequenced in its entirety. We transfect by electroporation, which is simple and efficient and we generally use 293-EBNA cells to obtain episomal plasmid replication.

2. Materials

2.1. Media for 293-EBNA Cells

1. Serum-free media; Dulbeccos MEM-nutrient mix F-12 (Gibco, Gaithersburg, MD) with final concentrations of 2 m*M* glutamine (Gibco) and 100 U/mL penicillin, 100 μg/mL streptomycin (Gibco).

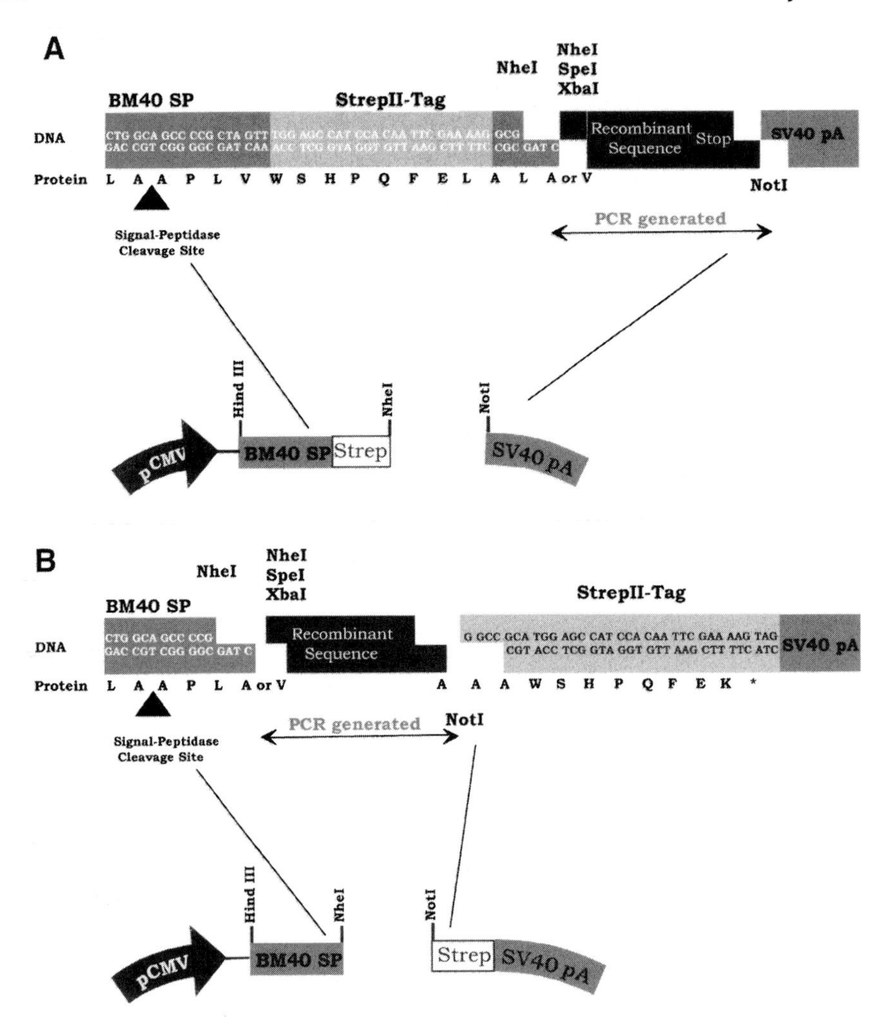

Fig. 1. Vectors for the expression of secreted proteins with N- and C-terminal Step tags in mammalian cells.

2. Growth media; serum-free media with 10% fetal calf serum (*not* heat-inactivated serum) and 175 μg/mL active G418 (Gibco).
3. Selection media; serum-free media with 10% fetal calf serum, 175 μg/mL active G418 and 0.5 μg/mL puromycin (Sigma, St. Louis, MO). (*See* **Note 1**.)

2.2. Materials for Purification of Protein

1. Protease inhibitors; X100 stock of NEM (N-ethylmaleimide, Fluka, Buchs, Switzerland) and PMSF (phenylmethylsulfonyl fluoride, Fluka) are dissolved together both at 100 m*M* in methanol and stored at –20°C.

2. Washing buffer; 100 m*M* Tris/HCl, 1 m*M* EDTA (pH 8.0). Elution buffer; washing buffer with 2.5 m*M* desthiobiotin (Sigma). Regeneration buffer; washing buffer with 1 m*M* HABA (4-hydroxyazobenzene–2-carboxylic acid) (Sigma). The above buffers are filtered through a 0.2-μm sterile filter prior to use.
3. StrepTactin (IBA, Göttingen) may be bought as free StrepTactin Sepharose or as a ready-to-use column (1 mL column volume) and is claimed to bind up to 350 nmol of biotin per mL gel matrix.

3. Methods

3.1. Transfection of 293-EBNA cells

1. 293 EBNA cells (Invitrogen, San Diego, CA) are cultured in growth media on 10-cm culture dishes at 37°C and in 5% CO_2 until 80% confluent.
2. The cells are trypsinized and dispersed by trituration. They are then pelleted at 150*g* for 3 min, resuspended in 5 mL of growth media, and counted (a 10-cm dish gives about 2×10^7 cells).
3. Cells are diluted to 6×10^5 per mL in growth media and 800 μL of cells are mixed with 10 μg of purified circular plasmid DNA, (Qiagen or equivalent) and left for 5 min at room temperature. A control electroporation with no DNA added to the cells is also carried out.
4. The cells are placed in a 4-mm electrode gap cuvette (Bio-Rad, Richmond, CA) and electroporated with the following settings:
   Capacitance 500 μF
   Voltage 230 V
5. After electroporation, the cells are left for 5 min to recover and then plated out on a 10 cm culture dish in growth medium.
6. Medium is changed daily and puromycin selection is initiated 48 h after transfection. After 6 d of selection, the negative control cells are dead, whereas the test plate is 60–70% confluent with obvious areas of clonal growth having been observed earlier.
7. The plate is trypsinized and the cells split on to two 10-cm dishes, which are allowed to grow for a further 2 d before the cells of one plate are cryopreserved in liquid nitrogen.
8. The other plate is allowed to grow for another 2 d until highly confluent. There should be no gaps between cells and the cell sheet should appear thick and undulating. As most of the contaminating proteins tend to be serum-derived, washing to remove the selection media should be thorough. The plates should be drained by being left at an angle for a minute to remove all residual medium. They are washed twice with PBS with care taken to remove all the PBS. Eight mL of serum free medium is added to the plate. The serum free media is replaced after 2, 4, and 6 d and the harvested media cleaned of debris by centrifugation and stored at –20°C until analyzed by SDS-PAGE and/or immunoblotting.

3.2. Purification of Expressed Recombinant Protein

If the protein is expressed, cells can be thawed and the culture scaled up. Generally, we grow the cells at this stage upon 5 to 20 dishes of 15-cm diameter and produce between 0.2 and 1 L of medium depending upon expression levels and experimental requirements.

1. Medium is harvested and stored at –20°C until required.
2. Upon thawing, the protease inhibitors NEM and PMSF are added and as both are sensitive to hydrolysis they are readded fresh every 24 h the protein is maintained above freezing. To further reduce degradation, EDTA is added to give a final concentration of 10 m*M*. All subsequent purification steps are carried out at 4°C.
3. Where large volumes are being used, the protein solution may be concentrated 5 to 10 times using an ultrafiltration cell (Amicon, Danvers, MA) with a membrane pore size chosen according to the molecular mass of the recombinant protein (protein binding to the membrane may be reduced by initially rinsing the membrane in 5% Tween-20 overnight and then rinsing it extensively with water before use).
4. The StrepTactin column is equilibrated with five column volumes of washing buffer passed through at 0.2 mL per min for a 1-mL column.
5. The protein is loaded on the column at the same flow rate and is recirculated over 12–24 h.
6. The column is washed with at least five column volumes of washing buffer at 0.2 mL per min.
7. Elution of the protein occurs upon the addition of three column volumes of elution buffer and 0.5-mL aliquots are collected and analyzed (**Fig. 2A,B**).
8. The column is cleaned using 15 column volumes of HABA-containing regeneration buffer. HABA has a yellow color and displaces the desthiobiotin. Upon interaction with avidin homologs, it has a color shift from yellow to red that can be used to indicate the removal of desthiobiotin from the column.
9. Finally, the column is washed free of the HABA using 10 column volumes of the wash buffer. The matrix should now loose its red pigmentation. The column can be stored at 4°C when maintained in wash buffer containing 0.02% w/v sodium azide.
10. A number of problems with recombinant protein purification may be encountered, and these are discussed in **Notes 3–5.**

4. Notes

1. Tags may not always be innocuous *(8)*. We have produced the CMVNstrep with protease sites for factor X, enterokinase, or thrombin to allow the removal of the Strep tag if it is felt to be a problem for later studies.
2. New batches of puromycin should be tested for their activity, to determine the minimum concentration able to give total cell death after 5 d of selection of the untransfected control.
3. Low or no expression of the protein may be caused by a number of causes.
   a. Frame shifts either owing to an error in choosing primers or an artifact in cloning.
   b. A PCR-induced mutation introducing a stop codon or leading to a major amino acid substitution affecting the structure of the expressed protein and leading to its intracellular degradation.

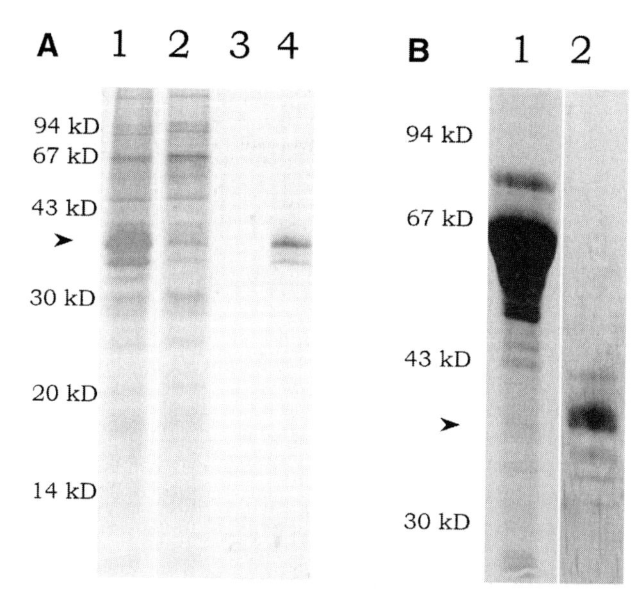

Fig. 2. Affinity purification of the IV domain from the laminin β_2 chain of transfected 293-EBNA cells. The protein carried an N-terminal Strep II tag and was chromatographed on a StrepTactin column as described in the text. The product occurs in different glycosylation forms. **(A)** Purification from serum-free medium: (lane 1) cell culture supernatant; (lane 2) flowthrough; (lane 3) wash; (lane 4) eluate. **(B)** Purification from medium containing 10% FCS: (lane 1) cell-culture supernatant; (lane 2) eluate. SDS-PAGE was performed on 15% polyacrylamide gels that were stained with Coomassie brilliant blue R250. All evident bands in the eluate were immunoreactive with an antibody specific for the laminin β_2 chain.

c. Incorrect estimation of the domain borders may lead to an incorrectly folded and nonsecreted protein.
d. Poor transfection efficiency with the survival of nontransfected cells gives low levels of the protein in the media. This may be corrected by hard (1:20) splitting of the cells and this is indicated when many dead cells are seen after routine passaging.
e. A low plasmid copy number in the 293-EBNA cells may produce low expression levels. Increasing the amount of puromycin in the selection media two- or threefold may raise yields of the recombinant protein.
f. The formation of insoluble deposits may occur where the secreted protein precipitates out of the media or is incorporated into a matrix around the cells or upon the plate. It is worthwhile immunoblotting the cell extract after the cells have been removed from the plate by prolonged washing with EDTA/PBS as well as trying to extract any proteins in the matrix laid on to the plate by treating it with SDS-PAGE sample buffer.

4. Expression of the protein as multiple bands when run on a SDS-PAGE gel may have numerous causes.
   a. Intra- or extracellular degradation of the protein may be occurring, the latter can be reduced by protease inhibitors, more frequent harvesting of media from the cells, or the use of different cell lines (e.g., COS cells) perhaps not expressing the protease.
   b. Variable levels of protein glycosylation may occur, this can be analyzed by treatment of the protein with N-glycosidase F, which should give an unglycosylated product of a single molecular mass if the differences are caused by N-linked carbohydrates. This may be caused by very high expression of the protein leading to exhaustion of the post-translational mechanisms of the cell. If it is a major hinderance, stable clones expressing at lower levels may reduce these problems.
5. The protein is expressed, but does not bind when added to the StrepTactin column.
   a. There has been insufficient cleaning of the column possibly owing to using low volumes of the regeneration and washing buffers.
   b. The medium was passed over the column at too great a speed for efficient binding.
   c. The tag is masked by folding of the protein, a tag placed on the opposite terminal may be more effective.

Acknowledgements

The authors gratefully thank the BMBF in the framework of the ZMMK (Centre for Molecular Medicine Cologne) and the Köln Fortune Fund of the Medical Faculty, University of Cologne for their financial assistance.

References

1. Bork, P., Downing, A. K., Kieffer, B., and Campbell, I. D. (1996) Structure and distribution of modules in extracellular proteins. *Rev. Biophys.* **29,** 119–167.
2. Pöschl, E., Fox, J. W., Block, D., Mayer, U., and Timpl, R. (1994) Two noncontiguous regions contribute to nidogen binding to a single EGF-like motif of the laminin γ1 chain. *EMBO. J.* **13,** 3741–3747.
3. Kohfeldt, E., Maurer, P., Vannahme, C., and Timpl, R. (1997) Properties of the extracellular calcium binding module of the proteoglycan testican. *FEBS. Lett.* **414,** 557–561.
4. Kohfeldt, E., Gohring, W., Mayer, U., Zweckstetter, M., Holak, T. A., Chu, M. L., and Timpl, R. (1996) Conversion of the Kunitz-type module of collagen VI into a highly active trypsin inhibitor by site-directed mutagenesis. *Eur. J. Biochem.* **238,** 333–340.
5. Schmidt, T. G. and Skerra, A. (1993) The random peptide library-assisted engineering of a C-terminal affinity peptide, useful for the detection and purification of a functional Ig Fv fragment. *Protein Eng.* **6,** 109–122.
6. Schmidt, T. G., Koepke, J., Frank, R., and Skerra, A. (1996) Molecular interaction between the Strep-tag affinity peptide and its cognate target, streptavidin. *J. Mol. Biol.* **255,** 753–766.

7. Voss, S. and Skerra, A. (1997) Mutagenesis of a flexible loop in streptavidin leads to higher affinity for the Strep-tag II peptide and improved performance in recombinant protein purification. *Protein Eng.* **10,** 975–982.
8. Ledent, P., Duez, C., Vanhove, M., Lejeune, A., Fonze, E., Charlier, P., et al. (1997) Unexpected influence of a C-terminal-fused His-tag on the processing of an enzyme and on the kinetic and folding parameters. *FEBS Lett.* **413,** 194–196.

II

Biophysics of Extracellular Matrix Proteins

6

Preparation of Isotopically Labeled Recombinant Fragments of Fibronectin for Functional and Structural Study by Heteronuclear Nuclear Magnetic Resonance Spectroscopy

Jeremy R. Bright, Andrew R. Pickford, Jennifer R. Potts, and Iain D. Campbell

1. Introduction

The collagens, proteoglycans, and glycoproteins of the extracellular matrix (ECM) are a diverse collection of macromolecules, heterogeneous in size, structure, and function. However, their primary structures have revealed that their polypeptide components are frequently comprised of a series of sequence repeats or "modules" *(1)*. This mosaic nature of ECM macromolecules permits their properties to be analyzed by a "dissection" strategy *(2)*, where protein fragments generated by proteolysis, chemical synthesis, or recombinant expression are employed in structural and functional investigations.

Fibronectin, an ubiquitous ECM glycoprotein, plays a major role in fundamental biological processes such as cell adhesion and migration, maintenance of normal cell morphology, cytoskeletal organization, and cell differentiation *(3,4)*. Fibronectin is constructed from three types of independently folding protein module (F1, F2, and F3) *(5)* and is found as a fibrillar network in the ECM where it interacts with other ECM components and provides anchorage sites for cell surface integrin receptors. It remains the focus of much structural investigation by both X-ray crystallography and nuclear magnetic resonance (NMR) spectroscopy *(6–12)*.

NMR spectroscopy is a powerful technique that can be used to determine the three-dimensional structure of small proteins and protein modules in solution and can also provide information on protein folding pathways, enzyme

From: *Methods in Molecular Biology, vol. 139: Extracellular Matrix Protocols*
Edited by: C. Streuli and M. Grant © Humana Press Inc., Totowa, NJ

mechanisms, and protein–ligand interactions. The introduction of isotope-labeling of proteins in conjunction with heteronuclear experiments has increased the upper limit for three-dimensional protein structure determination using NMR, from ~100 residues (for unlabeled proteins) to around 300 residues (for ^{15}N / ^{13}C-labeled proteins) ***(13,14)***.

Labeling of proteins with ^{15}N and/or ^{13}C allows the extension of NMR experiments into three and four dimensions, reducing the overlap of signals and facilitating the extraction of the interproton distance information used in structure calculations. Isotope labeling also allows additional information to be obtained, e.g., about protein backbone dynamics ***(15)***. These advantages can only be exploited if isotope-labeling can be performed cost-effectively, especially for labeling with the relatively expensive isotope ^{13}C.

Escherichia coli has been used extensively for the production of recombinant isotopically labeled proteins ***(16)*** and has been used to express recombinant fibronectin fragments, although the recombinant material often needs to be exposed to an involved refolding and purification procedure ***(17)***. We have found that *Pichia pastoris* can secrete recombinant fibronectin fragments very efficiently, without the need for refolding, but to date there are only two reports of isotope-labeling in this system ***(18–19C)***. We have developed a procedure for the cost-effective labeling of recombinant fibronectin fragments by growing *P. pastoris* strains in isotopically enriched media in a small fermentor.

The preparation of an isotopically labeled NMR sample is a multistep procedure. Typically, this involves the design, generation, and screening of recombinant clones, their growth in isotopically enriched media, purification of the recombinant product, and assessment of its purity and integrity by methods such as N-terminal peptide sequencing, mass spectrometry and 1-dimensional NMR. Coverage of all of these areas is beyond the scope of this chapter; here we describe a procedure for expression of isotope-labeled fibronectin modules and an initial purification step that we have found well suited to this system. Methodologies for the design, generation, and screening of *P. pastoris* strains use well-established molecular biological techniques and are described in detail elsewhere ***(20–24)***. Procedures for the final stages of protein purification have to be tailored to the individual protein and these procedures and those for NMR sample preparation are also described elsewhere ***(25,26)***.

The small fermentor used in this procedure is described in detail below; the strategy we employ uses media adapted from ***(27)*** and has been developed to make relatively small quantities of recombinant proteins reproducibly, using cost-effective amounts of labeled substrates (**Fig. 1**). However, it is a "first-shot" approach, which should be tried with unlabeled reagents to assess the performance of a given clone; the scale of the culture and/or the feed times can be adjusted to compensate for clones that perform poorly. Fermentation of

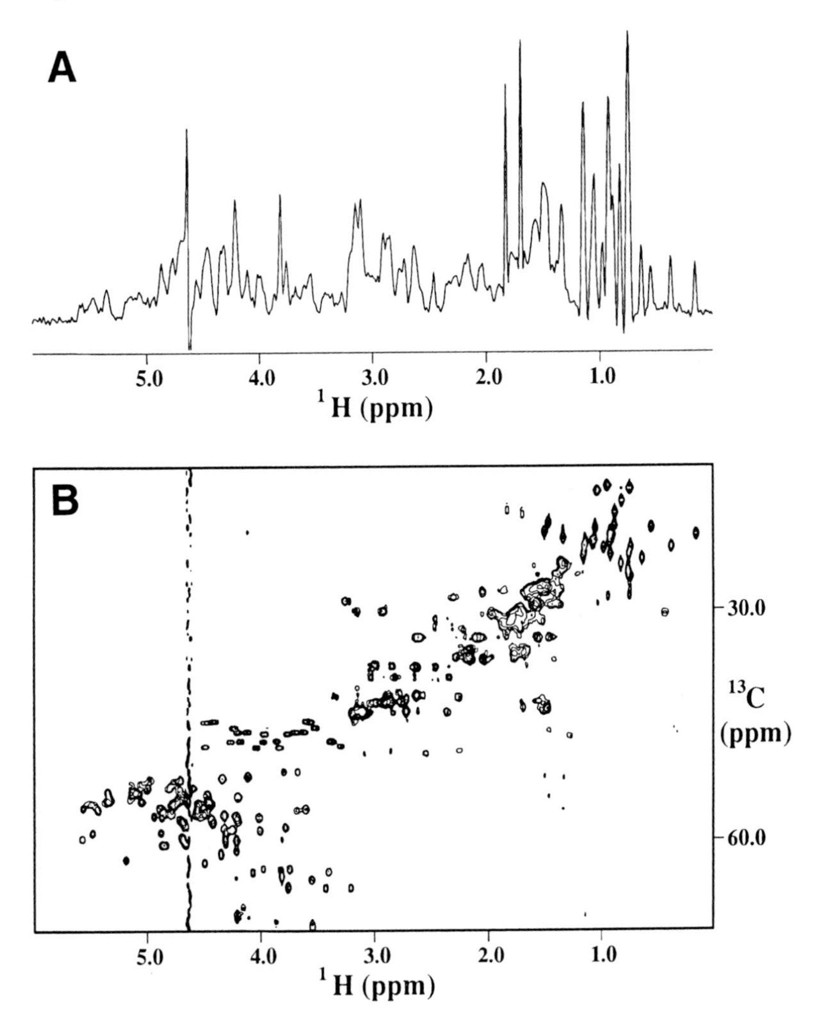

Fig. 1. ^{1}H–^{13}C heteronuclear single quantum correlation (HSQC) spectrum of a ^{15}N/^{13}C-labeled sample of the fourth and fifth fibronectin type 1 module pair *(8)*, gathered using a 0.5 m*M* (5.5 mg/mL) sample. Using our fermentation procedure, 2 g ^{15}N-ammonium sulphate, 1 g U-[^{13}C]-glucose, and 20 g ^{13}C methanol were sufficient to generate 60 mg pure ^{15}N / ^{13}C-labeled protein. The 1-dimensional projection **(A)** of the 2-D matrix, **(B)** shows how overlapping peaks are resolved by extending into the ^{13}C-dimension.

P. pastoris is not difficult, but we strongly advise anyone attempting it to consult a fermentation scientist for general information about safe handling and disposal; also there are variations in fermentation systems which may necessitate small adaptations to our protocol.

2. Materials

All unlabeled reagents were standard laboratory reagent grade; all solutions were prepared using milli-Q water. ^{15}N-ammonium sulphate (99 At%), U-[^{13}C]-glucose (99 At%) and ^{13}C-methanol (99 At%) were obtained from CK Gas Products, (Finchampstead, UK). The zeroing gel was obtained from Mettler-Toledo (Leicester, UK). Polypropylene glycol 1025 was obtained from Merck (Poole, UK). A 2-L glass fermentation vessel was obtained from Electrolab Ltd. (Tewkesbury, UK) with fittings for variable-speed impeller, temperature control, adjustable sterile air supply, air outlet fitted with a condenser, and with ports for pH, dissolved-oxygen (dO_2), and temperature probes, acid-, base-, and nutrient-feed lines, and for inoculation and sampling. The fermenter was operated using a controller (Electrolab Ltd.) with acid and base pumps for feedback pH control, feedback temperature control, impeller-speed control, and variable-speed nutrient-feed pump (**Fig. 2**). The chart recorder was obtained from Kipp & Zonen (St. Albans, UK).

2.1. Seed Culture

1. Biotin solution (per 100 mL): Dissolve 0.2 g D-biotin in 10 m*M* sodium hydroxide. Filter sterilize and store at 4°C for up to 3 mo.
2. Buffered Minimal Glucose (BMGc) pH 6.0 (per 100 mL): Dissolve 0.34 g yeast nitrogen base without amino acids or ammonium sulphate (Difco, UK), 1 g ammonium sulphate, 1 g D-glucose, 1.19 g potassium dihydrogen phosphate, 0.21 g di-potassium hydrogen phosphate, and 0.2 mL biotin solution in water. Filter sterilize and store at 4°C for up to 1 month. Use ^{15}N-ammonium sulphate and/or U-[^{13}C]-glucose instead as appropriate (*see* **Note 1**).

2.2. Preparation of the Vessel

1. Basal Salts Medium (BSM) pH 6.0 (per liter): Dissolve 0.93 g calcium sulphate, 14.9 g magnesium sulphate heptahydrate, 25.0 g potassium hydroxide (corrosive), and 26.7 mL orthophosphoric acid (85 wt % - corrosive) in water. Autoclave and adjust pH in the fermentation vessel (*see* **Note 2**).
2. BSM pH 3.0 (per liter): Dissolve 0.93 g calcium sulphate, 14.9 g magnesium sulphate heptahydrate, 18.2 g potassium sulphate, and 26.7 mL orthophosphoric acid (85 wt % - corrosive) in water. Autoclave and adjust pH in the fermentation vessel (*see* **Note 2**).
3. *Pichia* Trace Metals (PTM_1) (per liter): Dissolve 6 g copper (II) sulphate, 0.08 g sodium iodide, 3 g manganese sulphate monohydrate, 0.2 g sodium molybdate dihydrate, 0.02 g boric acid, 0.5 g cobalt chloride, 20.0 g zinc chloride, 65.0 g iron (II) sulphate, 5 mL sulfuric acid (98% - corrosive) in water. Store at room temperature for up to 1 yr; filter sterilize before use.
4. Ammonium sulphate solution: Prepare a 20 % (w/v) solution in water. Filter sterilize and store at room temperature for up to 1 yr. Use ^{15}N-ammonium sulphate as necessary (*see* **Note 3**).

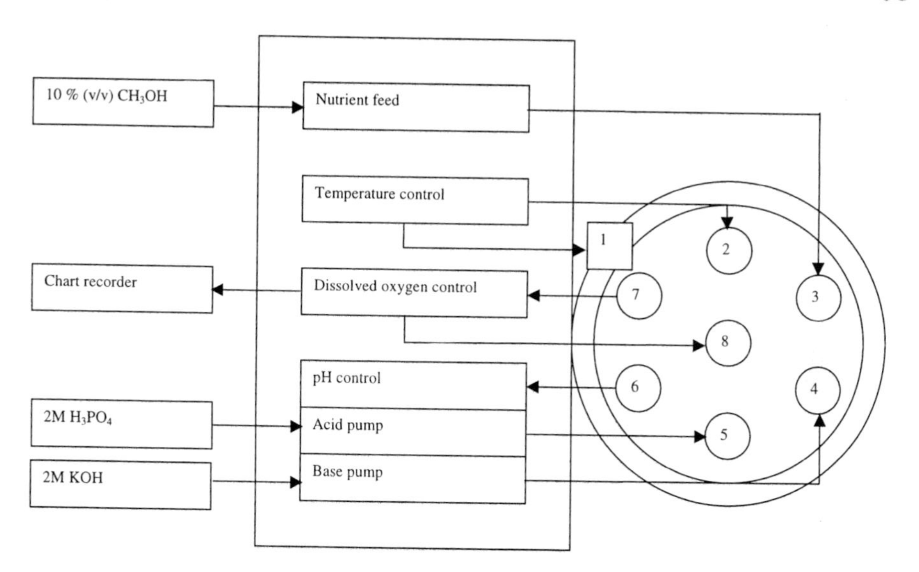

Fig. 2. Schematic representation of the fermentation vessel and controller. (1) temperature-controlled jacket; (2) temperature probe; (3) nutrient-feed port; (4) base-control port; (5) acid-control port; (6) pH probe; (7) dO_2 probe; (8) impeller.

5. D-glucose solution: Prepare a 20 % (w/v) solution in water. Filter sterilize and store at room temperature for up to 1 yr. Use U-[^{13}C]-glucose as necessary (*see* **Note 3**).
6. Antifoam solution: Autoclave polypropylene glycol 1025 and store at room temperature indefinitely (*see* **Note 4**).
7. Acid control solution: Prepare and autoclave 2 *M* orthophosphoric acid (corrosive) and store at room temperature indefinitely.
8. Base control solution: Prepare and autoclave 2 *M* potassium hydroxide (corrosive) and store at room temperature indefinitely.

2.3. Fermentation of P. pastoris

Methanol feed solution: Prepare fresh and filter sterilize 10% (v/v) methanol (toxic) in water.

2.4. Processing of Culture Supernatants

1. Citrate equilibration buffer pH 3.0 (per liter): Dissolve 3.15 g citric acid, 0.90 g trisodium citrate dihydrate, and 0.20 g sodium azide (highly toxic) in water.
2. Citrate wash buffer pH 5.0 (per liter): Dissolve 1.34 g citric acid, 3.82 g trisodium citrate dihydrate, and 0.20 g sodium azide (highly toxic) in water.
3. Citrate elution buffer pH 5.0 (per liter): Dissolve 1.34 g citric acid, 3.82 g trisodium citrate dihydrate, 58.44 g sodium chloride, and 0.20 g sodium azide (highly toxic) in water.

3. Methods

Media for the seed culture and the fermentation are defined such that replacement of the ammonium sulphate, glucose, and methanol, as appropriate, with their isotopically labeled analogs is sufficient to generate isotopically labeled recombinant material. Typical routes used to generate recombinant strains of *P. pastoris* can generate Mut^+ or Mut^s (methanol utilization positive or methanol utilization slow) phenotypes; the procedures described here pertain to strains with the Mut^+ phenotype.

3.1. Seed Culture

1. Thaw an aliquot of a *P. pastoris* strain that has been stored as a 20% (v/v) glycerol stock at –85°C.
2. Inoculate a 250 mL baffled flask containing 50 mL of BMGc with 1 mL of the glycerol stock and grow shaking at 30°C, for 16–24 h, until OD_{600} = 2–6. This will constitute the seed culture for a 0.5 l fermentation (*see* **Note 5**).

3.2. Preparation of the Vessel

1. Calibrate the pH probe, rinse it with deionized water and introduce it into one of the vessel's ports (*see* **Note 6**).
2. Test the dO_2 probe with zeroing gel, rinse it with deionized water, and introduce it into one of the vessel's ports (*see* **Note 7**).
3. Mask the external ends of all ports and lines with aluminium foil and autoclave tape, clamp any lines that will be in contact with the media (e.g., the air inlet and sampling line) and sterilize the vessel containing 500 mL of BSM.
4. Sterilize 3 empty aspirators for acid and base control and for nutrient feed.
5. Aseptically, fill the acid and base control aspirators with acid and base control solutions, respectively.
6. Connect the chart recorder and pH, dO_2, and temperature probes to the controller. Connect the temperature control facility (*see* **Note 8**).
7. Set the impeller speed to 150 rpm and introduce 2.5 mL D-glucose solution, 10 mL ammonium sulphate solution, 2.5 mL biotin solution, 2.5 mL PTM_1, and 0.1 mL antifoam solution via the inoculation port (*see* **Note 9**).
8. Aseptically, connect the aspirators for acid and base control to the vessel via the acid and base control pumps and allow the system to equilibrate to the desired temperature and pH (*see* **Note 9**).
9. Set the output from the oxygen probe to zero. Zero the chart recorder (*see* **Note 10**).
10. Connect the sterile airline, open the air outlet, and increase the impeller speed and air-flow rates to the maximum value to be used during the fermentation, typically 1000 rpm and 5 vol of air per volume of culture per min (vvm). After 10 min, calibrate the dO_2 probe at 100% and set the chart recorder to full scale deflection.

3.3. Fermentation of P. pastoris

1. Run the vessel initially at 2 vvm and at an impeller speed of 250 rpm, inoculate with the seed culture and start the chart recorder to monitor dO_2 levels. Take a

10-mL sample *via* the sample port for assay and determination of cell wet weight and/or OD_{600} (*see* **Note 11**).

2. Adjust the impeller speed during the course of the fermentation to maintain dO_2 above 20%.
3. When dO_2 "spikes," take a sample then add 5 mL 100% methanol (toxic) via the inoculation port, continuing to adjust the impeller speed to maintain dO_2 above 20% (*see* **Note 12**).
4. Aseptically, fill the nutrient-feed aspirator with the methanol feed solution and connect the aspirator to the vessel via the nutrient-feed pump.
5. When dO_2 spikes for the second time, start the methanol feed at 2.5 mL/h, increasing to 5 mL/h after 6 h. Keep dO_2 above 20% and take samples regularly. Add 0.1-mL aliquots of antifoam solution, as necessary (*see* **Note 13**).
6. To prevent methanol accumulating in the vessel, while at the same time maintaining optimal cell growth, the feed rate of methanol can be increased with time, but must remain limiting. Under conditions of limiting carbon source, the respiration rate rises and falls as methanol is constantly depleted then replenished in the vessel, this is signalled by oscillations in dO_2 levels (*see* **Note 14**).
7. Continue the fermentation until it has consumed at least 200 mL methanol feed solution (*see* **Note 15**).

3.4. Processing Culture Supernatants

This section of the procedure has to be tailored to the particular protein of interest, depending on its stability, its isoelectric point and any individual characteristics that may be exploited for affinity purification. In addition to the protein of interest, *P. pastoris* secretes a small number of proteins with isoelectric points in the range pH 3.0–5.0 and a nonproteinaceous high-molecular-weight species (possibly glycosaminoglycan), which is anionic under mildly acidic or neutral conditions of pH. With this in mind, and in the absence of an affinity purification step that can be exploited, we recommend an initial cation-exchange purification step to concentrate and partially purify the secreted protein of interest. However, if this protein is labile or insoluble under acidic conditions, other initial purification methods should be assessed.

1. Pellet the cells by centrifugation at 2000*g* for 15 min. Remove the supernatant and determine its volume.
2. For fermentations run at pH 6.0: Wash the cells by resuspending the cell pellet in a volume of 100 m*M* HCl equivalent to the original supernatant volume. Pellet the cells by centrifugation at 10,000*g* for 10 min. Remove the supernatant and pool with the original sample of supernatant (*see* **Note 16**).
3. For fermentations at pH 3.0: As for **Subheading 3.4.2.**, but wash the cells with distilled water (*see* **Note 16**).
4. Adjust the pooled supernatants to pH 3.0. Clarify by filtration through a 0.2-μm membrane (*see* **Note 17**) and load onto a column of SP-sepharose, pre-equilibrated with 5 column volumes of citrate equilibration buffer.

5. Wash the SP-sepharose with citrate wash buffer, monitoring A_{280} of the eluate, until a flat baseline is obtained (*see* **Note 18**).
6. Elute bound proteins by washing the SP-sepharose with a linear gradient of 0–100% citrate elution buffer over 5 column volumes, collecting fractions, and monitoring A_{280} (*see* **Note 18**).

4. Notes

1. *P. pastoris* growth media commonly contain glycerol instead of glucose as the carbon source. Glucose is merely chosen because of the cost and availability of its ^{13}C isotopically labeled analog.
2. Selection of the growth and induction pH of *P. pastoris* can help to overcome proteolysis and insolubility of the recombinant product. The optimum value should be determined in small-scale cultures prior to fermentation. Final pH adjustment is performed in the vessel because precipitation of inorganic salts during the sterilization process can occur if the pH of BSM >3.5.
3. Ammonium sulphate and D-glucose can be sterilized by autoclaving. Filter sterilization is used instead as a precaution against accidental losses of the isotopically labeled compounds.
4. Different molecular-weight preparations of polypropylene glycol are available; we routinely use 1025. The disadvantage of polypropylene glycol as an antifoam reagent is that its presence, even at low concentrations, can significantly reduce flow rates through certain ultrafiltration membranes.
5. The volume of seed culture can be changed according to the scale of fermentation. Using 10% of the fermentation volume as the seed culture works well.
6. The performance of gel-filled autoclavable pH probes will deteriorate with repeated sterilization, but can be prolonged by periodic cleaning as recommended by the manufacturer. Probes should be stored in 3 *M* KCl.
7. We have only used polarographic dissolved oxygen probes; information on the use of galvanic oxygen probes should be obtained from the manufacturer. Polarographic oxygen probes require routine maintenance to prolong their working life and these details should also be obtained from the manufacturer. Probes should be stored wet.
8. Facilities for temperature control vary; our vessel has an external heater and an internal cold finger, other glass vessels may have a water jacket connected to an external thermostatically controlled bath.
9. As the vessel reaches its target pH and particularly on addition of PTM_1 salts, some inorganic salts will precipitate. This does not affect cell growth.
10. This operation will depend on the design of the controller.
11. Regular sampling is important to monitor the growth of the fermentation, allows assay for any recombinant product, and permits microscopic inspection of the culture to screen for contaminating microorganisms. The precipitated inorganic salts in the sample can interfere with these tests, but can be dissolved by the addition of an equal volume of 100 m*M* HCl without leading to cell lysis. If the recombinant product is acid labile, then a sample of supernatant should be removed for assay before addition of the HCl.

12. When a culture is growing normally and consumes all of the carbon source, respiration will slow quite dramatically causing the dO_2 level to rise quite sharply or "spike." Addition of more carbon source should lead to an almost instantaneous rise in respiration signaled by a concomitant fall in dO_2. Many fermentation controllers have automated feedback control, which adjusts the impeller speed to maintain the dO_2 at a preset level. This is very useful when the system is left unattended for long periods of time, but can mask dO_2 spikes.
13. The addition of antifoam is usually necessary, but when growing *P. pastoris* at relatively low cell densities only small quantities are required. If the methanol feed is continued for longer periods, more antifoam solution may be needed. Alternatively, if available, antifoam feedback control may be used.
14. A useful guide to whether the methanol is limiting is to switch off the methanol feed and measure the time taken for dO_2 to rise by 10%. If this time is <1 min, the carbon source is limiting.
15. The actual amount of methanol feed solution required to generate sufficient recombinant material must be determined empirically. We have never used less than 200 mL 10% (v/v) methanol.
16. The purpose of washing the cells is threefold: i) to increase the yield of recombinant protein; ii) to dilute the supernatant and therefore to reduce its ionic strength prior to ion-exchange chromatography; and iii) to reduce the pH to ~3.0 for cation exchange.
17. We use an Amicon hollow-fiber cartridge with an exclusion limit of 0.2 μm to clarify the supernatant, but any low protein-binding 0.2-μm membrane filtration device could be used. Alternatively, centrifugation for 20 min at > 20,000*g* could be used.
18. Washing the SP-sepharose at pH 5 should remove most of the nonrecombinant *P. pastoris* proteins from the column, but may also remove the recombinant protein if its p*I* <6. If this occurs, washing and eluting at a slightly lower pH may be an option.

Acknowledgments

The authors thank the Wellcome Trust for financial support.

References

1. Bork, P., Downing, A. K., Kieffer, B., and Campbell, I. D. (1996) Structure and distribution of modules in extracellular proteins. *Quart. Rev. Biophys.* **29,** 119–167.
2. Baron, M., Norman, D. G., and Campbell, I. D. (1991) Protein modules *TIBS* **16,** 13–17.
3. Hynes, R. O. (1990) Fibronectins, (Rich, A., ed.), Springer-Verlag, Berlin.
4. Mosher, D. F. (1993) Assembly of fibronectin into extracellular matrix. *Curr. Opin. Struct. Biol.* **3,** 214–222.
5. Petersen, T. E., Thørgersen, H. C., Skorstengaard, K., Vibe-Pedersen, K., Sahl, P., Sottrup-Jensen, L., and Magnusson, S. (1983) Partial primary structure of bovine plasma fibronectin: three types of internal homology. *Proc. Nat. Acad. Sci. USA* **80,** 137–141.

6. Baron, M., Norman, D., Willis, A., and Campbell, I. D. (1990) Structure of the fibronectin type 1 module. *Nature* **345,** 642–646.
7. Main, A. L., Harvey, T. S., Baron, M., Boyd, J., and Campbell, I. D. (1992) The three-dimensional structure of the tenth type III module of fibronectin: an insight into RGD-mediated interactions. *Cell* **71,** 671–678.
8. Williams, M. J., Phan, I., Harvey, T. S., Rostagno, A., Gold, L. I., and Campbell, I. D. (1994) Solution structure of a pair of fibronectin type 1 modules with fibrin binding activity. *J. Mol. Biol.* **235,** 1302–1311.
9. Potts, J. R., Phan, I., Williams, M. J., and Campbell, I. D. (1995) High resolution structural studies of the factor XIIIa crosslinking site and the first type 1 module of fibronectin. *Nat. Struct. Biol.* **2,** 946–950.
10. Leahy, D. J., Aukhil, I., and Erickson, H. P. (1996) 2. 0 Å crystal structure of a four-domain segment of human fibronectin encompassing the RGD loop and synergy region. *Cell* **84,** 155–164.
11. Pickford, A. R., Potts, J. R., Bright, J. R., Phan, I., and Campbell, I. D. (1997) Solution structure of a type 2 module from fibronectin: implications for the structure and function of the gelatin-binding domain. *Structure* **5,** 359–370.
12. Sticht, H., Pickford, A. R., Potts, J. R., and Campbell, I. D. (1998) Solution structure of the glycosylated second type 2 module of fibronectin. *J. Mol. Biol.* **276,** 177–187.
13. Clore, G. M. and Gronenborn, A. M. (1991) Structures of larger proteins in solution: three- and four-dimensional heteronuclear NMR spectroscopy. *Science* **252,** 1390–1399.
14. Martin, J. R., Mulder, F. A. A., Karimi-Nejad, Y., van der Zwan, J., Mariani, M., Schipper, D., and Boelens, R. (1997) The solution structure of a serine protease PB92 from *Bacillus alcalophilus* presents a rigid fold with a flexible substrate-binding site. *Structure* **5,** 521–532.
15. Carr, P. A., Erickson, H. P., and Palmer, A. G. III (1997) Backbone dynamics of homologous fibronectin type III cell adhesion domains from fibronectin and tenascin. *Structure* **5,** 949–959.
16. Markley, J. L. and Kainosho, M. (1993) Stable isotope labelling and resonance assignments in larger proteins, in *NMR of Macromolecules—a practical approach* (Roberts, G. C. K., ed.), IRL Press at Oxford University Press, Oxford, pp. 101–152.
17. Skorstengaard, K., Holtet, T. L., Etzerodt, M., and Thørgersen, H. C. (1994) Collagen-binding recombinant fibronectin fragments containing type II domains. *FEBS Letts.* **343,** 47–50.
18. Laroche, Y., Storme,V., De Meutter, J., Messens, J., and Lauwereys, M. (1994) High level secretion and very efficient isotopic labelling of tick anticoagulant peptide (TAP) expressed in the methylotrophic yeast *Pichia pastoris*. *Bio/Technology* **12,** 1119–1124.
19. Wiles, A. P., Shaw, G., Bright, J., Perczel, A., Campbell, I. D., and Barlow, P. N. (1997) NMR studies of a viral protein that mimics the regulators of complement activation. *J. Mol. Biol.* **272,** 253–265.

19a. McAlister, M. S., Davis, B., Pfuhl, M., and Driscoll, P. C. (1998) NMR analysis of the N-terminal SRCR domain of human CD5: engineering of a glycoprotein for superior characteristics in NMR experiments. *Protein Eng.* **11,** 847–853.
19b. Wood, M. J. and Komives, E. A. (1999) Production of large quantities of isotopically labeled protein in *Pichia pastoris by fermentation. J. Biomol. NMR* **13,** 149–159.
19c. Johnson, T. M., Holaday, S. K., Sun, Y., Subramaniam, P. S., Johnson, H. M., and Krishna, N. R. (1999) Expression, purification and characteriszation of interferon-tau produced in *Pichia pastoris* grown in mininal medium. *J. Interferon Cytokine Res.* **19,** 631–636.
20. Cregg, J. M., Barringer, K. J., Hessler, A. Y., and Madden, K. R. (1985) *Pichia pastoris* as a host system for transformations. *Mol. Cell. Biol.* **5,** 3376–3385.
21. Cregg, J. M., Vedvick, T. S., and Raschke, W. C. (1993) Recent advances in the expression of foreign genes in *Pichia pastoris. Bio/Technology* **11,** 905–910.
22. Scorer, C. A., Clare, J. J., McCombie, W. R., Romanos, M. A., and Sreekrishna, K. (1994) Rapid selection using G418 of high copy number transformants of *Pichia pastoris* for high-level foreign gene expression. *Bio/Technology* **12,** 181–184.
23. Romanos, M. A. (1995) Advances in the use of *Pichia pastoris* for high-level gene expression. *Curr. Opin. Biotechnol.* **6,** 527–533.
24. Invitrogen Corp., San Diego, CA, USA. *Pichia* Expression System: Manual of methods for expression of recombinant proteins in *Pichia pastoris*, Version G, http://www. invitrogen. com/manuals. html.
25. Harris, E. L. V. and Angal, S., eds. (1989) *Protein Purification Methods—A Practical Approach.* IRL Press at Oxford University Press, Oxford and New York.
26. Primrose, W. U. (1993) Sample Preparation in *NMR of Macromolecules—A Practical Approach* (Roberts, G. C. K. ed.), IRL Press at Oxford University Press, Oxford, pp. 7–34.
27. Invitrogen Corporation, San Diego, USA. *Pichia* Fermentation Process Guidelines, http://www. invitrogen. com/manuals. html.

7

Reconstitution of Functional Integrin into Phospholipid Vesicles and Planar Lipid Bilayers

Eva-Maria Erb and Jürgen Engel

1. Introduction

Integrins are heterodimeric cell-surface receptors involved in a variety of functions such as binding to the ECM, regulation of the cellular organization, cell migration, proliferation, differentiation, and gene expression *(1,2)*. Integrins in their native environment can diffuse in the plane of the membrane, and this mobility is required for processes like the assembly of integrins to focal contacts or cell migration *(3,4)*.

The integrin aIIIbb3 was used in the present study as a model integrin because of the relative ease of isolation. It is the main receptor on blood platelets and plays an important role in the coagulation process. The properties of the purified receptor, e.g., ligand binding or receptor mobility, were investigated using integrin αIIbβ3 reconstituted into phospholipid vesicles and supported planar bilayers. As compared to studies with whole cells, both systems allow an investigation of the receptor without the complex influences of cytoplasmic components and other membrane proteins, but with the receptor still embedded in a lipid membrane. An advantage of the bilayer system compared to solid-phase assays is the receptors' mobility and native state. A thin water layer of about 30 Å between the bilayer and the support *(5)* is important for the lateral mobility of lipids and proteins in the plane of the membrane. Proteins adsorbed on lipid coated surfaces are often highly restricted in their lateral motion *(6)* and when directly adsorbed on plastic, proteins are immobilized and the native state may be lost by adsorption *(7)*. Furthermore, the surface concentration of the receptor may be controlled in the vesicle and bilayer system.

From: *Methods in Molecular Biology, vol. 139: Extracellular Matrix Protocols*
Edited by: C. Streuli and M. Grant © Humana Press Inc., Totowa, NJ

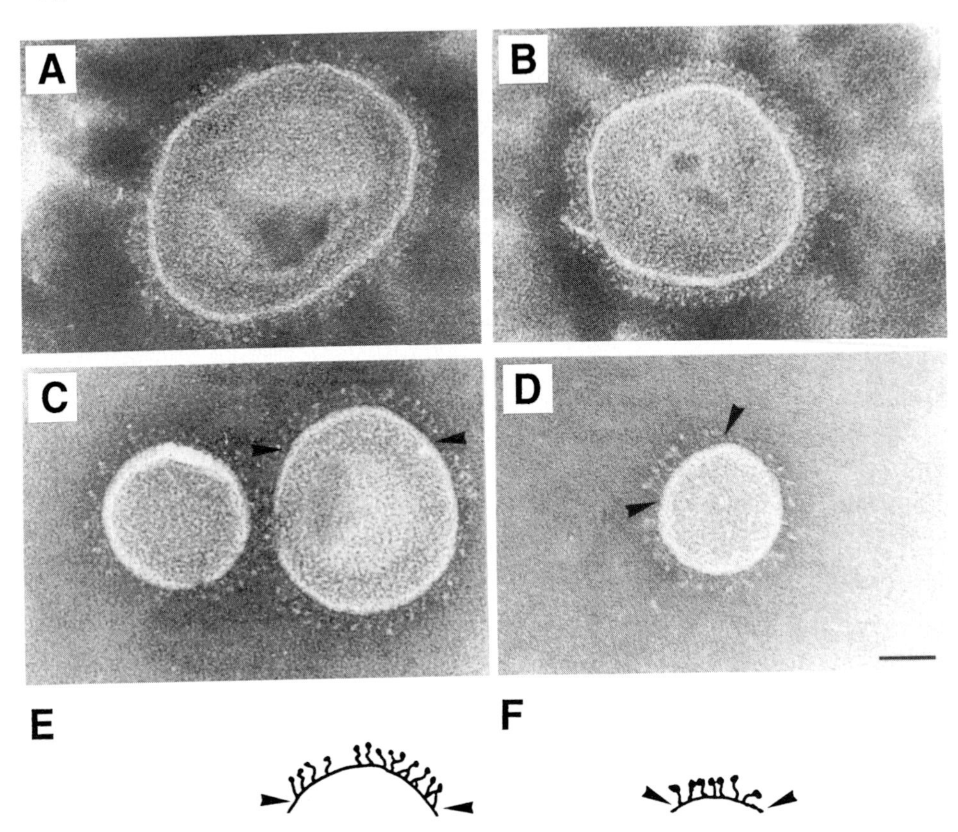

Fig. 1. Negatively stained electron micrographs of integrin αIIbβ3 incorporated into DMPC/DMPG vesicles. **A** and **B**: vesicles with high surface density of the integrin αIIbβ3, **C** and **D**: with low surface density of the integrin αIIbβ3. The bar represents 100 nm. Schematic drawings of selected regions in **C** and **D** represent the integrin organization within the membranes of vesicles (**E** and **F**).

Several methods for the reconstitution of integrins into phospholipid vesicles are described *(8–10)*. Müller et al. *(10)* were the first to use the method of detergent removal by BIO-Beads *(11)* to reconstitute integrin αIIIbβ3. It is important to analyze the vesicles after the reconstitution by electron microscopy in addition to the quantification of protein *(12)* and lipid *(13)* to distinguish between real protein incorporation into the vesicles or protein attachment to the vesicle surface. Electron micrographs of negatively stained specimen show integrin incorporation into phospholipid vesicles at different receptor densities (**Fig. 1A–D**). At a lower integrin density, well-separated integrin molecules can clearly be distinguished, indicating a nonclustered monodispersed distribution of the integrin. **Figure 2A** shows the schematic drawing

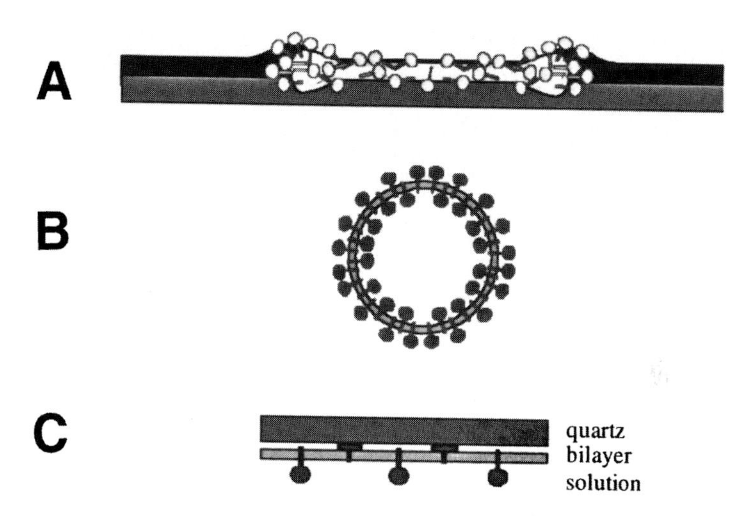

Fig. 2. Schematic drawing of negatively stained and hydrated specimens and distribution of integrins in a supported bilayer. (**A**) negatively stained specimen, (**B**) hydrated vesicle, (**C**) orientation of integrins in a quartz supported bilayer with their extracellular domains adsorbed to the quartz or in solution.

of a negatively stained specimen. The vesicles adsorbed on the grid became flattened and dehydrated. With this technique, only the outside of the vesicle can be visualized.

The vesicles with reconstituted integrin were further used to generate supported planar bilayers by vesicle fusion to quartz slides *(14)*. Using fluorescently labeled integrin αIIbβ3, the time course of vesicle fusion to quartz slides was monitored by total internal reflection fluorescence microscopy, TIRFM (**Fig. 3**). This method allows excitation of fluorophores, which are in close proximity to the quartz buffer interface (distance less than 100 nm) *(15)*. The integrins are homogeneously distributed in the plane of the supported bilayer.

Fluorescence recovery after photobleaching (FRAP) experiments with FITC-labeled αIIbβ3 incorporated into supported planar bilayers of DMPG/DMPC were performed to determine the lateral mobility of αIIbβ3 incorporated into supported planar bilayers. As a control, the mobility of the fluorescently labeled lipid NBD-PE in the same lipid mixture was determined. DMPG/DMPC vesicles (molar ratio 1:1) containing 0.5 mol% NBD-PE were prepared by the extrusion method according to *(16)*. By fusion to quartz slides, supported planar bilayers with homogeneously distributed labeled-NBD PE were generated. FRAP experiments were performed according to Kalb and Tamm *(17)* and for NBD-PE a diffusion coefficient $D = (4.4 \pm 0.40) \times 10^{-8}$ cm^2/s with a mobile fraction of $F = (93 \pm 7)\%$ was calculated. These values are in

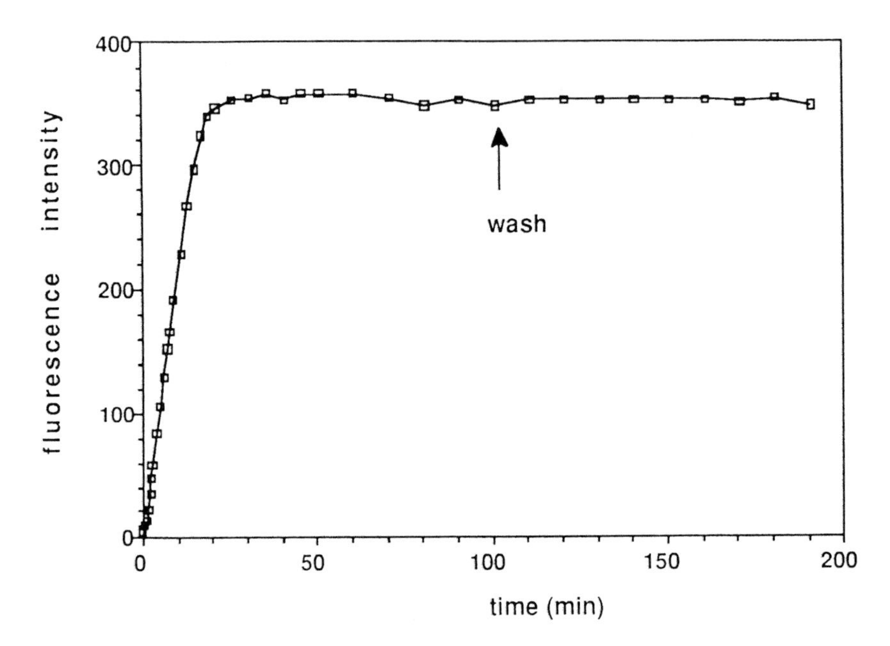

Fig. 3. Kinetics of formation of supported planar bilayers by fusion of DMPG/DMPC vesicles containing integrin αIIbβ3. Integrin was fluorescently labeled with FITC prior to incorporation into DMPG/DMPC vesicles. The cell was filled with Buffer B and vesicle suspension (about 50 nmol of lipid) was injected into the measuring cell containing vesicle fusion buffer to let the vesicles fuse onto a quartz slide. Fusion kinetics were followed by measuring the increase of fluorescence intensity by TIRFM. After 100 min the cell was washed with buffer to remove excess of vesicles. No change in fluorescence intensity was observed after the washing step (arrow).

close agreement with earlier data obtained earlier ***(18,19)*** and clearly indicate the intactness of the membrane with free mobility of the fluorescently labeled lipid in both leaflets of the supported bilayer.

The lateral diffusion coefficient of the FITC-labeled αIIbβ3 was found to be only six times smaller than that of the lipid. The value of 0.70×10^{-8} cm^2/s corresponds well with values found for other monodispersed membrane-spanning proteins at comparable surface concentrations ***(20)***. FRAP also revealed that about 50% of the integrin population was mobile (**Fig. 4**) and the immobility of the remaining population is consistent with integrins oriented with their large extracellular part towards the quartz support of the bilayer, allowing attachment of these domains to the support. In the mobile population, the small size of the cytoplasmic domains prevented the integrin from sticking to the support ***(21)***. **Figure 2C** depicts a schematic drawing of the orientation of the integrins in supported planar bilayers.

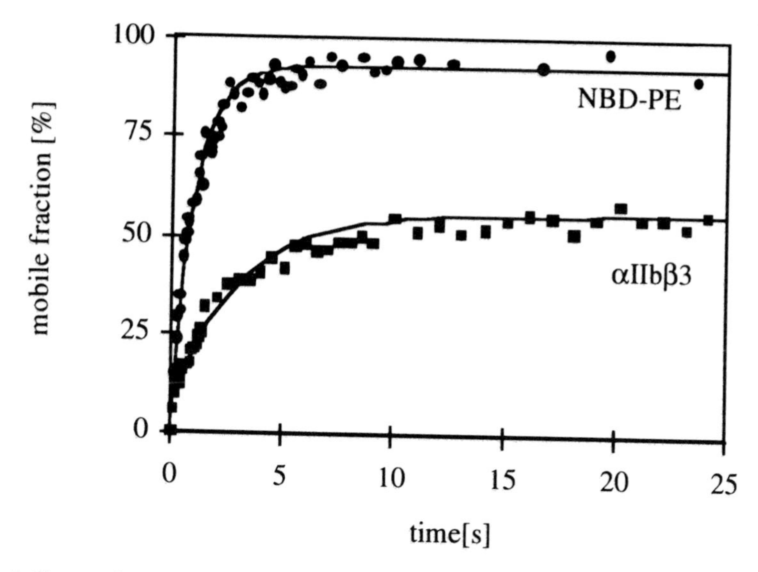

Fig. 4. Determination of the lateral diffusion coefficients of FITC-labeled αIIbβ3 and NBD-labeled PE incorporated into DMPG/DMPC bilayers. The time course of fluorescence recovery after photobleaching in supported planar DMPG/DMPC bilayers containing 0.5 mol% NBD-labeled PE (●) and FITC-labeled αIIbβ3 (■) were fitted to single exponential curves. All values are determined at 33°C.

Integrin αIIbβ3 reconstituted into supported planar bilayers retained its activity and ligand binding can be monitored in real time by TIRFM on condition that the ligand is fluorescently labeled. **Figure 5** shows the association and dissociation kinetics of TRITC-labeled GRGDSPC-peptide to αIIbβ3 incorporated into a DMPC/DMPG bilayer. The binding was fully reversible as shown by competitive inhibition with unlabeled GRGDS-peptide. A second addition of labeled GRGDSPC-peptide yielded again the same fluorescence intensity as before, demonstrating the stability of the membrane.

2. Materials

2.1. Fluorescence Labeling of Proteins

1. 0.2 *M* $NaHCO_3$ in distilled water.
2. FITC or TRITC dissolved in dry DMSO to a final concentration of 5 m*M*, freshly prepared prior to labeling (*see* **Note 1**).
3. Buffer A: 20 m*M* Tris-HCl, pH 7.4, 50 m*M* NaCl, 1 m*M* $CaCl_2$.
4. Buffer B: Dissolve 45.4 µL Triton X-100 in Buffer A, final volume 50 mL, final concentration 0.1% (w/v).
5. Sephadex G-25 M column, (prepacked, Pharmacia Biotechnology, Uppsala, Sweden) equilibrated with Buffer B.

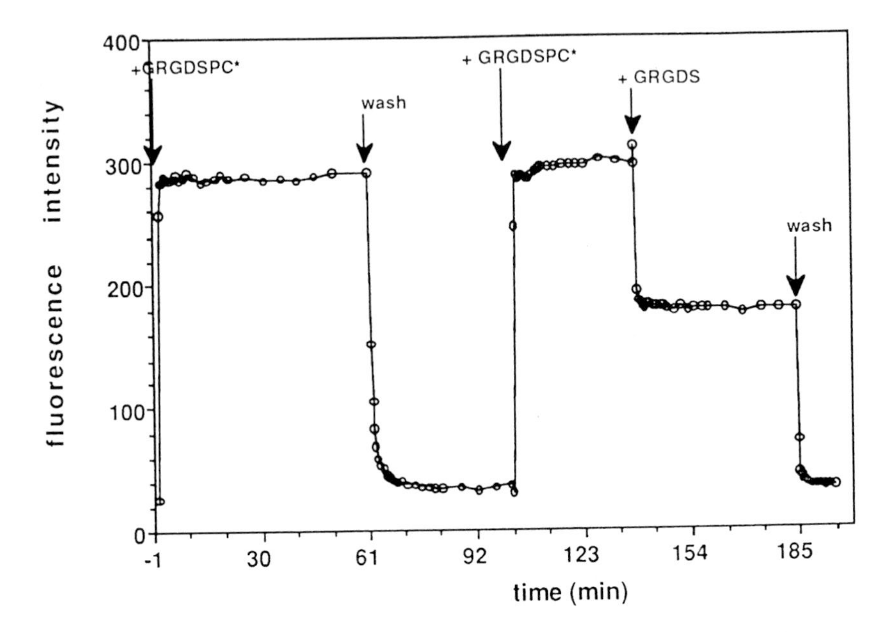

Fig. 5. Association and dissociation kinetics of peptide binding to αIIbβ3 in bilayers by TIRFM. Bilayers with αIIbβ3 in DMPG/DMPC vesicles were formed by vesicle fusion. TRITC-labeled peptide GRGDSPC was added at a concentration of 3300 n*M* and association kinetics were followed by measuring the time course of the fluorescence intensity (°) at an excitation of 514 nm. After reaching the equilibrium, the cell was washed with Buffer C (4 mL/min) while monitoring the time-course of fluorescence intensity. TRITC-labeled GRGDSPC-peptide could be dissociated by addition of a 100-fold molar excess of unlabeled GRGDS-peptide. The background signal of about 180 relative fluorescence units was caused by labeled peptide which stayed free in solution and could be abolished by washing with buffer.

2.2. Vesicle Preparation

1. Lipid stock solution: lyophilized phospholipids DMPG and DMPC (Avanti Polar Lipids, Alabaster, AL) were dissolved in chloroform/methanol 2:1 at a concentration of about 15 m*M*. Precooled solvent (30 min at –20°C) is easier to pipet (*see* **Note 2**).
2. 2 *M* (w/v) Tris-HCl, adjust pH to 7.4 with concentrated HCl.
3. 5 *M* (w/v) NaCl, warm the solution to facilitate dissolvation.
4. Buffer A: 20 m*M* Tris-HCl pH 7.4, 50 m*M* NaCl, 1 m*M* $CaCl_2$.
5. Buffer B: Dissolve 45.4 µL Triton X-100 in buffer A, final volume 50 mL, final concentration 0.1% (w/v) (*see* **Note 3**).
6. About 500 mg of BIO-Beads SM-2 (Bio-Rad, Richmond, CA) were first washed with 50 mL methanol and then with 500 mL distilled water. The wet beads were collected on a sintered glass funnel (*see* **Note 4**).

7. Four-step sucrose gradient with 0.4 M, 0.8 M, 1.2 *M* and 2 *M* sucrose: prepare stock solutions of 0.4 M, 0.8 M, 1.2 *M*, and 2 *M* sucrose in Buffer A. Take a clear 4-mL centrifugal tube and add 850 µL 2 *M* sucrose, then overlay with 850 µL 1.2 *M* and 850 µL 0.8 M, followed by 600 µL of 0.4 *M* sucrose.

2.3. Vesicle Analysis

2.3.1. Protein Determination

1. BSA standard, 0.2 mg/ml
2. 0.15% Na-deoxycholate: dissolve 15 mg Na-deoxycholate in 10 mL distilled H_2O, store at –20°C.
3. 72–100 % (w/v) TCA: Dissolve 100 g TCA (72–100%) in distilled water to a final volume of 100 mL.
4. 1% (w/v) $CuSO_4$, store at 4°C.
5. 2% (w/v) Na-K-tartrate, store at 4°C.
6. 10% (w/v) SDS, store at room temperature.
7. 0.8 *M* NaOH, store at room temperature.
8. Solution A: Dissolve 200 µg Na_2CO_3 in 1600 µL H_2O, add 200 µL 1% $CuSO_4$ and 200 µL 2% Na-K-tartrate, add 2 mL 10% SDS, 2 mL 0.8M NaOH, and 10 mL H_2O.
9. Solution B: Dilute Folin-Ciocalteau-reagent (Merck, Darmstadt, Germany) 1:6 with distilled water.

2.3.2. Phosphorus Determination

1. Perchloric acid, not exceeding 72%.
2. 10% ascorbic acid, freshly prepared and stored at –20°C at the latest for 1 mo.
3. Molybdate reagent: To 2.2 g $(NH_4)_6Mo_7O_{24}$ *4 H_2O add 14.3 mL H_2SO_4, 95–98% and then slowly add distilled water to 1000 mL. Prevent from exposure to light.
4. X100 phosphate standard, 40 m*M*,: To 712.0 mg Na_2HPO_4 *2H_2O add distilled water to a final volume of 100 mL. Dilute to X1 phosphate standard, 400 µ*M*, standard, store at –20°C.
5. Phosphate-free, disposable glass tubes (borate glass).

2.3.3. Electron Microscopy

1. 2% (v/v) Na-phosphotungstate in distilled water, adjusted to pH 7.0, sterile filtered through a 0.22-µm filter prior to use.
2. Collodion carbon grids, glow discharged for 30–45 s in a glow discharger.

2.4. Vesicle Fusion

1. Buffer C: 20 m*M* Tris-HCl pH 7.4, 150 m*M* NaCl, 1 m*M* $CaCl_2$, 1 m*M* $MgCl_2$, 1 m*M* $MnCl_2$. Dissolve $MnCl_2$ directly before the buffer is used (*see* **Note 5**).
2. Detergent for cleaning the quartz slides: Linbro 7X-PF (ICN Flow Biochemicals, Costa Mesa, CA), diluted 1:6 with distilled water, can be used five times for washing.
3. Quartz slides.

3. Methods

3.1. Fluorescence Labeling of Proteins

1. Integrin αIIbβ3, dissolved in buffer B, is adjusted to pH 8.0 by addition of 0.2 *M* $NaHCO_3$.
2. Add a 50-fold molar excess of FITC or TRITC to the protein, vortex shortly.
3. Incubate the mixture for 30 min in the dark at room temperature.
4. Remove excess of FITC and TRITC by gelfiltration on a Sephadex G-25 *M* column, equilibrated with Buffer B.
5. Measure the OD of the FITC- or TRITC-labeled protein at 493 nm ($\varepsilon = 153.8\ cm^2/mg$) or 555 nm ($\varepsilon = 151.5\ cm^2/mg$), respectively. The degree of labeling can be calculated according to the following equation: $c = OD/\varepsilon * M * d$. The degree of labeling is between 1–2 FITC or TRITC groups per integrin.

3.2. Vesicle Preparation

1. Take an aliquot corresponding to 435 nmol of each lipid solution using glass micropipets.
2. Dry lipids first under an N_2-stream, then in high vacuum for 2 h.
3. Add 930 µL Buffer B to the lipid film and vortex the solution until the lipid film is completely solubilized. Transfer to a 1.5-mL Eppendorf tube.
4. Add 70 µL integrin αIIbβ3 dissolved in Buffer B (3 mg/mL) to the lipid solution.
5. Shake the mixture for 120 min at 37°C.
6. Add 50 mg of washed BIO-Beads and shake for 210 min (*see* **Note 6**).
7. Spin briefly to precipitate the beads and transfer the supernatant into a new tube. Add a second 50 mg of BIO-beads (*see* **Note 7**).
8. Spin again, remove the supernatant, it should now be opalescent. Load the supernatant on top of the sucrose gradient and fill the tube with Buffer A.
9. Use a TST60 (Beckman, Fullerton, CA) or equivalent vertical rotor and centrifuge at 4°C and 275,000*g* for 24 h.
10. The vesicle-containing band can be seen as a white band in the 0.8 *M* sucrose step. Remove the upper part of the gradient carefully with a Pasteur pipet. Collect the white band from the gradient, try to keep the volume small, normally between 250–400 µL.
11. Dialyze the vesicles against Buffer B for 2–3 d with 2 buffer changes per day.
12. Analyze the vesicles by protein and lipid determination and by EM.

3.3. Vesicle Analysis

3.3.1. Protein Determination

This method allows the determination of protein even at a high lipid/protein ratio.

1. Ideal protein quantity in the sample: 10–20 µg.
2. Make a calibration curve consisting of 0, 20, 40, 60, 80, 100 µL of the BSA standard.
3. Add water to 1000 mL.

4. Add 100 µL TCA. Leave for 1 h at room temperature (*see* **Note 8**).
5. Spin 10 min at full speed in an Eppendorf centrifuge, aspirate supernatant carefully.
6. Turn tubes and dry on paper.
7. Add 750 µL solution A and vortex to resuspend the precipitate.
8. Incubate for exactly 10 min per tube and stop the reaction by adding 187.5 µL of solution B (*see* **Note 9**).
9. Leave for 30 min at room temperature and measure the absorption at 750 nm.
10. The yield of reconstituted protein is normally between 25% and 40%.

3.3.2. Lipid Analysis

1. The amount of lipid in the sample should be around 25 nmol.
2. Make a calibration curve consisting of 0, 25, 50, 75, 100, 125 µL of the 400 µ*M* phosphate standard.
3. Add 0.2 mL 68–72% perchloric acid.
4. Heat 30 min to 160°C, then vortex the solution (could have a brown color) and increase the temperature to 180°C for additional 90 min (solution should now be colorless).
5. Allow the tubes to cool down and add 0.2 mL of the molybdate reagent, vortex and add 0.25 mL of the 10% ascorbic acid and vortex again.
6. Immerse the tubes in boiling water for precisely 10 min.
7. Stop the reaction by cooling the tubes in ice water for some minutes, remove the ice and allow the tubes to equilibrate to room temperature.
8. Measure the absorption at 812 nm using zero of calibration as reference (*see* **Note 10**).
9. The yield of reconstituted lipid is normally between 40–60% (*see* **Note 11**).

3.3.3. Electron Microscopy of Negatively Stained Specimen

1. Dilute the vesicle suspension 1:5 with Buffer A.
2. Glow discharge collodion carbon grids for 30–45 s.
3. Put the grid into a pair of tweezers and add 10 µL of the diluted vesicle suspension.
4. Incubate for 3–4 min.
5. Add 5 µL stain, incubate for 30 s (*see* **Note 12**).
6. Blot the fluid from the grid, add another 10 µL stain.
7. Incubate for 3–4 min, the grid covered to prevent drying.
8. Blot the solution and let the grid dry completely.

3.4. Formation of Supported Planar Bilayers

1. To clean the quartz slides, put them on a rack so that the detergent can reach both sides of the slide.
2. Put the quartz slides in a 300-mL beaker, add diluted Linbro 7X-PF until they are completely covered and boil the quartz slides for 20 min.
3. Put the hot beaker in a bath sonicater and sonicate for 30 min.
4. Remove the detergent and rinse the slides for 30 min with distilled water.
5. Dry the slides for 1 h at 150°C.

6. Clean the slides in an argon plasma cleaner prior to vesicle fusion: Put a slide in the plasma cleaner, close and evacuate the chamber, flush the chamber carefully with argon, (the vacuum should be kept at low pressure), apply high voltage. The purple color of the plasma is now visible. Leave the slide inside for 15 min.
7. Take the slide out of the plasma cleaner and put it into the analysis chamber. Fill the chamber with Buffer C and inject about 50 nmol of the vesicle suspension. Stir the buffer during the fusion process, which is finished after about 30 min. Wash the analysis chamber with Buffer C to remove excess of vesicles. The fluorescence signal should be stable after the washing step (*see* **Note 13**).

4. Notes

1. Prevent DMSO from absorbing water, this will reduce the amount of active fluorophores before the labeling reaction is started.
2. The reconstitution was also performed using other lipids/lipid mixtures as POPC, POPC/POPS (molar ratio 75:25), POPC/POPG (molar ratio 50:50), PC/PE (molar ratio 70:30), which did in neither case result in a good integrin reconstitution. With pure DMPC the result was comparable to the equimolar mixture of DMPG/DMPC.
3. When 100 m*M* octyl β-D-glucopyranoside or 30 m*M* octyl-POE were used to solubilize lipid and protein no reconstitution could be seen in the electron microscope.
4. Washed BIO-Beads were stored in 20% EtOH at 4°C and washed again with H_2O prior to use. The beads were not allowed to dry.
5. Solutions of $MnCl_2$ are not very stable because of oxidation to MnO_2 and should always be freshly prepared.
6. For control measurements, vesicles without protein were prepared using the same reconstitution protocol, but the incubation time for the first addition of BIO-Beads was reduced to 120 min. For reconstitution of fluorescently labeled integrin the reconstitution protocol was used with the following modifications: the integrin-lipid mixture was incubated twice with only 25 mg of BIO-Beads for 90 min and 60 min, respectively.
7. The supernatant can be easily removed from the beads with a plastic pipet tip that was squeezed with a pair of tweezers, so that you can only take up the fluid, not the beads.
8. The precipitation can also be done at –20°C for 15 min.
9. The first reaction is not an end point reaction and must, therefore, be timed accurately. 15 s between tubes should be sufficient to resuspend the precipitate.
10. The absorbance of the reference in the phosphate determination should be 0.012-0.020. If higher, the molybdate solution was too old. The absorbance should be about 0.1–0.125/ 20 nmol of phosphate. Using linear regression, the regression coefficient is generally better than 0.997.
11. Calculate the molar protein to lipid ratio, for a good preparation it should be in between 1:1600 to 1:1200.
12. Use only stain at neutral pH and do not stain with conventional uranyl acetate or uranyl formiate, which would destroy the vesicles.

13. The membranes were stable for at least 48 h and could not be damaged by washing at high flow rates or stirring the buffer at high speed. To prevent evaporation of water, the inlet and outlet of the analysis chamber was sealed with parafilm.

References

1. Hynes, R. O. (1987) Integrins: a family of cell surface receptors. *Cell* **48,** 549–554.
2. Hynes, R. O. (1992) Integrins: versatility, modulation and signaling in cell adhesion. *Cell* **69,** 11–25.
3. Duband, J.-L., Nuckolls, G. H., Ishihara, A., Hasegawa, T., Yamada, K. M., Thiery, J. P., and Jacobson, K. (1988) Fibronectin receptor exhibits high lateral mobility in embryonic locomoting cells but is immobile in focal contacts and fibrillar streaks in stationary cells. *J. Cell Biol.* **107,** 1385–1396.
4. Schmidt, C. E., Horwitz, A. F., Lauffenburger, D. A., and Sheetz, M. P. (1995) Integrin-cytoskeletal interactions in migrating fibroblsts are dynamic, asymmetric and regulated. *J.Cell Biol.* **123,** 977–991.
5. Johnson, S. J., Bayerl, T. M., McDermott, D. C., Adam, G. W., Rennie, A. R., Thomas, R. K., and Sackmann, E. (1991) Structure of an adsorbed dimyristoylphosphatidylcholine bilayer measured with specular reflection of neutrons. *Biophys. J.* **59,** 289–294.
6. Pachence, J. M., Amador, S., Maniara, G., Vanderkooi, J., Dutton, P. L., and Blasie, J. K. (1990) Orientation and lateral mobility of cytochrome c on the surface of ultrathin lipid multilayer films. *Biophys. J.* **58,** 379–389.
7. Tilton, R. D., Gast, A. P., and Robertson, C. R. (1990) Surface diffusion of interacting proteins. Effect of concentration on the lateral mobility of adsorbed bovine serum albumin. *Biophys. J.* **58,** 1321–1326.
8. Müller, B., Zerwes, H.-G., Tangemann, K., Peter, J., and Engel, J. (1993) Two-step binding mechanism of fibrinogen to αIIbβ3 integrin reconstituted into planar lipid bilayers. *J. Biol. Chem.* **268,** 6800–6808.
9. Parise, L. V. and Philips, D. R. (1985) Reconstitution of the purified fibrinogen-receptor. *J. Biol. Chem.* **260,** 10,698–10,707.
10. Pytela, R., Pirschbacher, M. D., and Ruoslahti, E. (1985) Platelet membrane glycoprotein IIbIIIa; Member of a family of arg-gly-asp-specific adhesion receptors. *Cell* **40,** 191–198.
11. Holloway, P. W. (1973) A simple procedure for removal of triton X-100 from protein samples. *Analyt. Biochem.* **53,** 304–308.
12. Peterson, G. L. (1977) A simplification of the protein assay method of Lowry et al. which is more generally applicable. *Anal. Biochem.* **83,** 346–356.
13. Böttcher, C. J. F., Van Gent, C. M., and Fries, C. (1961) A rapid and sensitive sub-micro phosphorus determination. *Anal. Chim. Acta* **24,** 203–204.
14. Brian, A. A., and McConnell, H. M. (1984) Allogenic stimunlation of cytotoxic T cells by supported planar membranes. *Proc. Natl. Acad. Sci. USA* **81,** 6159–6163.
15. Tamm, L. K. and Kalb, E. (1993). Microspectofluorometry on supported planar membranes, in Molecular Luminescence Spectroscopy, Part **3,** Chemical Analysis Series (Schulman, S. G., ed.), J Wiley, New York, pp. 253–305.

16. Hope, M. J., Bally, M. B., Webb, G., and Cullis, P. R. (1985) Production of large unilamellar vesicles by a rapid extrusion procedure. Characterization of size distribution, trapped volume and ability to maintain a membrane potential. *Biochim. Biophys. Acta* **812,** 55–65.
17. Kalb, E. and Tamm, L. K. (1992) Incorporation of cytochrom b5 into supported phospholipid bilayers by vesicle fusion to supported monolayers. *Thin Solid Films* **210/211,** 763–765.
18. Kalb, E., Frey, S., and Tamm, L. K. (1992) Formation of supported planar bilayers by fusion of vesicles to supported phospholipid monolayers. *Biochim. Biophys. Acta* **1103,** 307–316.
19. Tamm, L. K. (1988) Lateral diffusion and fluorescence microscope studies on a monoclonal antibody specifically bound to supported phospholipid bilayers. *Biochemistry* **27,** 1450–1457.
20. Vaz, W. L. C., Goodsaid-Zalduondo, F., and Jacobson, K. (1984) Lateral diffusion of lipids and proteins in bilayer membranes. *FEBS Lett.* **174,** 199–207.
21. Erb, E.-M., Tangemann, K., Bohrmann, B., Müller, B., and Engel, J. (1997) Integrin αIIbβ3 reconstituted into Lipid Bilayers is nonclustered in its activated state but clusters after fibrinogen binding. *Biochemistry* **36,** 7395–7402.

8

Confocal-FRAP Analysis of ECM Molecular Interactions

Timothy Hardingham and Philip Gribbon

1. Introduction

Extracellular matrices (ECM) contain a mixture of fibrillar and nonfibrillar macromolecular components, which together form a composite structure *(1–3)*. It is the ECM that defines the architecture, the form, and the biomechanical properties of different tissues *(4,5)*. Among the nonfibrillar macromolecules, the highly charged proteoglycans and hyaluronan are major components that occur at high concentration and greatly influence the movement of solutes and water between the tissue and the circulation, and control the access to cells of nutrients, metabolites, growth factors, and chemokines *(6,7)*. This local environmental regulation may have important consequences on cellular functions, especially in tissues with large dense ECMs, such as articular cartilage.

Studies on the physical properties of proteoglycans and glycosaminoglycans have commonly included rheological, light scattering, and ultracentrifugation analyses. These are carried out under "ideal" conditions, which are at low concentration in order to minimize the effects of intermolecular interactions *(8–11)*. However, to understand their properties at the high concentrations found physiologically, these methods are of limited application. An alternative experimental approach has been developed using fluorescence recovery after photobleaching (FRAP) *(12)*. It was originally used for investigating cell membrane dynamics, but in an adapted procedure with a confocal microscope, *(13–15)*, it has now been applied to study intermolecular interactions and network formation of biopolymers in solution *(16)*. It was used to characterize the intermolecular interactions of aggrecan, the large aggregating proteoglycan containing many chondroitin sulphate and keratan sulphate

From: *Methods in Molecular Biology, vol. 139: Extracellular Matrix Protocols*
Edited by: C. Streuli and M. Grant © Humana Press Inc., Totowa, NJ

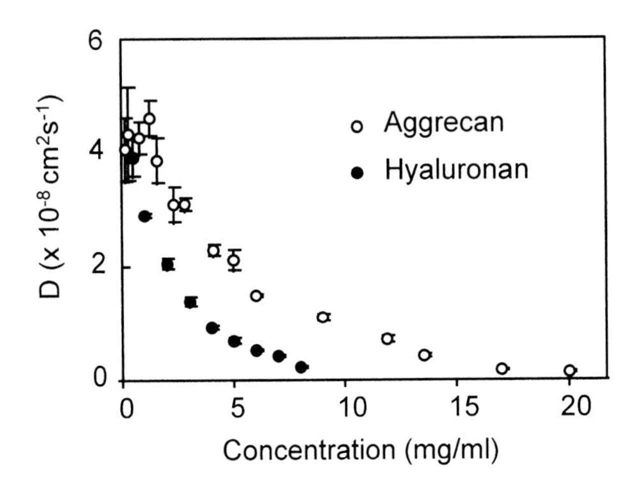

Fig. 1. Confocal FRAP measurements of the lateral translational selfdiffusion coefficient of 2500 kDa aggrecan and 870 kDa hyaluronan, as a function of concentration. These data show aggrecan forms a more dynamic network than hyaluronan, because of its more compact branched structure.

chains. Comparison of the self diffusion of aggrecan and hyaluronan shows a contrast between the extended stiffened random coil behavior of hyaluronan with the more compact conformation of aggrecan (**Fig. 1**). Investigations of the permeability of aggrecan networks to FITC-dextran solutes demonstrate the effects of size-selective molecular sieving (**Fig. 2**) with the mobility of a 2000 kDa probe being reduced much more than a 167 kDa probe as the concentration of aggrecan rises.

Experimentally, confocal-FRAP involves viewing a solution of a fluorescently labeled molecule with a confocal microscope, locally creating a bleached area in the solution, followed by observing the subsequent redistribution of fluorescence. The bleaching can be either continuous or, as described below, it can be achieved rapidly by brief high-power laser illumination. In simple solutions of monodisperse macromolecules, the measurement of the recovery of fluorescence yields a long-time translational lateral self diffusion coefficient *(17)*. This describes the movement of the macromolecule through a matrix of like macromolecules. Importantly, the technique measures self diffusion coefficients under equilibrium conditions, which in the absence of shear, or flow, favors any weak intermolecular interactions that may stabilize the network. The technique is also well suited to measure the mobility of a labeled probe molecule within a macromolecular network in solution. This measures the lateral diffusion coefficient of the tracer from which the effective pore dimensions of the matrix can be calculated. It thus enables the effects of com-

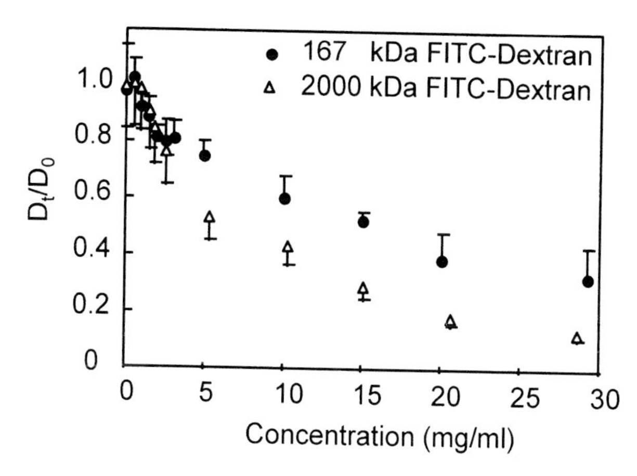

Fig. 2. Relative lateral tracer diffusion coefficients of FITC-dextrans in aggrecan (2000 kDa) (Gribbon and Hardingham, unpublished data). The relative lateral tracer-diffusion coefficient is the ratio of the tracer-diffusion coefficient at a finite concentration (D_t) to that at zero concentration (D_0). These data show aggrecan networks influence solute transport by acting as size-selective molecular sieves.

plex mixtures of macromolecules on the characteristics of the matrix to be determined by measurements of a single-labeled probe.

In its original development, the nonconfocal FRAP technique was applied to measure solute mobility in tissues and within single cells *(18)*. Experimentally, this provided a number of constraints on the choice of bleach mode, geometry, and recovery analysis, which were also limited by the fluorophore concentration and the time-scale of diffusion. Fortunately, with homogenous solutions of biological macromolecules there are fewer experimental constraints and the technique is well suited to a confocal microscope. Using the scanning facility of a confocal microscope, the bleach dimension can be adjusted to give bleach-recovery times suited to the macromolecule being investigated. ECM macromolecules frequently have a high molecular mass and relatively low translational-diffusion coefficients. But with appropriate manipulation of the bleach and recovery conditions, the translational diffusion of both low-molecular-weight probes and of concentrated- macromolecules-forming networks are open to investigation.

In this chapter, the practical use of an unmodified commercial confocal microscope for determining the diffusion properties of macromolecular ECM components and of matrix-probe molecules will be outlined. The preparation and characterization of aggrecan and hyaluronan solutions will be used as examples, although the technique can be generally applied to the investigation of a wide range of polymer–polymer and polymer–solute interactions.

2. Materials

2.1. Protein and Carbohydrate Labeling

1. Protein labeling buffer, 0.2 *M*: $NaHCO_3$ (1.48 g/100 mL), Na_2CO_3 (0.24 g/100 mL), pH 9.0. Filter 0.2-mm before use. Should be prepared fresh, keeps at 4°C for 2 wk.
2. Phosphate-buffered saline (PBS): 8 g NaCl, 0.2 g KCl, 1.44 g Na_2HPO_4. $2H_2O$, 0.2 g KH_2PO_4, made up to 1 L, pH 7.4. Filter 0.2-µm before use, keep at 4°C.
3. Carbohydrate (hyaluronan) labeling buffer, 0.2 *M*: $Na_2B_4O_7.10H_2O$ (0.572 g/100 mL), boric acid (0.866 g/100 mL), pH. 8.0. Filter 0.2-µm before use.
4. FITC.
5. Hyaluronan (HA).
6. CNBr.
7. Hi-trap columns.
8. Fluoresceinamine.
9. Centriplus (Amicon, Danvers, MA).
10. Dialysis tubing.

2.2. Photobleaching Experiments

1. Upright confocal laser scanner and fluorescence microscope: the authors use a 100 mW Argon-Ion laser (*see* **Note 1**) with a MRC-1000 scanner (Bio-Rad, Hemel Hempstead, UK) attached to a Optiphot microscope (Nikon [Membrane filtration Products, San Antonio, TX]) (**Fig. 3**). The chosen scan head needs to include a motorized intensity filter changer and a fluorescein compatible filter set. With the MRC-1000, data acquisition and analysis are controlled by a simple programming language (MPL) for which the authors have written appropriate "macro" procedures. The majority of confocal microscopes include similar facilities for user application development.
2. Microscope objectives of numerical aperture <0.6 (magnification: ×10, ×20). A low numerical aperture is most appropriate, as it gives a more uniform and parallel bleach volume, allowing the results to be calculated using a 2-dimensional analysis ***(15)***.
3. Temperature-controlled microscope stage capable of operating between 5 and 60°C at ± 0.1°C, for example Linkam Instruments PE 60 (Linkam Ltd, UK).
4. Cavity microscope slides, 12-mm cavity diameter, 30-µL volume (Scientific Lab Supplies, Nottingham, UK). Slides should be scrupulously cleaned and rinsed with deionized water before use.

3. Methods

3.1. Sample Preparation and Fluorescence Labeling

Glycosaminoglycans, including chondroitin sulphate and hyaluronan, are available commercially (e.g., Sigma Corp., St. Louis, MO, Seikagaku Corp., Tokyo, Japan) from several tissue sources (e.g., bacteria, cock's comb, umbilical cord), although the molecular weight and degree of protein contamination

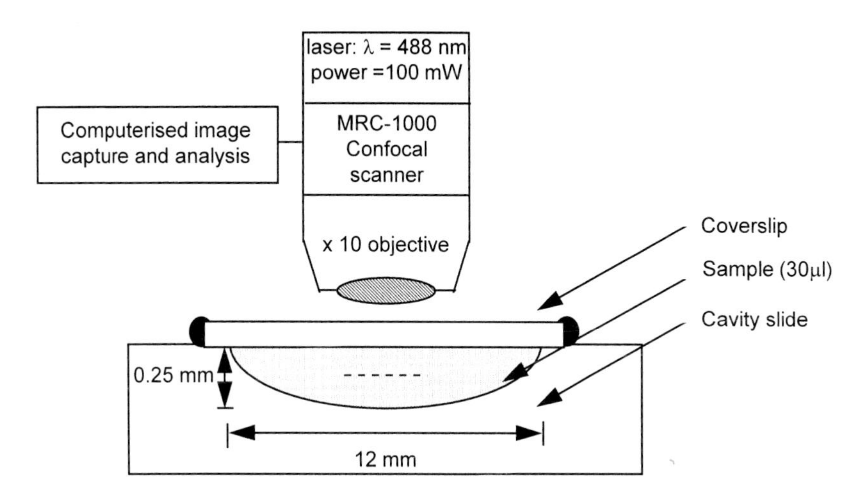

Fig. 3. Experimental arrangement for confocal FRAP. Sample (30 μL) is contained in a cavity microscope slide with a sealed coverslip. Square bleaches are generated by scanning at high laser power through the volume of the sample. Fluorescence recovery is observed at low laser power in a focal plane (- - - -) in the centre of the slide, midway between coverslip and slide.

vary greatly. The pharmaceutical grade material is least contaminated (<0.1 % protein). For tracer experiments, commercial protein preparations sold as "molecular weight standards" can usually be fluorescein-isothiocyanate (FITC) labeled and used without further purification (e.g., soy bean trypsin inhibitor and bovine serum albumin from Sigma). However, many commercially available FITC-labeled proteins are excessively substituted (>10 mol FITC per mol protein) and should be avoided as their diffusion properties may differ significantly from those of the native protein. FITC-dextrans are useful, uncharged, matrix probes and are available in many different sizes from molecular weight 4 kDa to 2500 kDa (Sigma). At present, commercial sources of aggrecan are expensive, but it can be prepared by extraction of cartilage and stored dry or in solution at –20°C ***(19)***. The labeling protocols outlined for proteins and carbohydrates such as hyaluronan are generally applicable, although for molecules, such as aggrecan, the protein content is only ~10% by weight, so correspondingly reduced amounts of FITC are sufficient for labeling. Concentrated solutions of aggrecan and HA are viscous, so care should be taken to minimize handling losses.

3.1.1. Protein Labeling

1. Dissolve 10 mg of protein in 5 mL of carbonate labeling buffer (pH 9.0) by mixing with gentle rotation for 24 h at 4°C (48 h for aggrecan solutions).

2. Add 0.1 mg of FITC to 1 mL of labeling buffer. Immediately add the cold FITC solution dropwise until the final molar ratio of FITC to protein is 5:1 and incubate overnight at 4°C with gentle rotation.
3. Remove unconjugated FITC by exhaustive dialysis against deionized water at 4°C (*see* **Note 2**) and freeze-dry to constant mass.

3.1.2. Hyaluronan Labeling

1. Dissolve 2 mg of HA in 1 mL of deionized water for 48 h at 4°C with gentle stirring. Add 0.2 mL of CNBr solution (10 mg/ml in H_2O, make up just before use, *caution—toxic and hazardous*) and leave for 5 min (*see* **Note 3**).
2. Isolate activated HA using 2 x Hi-Trap columns (Pharmacia, Uppsala, Sweden) in series, elute with 0.2 *M* borate buffer and collect 1 mL fractions. Add 2 mg of fluoresceinamine (FA) to the activated GAG (fractions 4 - 6) and incubate overnight at 4°C with gentle rotation.
3. Remove salt and free FA by ultrafiltration at 4°C (Centriplus, Amicon, centrifuge 3000*g*, 10 × washes with 10 mL H_2O) and freeze-dry to constant mass. Alternatively, free FA and salt may be removed by exhaustive dialysis, although this takes considerably longer, especially for viscous hyaluronan solutions.

3.1.3. Preparation of Samples for Confocal FRAP Analysis

Fluoresceine concentration in final solutions for analysis should be kept below 1 m*M* to avoid fluorescence self quenching effects. For solution studies below 10 m*M* ionic strength, or below pH 6.0, the fluorescence signal is greatly reduced, thus producing noisy data at low concentrations. More heavily labeled material may then be necessary ***(20)***.

1. Before using labeled samples, it is advisable to check for the presence of unconjugated fluorescein. Load 0.2 mg sample in 0.5 mL on a Hi-Trap column, elute with PBS, and take 1-mL fractions. No fluorescence should be evident after fraction 4 (*see* **Note 4**).
2. Make up solutions of labeled or unlabeled matrix macromolecules by weight and equilibrate at 4°C for 48 h. In tracer studies, add FITC-labeled globular protein probes (e.g., FITC-bovine serum albumin and FITC-soy bean trypsin inhibitor), or polymer probes (e.g., FITC-dextrans) at <0.1 mg/ml and equilibrate for a further 24 h at 4°C.
3. Pipet 30 µL of sample onto a cavity slide and seal under a 20-mm diameter circular coverslip using nail varnish. Allow 10 min to dry (**Fig. 3**).
4. Place sealed cavity slide on thermally regulated microscope stage and allow 15 min to reach thermal equilibrium. Care should be taken to avoid condensation in low temperature studies.
5. Adjust the microscope stage height so the objective focal plane is midway between the coverslip and the lower interface between the solution and the slide, this is usually ~120 µm below the coverslip.

3.2. Microscope Setup and Data Acquisition

In a confocal-FRAP experiment, a prebleach image is acquired, the sample is bleached, and a recovery image series is collected and analyzed. Scanning confocal microscopes excite fluorescence with a laser and detect emitted light with a photomultiplier (PMT). The final image intensity is a complex function of the sample concentration, laser intensity, scan rate, optical path, confocal pinhole size, and PMT gain. These parameters must be set so that imaging of the sample is within the linear region of PMT response for imaging prebleach and during the recovery. The essential elements of a protocol, based on a square bleach and a moments recovery analysis will be given. This protocol, though general, does require that the switching between microscope configurations is directed by software, a facility which is available on all recent generations of confocal scanners. The analysis is not dependent on the shape of the bleach as long as it is symmetrical about the bleach center, and whereas a circular bleach may give the optimum geometry, it is much easier and faster to generate a square bleach because of the scanning action of a confocal microscope.

1. Determine instrument settings for bleach and prebleach/recovery and save as separate set-up configurations to be accessed by the bleaching software (*see* **Note 5**). Typically, for prebleach/recovery the scan time = 2 s, laser intensity = 1% of maximum, PMT gain = 1000 V, iris size = 75% of maximum. Typical bleach settings are scan time = 2 s, laser intensity = 100%, PMT gain = 0, iris size = minimum (*see* **Note 6**).
2. At the start of the bleaching and data acquisition program, a square region for bleaching is defined at the center of the field of view. In general, the more mobile species the larger the bleach and the faster the image scan rate required.
3. The microscope is set and one or more prebleach images are scanned, centered on the proposed bleach. Scan dimensions are ×4 the bleach width, optical zoom = 1. Several prebleach images may be averaged to reduce noise.
4. The microscope switches to bleach configuration and sample bleaching is performed at 100% laser power (*see* **Note 7**). The bleach scan area is ×16 smaller in area than the prebleach area, although the scan time is the same. The total energy delivered to the bleach scan area is therefore ×1600 that conveyed to the same area during a single prebleach or recovery scan.
5. The microscope resets to the prebleach setup, and a series of recovery images (up to 50) are collected. For slowly moving species, two or more images should be taken and averaged in real time at each scan time-point (Kalman type averaging is suitable). More mobile species require faster data acquisition, the rate of image acquisition can be increased by saving images direct to the computer's random access memory (use a software utility such as MS RAM-disk) rather than the hard drive (*see* **Note 8**).
6. Final images should be taken at two long time points and compared to check that the recovery is complete. This is important as it enables the mobile fraction to be calculated.

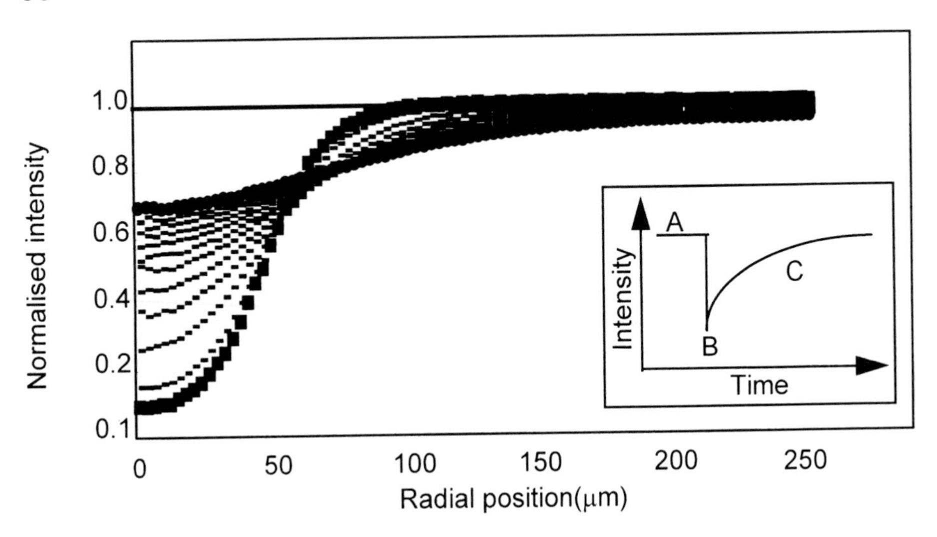

Fig. 4. Time series of calculated radially averaged fluorescence-intensity profiles (normalized) for a confocal-FRAP experiment with FITC-dextran (260 kDa) at 0.1 mg/mL. First postbleach profile is at 3 s (■). Subsequent profiles are at 6 s intervals (–). The final profile is at 72 s postbleach (●). Mobile fraction = 1.00. The inset is a schematic of a classical FRAP experiment, showing total fluorescence intensity solely within the bleached region as a function of time, during prebleach **(A)**, bleach **(B)**, and recovery **(C)**, stages.

7. The image series is transferred to the hard drive and the RAM-disk cleared. A file containing the essential experimental parameters including magnification, laser illumination, and so on, should be saved with each data series. Image processing and analysis are ideally carried out on a separate computer.
8. Less than 0.01% of the total sample volume is typically included in the bleach. This allows a number of repeat experiments to be made on a single sample. The authors normally perform six measurements per sample.

3.3. Data Analysis

The data analysis strategy for confocal-FRAP is based on the variance analysis of Kubitscheck et al. ***(13)***. The redistribution of fluorescence is measured over the whole imaged area, which includes the area inside and immediately outside the bleached region (**Fig. 4**) (*see* **Note 9**). This increases the data available for analysis, compared to conventional integral analyses, which consider the recovery of fluorescence only in the bleached region (Fig. 4 inset). The straightforward computing routine developed by the authors combines the image-analysis tools contained within the MRC-1000 Microscope Programming Language (MPL) and the data analysis capabilities found in Excel.

1. Each recovery series is checked to ensure that no convective motion of the solution has occurred. The authors use a subroutine in MPL that determines any lateral displacement of the center of the bleach using a center of mass calculation.
2. The effects of nonuniformity in illumination and detector response are corrected by dividing each recovery image by the prebleach image to produce a floating-point ratio image.
3. Pixels contained within three bleach widths of the center are quantified and a normalized radial intensity distribution is determined for each image, with the bleach center set as the origin (**Fig. 4**). The average intensity at the image periphery (between three and four bleach widths from the center) is set as the mean background intensity.
4. Intensity profile data are passed from MPL to Excel. The radial distribution of bleached fluorophores is determined by subtracting radial intensity values from the image background intensity.
5. The variance of the radial distribution of bleached fluorophores is calculated and plotted as a function of time (*see* **ref.** *[13]* for background details of how to calculate the variance values). For well-defined populations of monomeric diffusants, a linear relationship exists between variance and time. The lateral translational diffusion coefficient is calculated from the gradient determined from a linear least-squares fit to the data.
6. A nonlinear dependence of variance upon time can result from either an underestimate of the apparent background intensity as photobleached components reach the edge of the field of view, or, the presence of multiple independently diffusing components. The first problem is rectified by excluding data from longer time-points. Multiple components having greatly differing diffusion coefficients show a characteristic multicomponent recovery of total intensity in the bleached area. Typically, there is fast initial recovery as the smaller components redistribute, followed by a slower recovery phase as the mobility of larger species dominates. It is difficult to resolve components if the difference in diffusion coefficients is not large.

4. Notes

1. Argon-ion lasers are preferable to krypton-argon for this technique as they have superior lifetimes, higher beam power, and lower maintenance costs.
2. Dialysis sacs should be double sealed for aggrecan and hyaluronan solutions as they generate high osmotic pressures when dialyzed against water.
3. Allowing CNBr activation of carbohydrate chains to proceed for longer than 5 min results in depolymerization.
4. The fluorescence concentration in Hi-Trap fractions can be checked quickly using a standard UV gel transilluminator.
5. Higher scan rates with larger pixel dimensions are needed to monitor the recovery of very mobile species, therefore, a compromise is needed between increased temporal resolution and decreased spatial resolution.
6. The gain and aperture of the photomultiplier tube must be set to zero during bleaching to prevent damage to the tube.

7. Owing to the high inertia of the scan head mirror assemblies found in some confocal microscopes, the scan-beam dwell time can be longer at the edge of a scan than the center, leading to uneven bleaches, especially of rectangles. Before purchasing a confocal microscope for confocal-FRAP, the user should ensure that this problem does not occur. The MRC-1000 instrument is not prone to this effect.
8. To allow the collection of extended image series, a minimum computer RAM size of 32 Mb is recommended.
9. At short displacements from the bleach center (<10 pixels) the total signal is low and also pixelation means radial positions cannot be accurately defined. However, the variance calculation is insensitive to bleached fluorophore concentration near the bleach center, and these effects do not constitute significant errors. Conversely, variance calculations are very sensitive to bleached fluorophore concentrations distant from the bleach center. The accurate measurement of fluorescence intensities at such positions is thus very important.

Acknowledgments

We are grateful to The Wellcome Trust for support.

References

1. Winlove, C. P. and Parker, K. H. P. (1995) The physiological functions of extracellular matrix macromolecules, in *Interstitium, Connective Tissue and Lymphatics* (Reed, R. K., McHale, N. G., Bert, J. L., Winlove, C. P., and Laine, G. A., eds.), Portland, London, UK, pp. 137–165.
2. Comper, W. D. and Laurent, T. C. (1978) Physiological functions of connective tissue polysaccharides. *Physiol. Rev.* **58,** 255–316.
3. Comper, W. D., ed. (1996) *Extracellular Matrix.* Harwood, Amsterdam, The Netherlands.
4. Maroudas, A. (1976) Balance between swelling pressure and collagen tension in normal and degenerate cartilage. *Nature* **260,** 808–809.
5. Grodzinsky, A. J. (1983) Electromechanical and physicochemical properties of connective tissue. *CRC Critical Rev. Bioeng.* **14,** 133–199.
6. Hardingham, T. E. and Fosang, A. (1992) Proteoglycans: many forms and many functions. *FASEB. J.* **6,** 861–870.
7. Urban, J. P. G., Holm, S., and Maroudas, A. (1982) Nutrition of the intervertebral disk: Effect of fluid flow on solute transport. *Clin. Orthop.* **170,** 293–302.
8. Hardingham, T. E., Muir, H., Kwan, M. K., Lai, W. M., and Mow, V. C. (1987) Viscoelastic properties of proteoglycan solutions with varying proportions present as aggregates. *J. Orthop. Res.* **5,** 36–46.
9. Li, X. and Reed. W. F. (1991) Polyelectrolyte properties of proteoglycan monomers. *J. Chem. Phys.* **94,** 4658–4580.
10. Sheehan, J. K. Arundel, C., and Phelps, C. F. (1983) Effects of the cations sodium, potassium and calcium on the interaction of hyaluronate chains: a light scattering and viscometric study. *Int. J. Biol. Macromol.* **5,** 222–228.
11. Harper, G. S., Comper, W. D. Preston, B. N., and Daivies, P. (1985) Concentration dependence of proteoglycan diffusion. *Biopolymers* **24,** 2165–2173.

12. Axelrod, D., Koppel, D. E., Schlessinger, J. Elson, E., and Webb, W. W. (1976) Mobility measurements by analysis of fluorescence photobleaching recovery kinetics. *Biophys. J.* **16,** 1055–1069.
13. Kubitscheck, H., Wedekind, P., and Peters R. (1994) Lateral diffusion measurements at high spatial resolution by scanning microphotolysis in a confocal microscope. *Biophys. J.* **67,** 946–965.
14. Bayley, P. M. and Clough, B. (1995) Application of optical microscopy to cellular-dynamics: studies of fluorescence photobleaching (FRAP) of erythrocyte-membrane proteins using the confocal microscope. *J. Trace. Microprobe T.* **13,** 209–216.
15. Blonk, J. C. G., Don, A., Van Aalst, H., and Birmingham, J. J. (1993) Fluorescence photobleaching in the confocal scanning light microscope. *J. Microsc.* **169,** 363–374.
16. Gribbon, P. and Hardingham, T. E. (1998) Macromolecular diffusion of biological polymers measured by confocal fluorescence recovery after photobleaching. *Biophys. J.* **75,** 1032–1039.
17. Imhof, A., Van Blaadren, A., Maret, G., Mellema, J., and Dhont. J. K. G. (1994) A comparison between the long time self diffusion of and low shear viscosity of concentrated dispersions of charged colloidal silica spheres. *J. Chem Phys.* **100,** 2170–2181.
18. Peters, R., Brunger, A. and Schulten, K. (1981) Continuous fluorescence microphotolysis: a sensitive method for the study of diffusion processes in single cells. *Proc. Natl. Acad. Sci. USA* **78,** 962–966.
19. Hardingham, T. E., Ewins, R. J. F., and Muir, H. (1976) Cartilage proteoglycans: structure and heterogeneity of the protein core and the effects of specific protein modifications on the binding to hyaluronate. *Biochem. J.* **157,** 127–143.
20. Gribbon, P., Heng, B. C., and Hardingham, T. E. (1999) The molecular basis of the solution properties of hyaluronan investigated by confocal fluorescence after photobleaching. *Biophy. J.* **77,** 2210–2216.

9

Electron Cryomicroscopy of Fibrillar Collagens

Roger S. Meadows, David F. Holmes, Chris J. Gilpin, and Karl E. Kadler

1. Introduction

Collagen fibrils in tissue are generally heterotypic with more than one type of collagen molecule incorporated into the fibril structure. Furthermore specific macromolecules are bound onto the fibril surface influencing both the assembly and the interaction of the fibril with the surrounding matrix. The electron cryomicroscopy procedures described here form part of a program of work to determine the structure and assembly of tissue fibrils containing surface-associated components.

Collagen fibrils show a characteristic axial periodicity of ~67 nm, which is most apparent in the gap/overlap structure of fibrils that have been prepared for transmission electron microscopy by negative staining. The details of this pattern can be related to the axial distribution of amino acids in the collagen triple helix when each residue is scored for "bulkiness" *(1)*. The negative stain distribution does not, however, allow the scattering mass of the proteins to be determined nor is it possible to distinguish between contributions from internal and surface structure in single projections. We have used dark-field scanning transmission electron microscopy (STEM) to mass-map the unstained collagen fibril allowing the mass per unit length to be measured along the fibril axis *(2)*. In collagen fibrils made in vitro from type I collagen (in the absence of other macromolecule), it is possible to determine the number of collagen molecules in the fibril cross-section *(3)*. Studies of the axial mass distribution by STEM within a single *D*-period are limited, however, by differential shrinkage brought about by dehydration of the fibrils *(2)*. Consequently, fine structure below 10 nm is lost.

Two techniques to overcome these problems are detailed in this chapter. The first is electron cryomicroscopy of unstained collagen fibrils in vitreous

From: *Methods in Molecular Biology, vol. 139: Extracellular Matrix Protocols*
Edited by: C. Streuli and M. Grant © Humana Press Inc., Totowa, NJ

(amorphous) ice (**Fig. 1**). The vitreous ice technique has been described in detail elsewhere (*see [4]* for review), and a complete description is beyond the scope of this chapter. In brief, molecules rapidly frozen in a thin (10–100 nm) ice layer can be imaged unstained in a suitable transmission instrument operated in low-dose mode (1–10 electrons $Å^{-2}$). The technique preserves the native structure of the specimen to near-atomic resolution. Images recorded under these conditions have a low signal-to-noise ratio and generally require image processing and averaging procedures to extract high-resolution information. The second technique is aimed at acquiring structural information about the surface of hydrated collagen fibrils. This consists of high resolution shadowing of freeze-etched fibrils with retained bound water (**Fig. 2**). Raspanti et al. *(5)* have compared different methods for studying the surfaces of collagen fibrils and have found that the highest resolution information is obtained by freeze-etching and rotary shadowing of a freeze-fractured tendon sample. To examine collagen fibrils in suspension, originating either from in vitro collagen fibril assembly systems or tissue, we have developed a freeze-etching and shadowing (mica ice-wedge) procedure to cryopreserve collagen fibrils in a hydrated state. The mica ice wedge procedure is related to the quick-freeze deep-etch method described by John Heuser (for review *see* **ref.** *[6]*).

2. Materials

2.1. Vitreous-Ice Procedure

1. Chloroform (BDH, London, UK).
2. Formvar (BDH).
3. Copper and nickel grids (200 and 400) (Agar Scientific, Stansted, UK).
4. 6-mm carbon rods (Agar Scientific).
5. Nanotech coating unit.
6. Ultrapure water (Purite Stillplus HP).
7. Glass slides (Blue star).
8. Glycerol (BDH).
9. Fine forceps (Agar Scientific).
10. Acetone (BDH).
11. 50-Grade filter paper (Whatman, Clifton, NJ).
12. Ultrasonicator bath (Agar Scientific).
13. 3-in-1 oil (E. R. Howard Ltd., Stowmarket).
14. Dispersed collagen fibrils from bovine skin *(7)*.
15. Ethane (BOC, Trafford Park, Manchester, UK).
16. Controlled Environment Vitrification System (CEVS, provided by Professor Ishi Talmon, ishi@tx.technion.ac.il: http://www.technion.ac.il/technion/chem-eng/talmon/) (see also **ref.** *[7]*).

2.2. Mica Ice-Wedge Procedure

1. Mica sheets 75 mm × 25 mm × 0.15 mm thick (Agar Scientific).
2. Rotary cutter (Myers precision cutter).

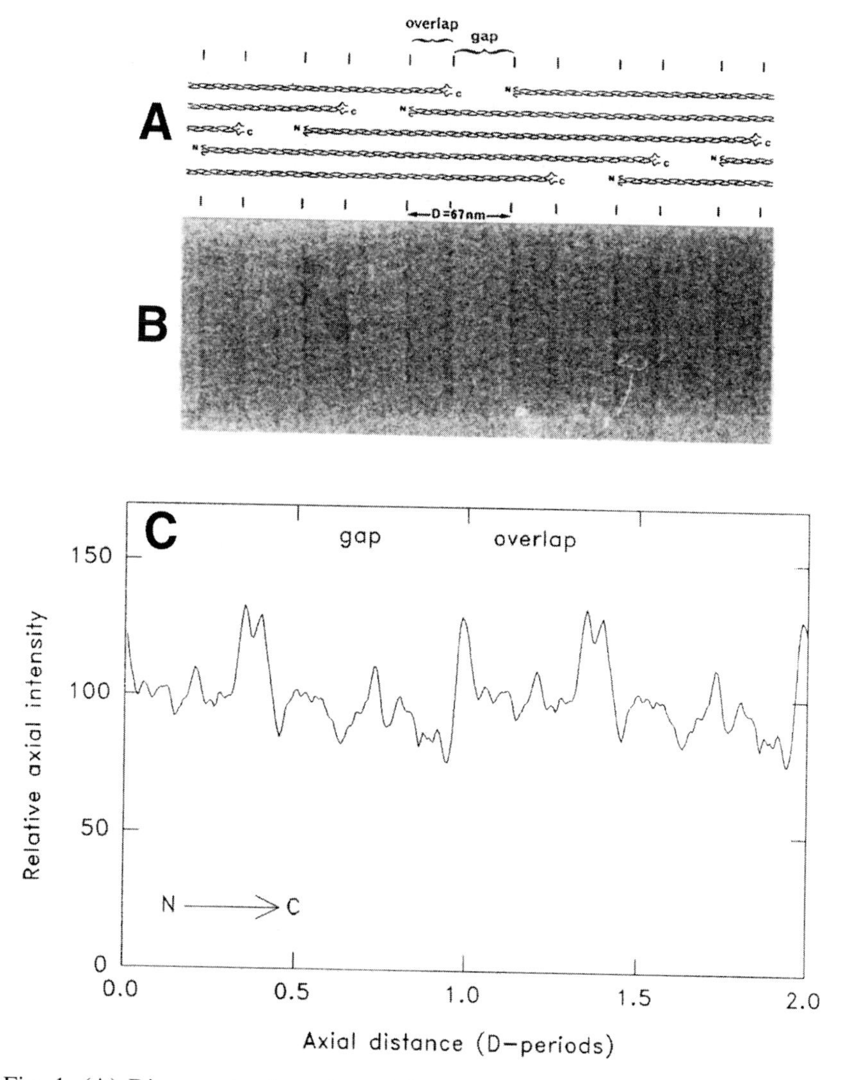

Fig. 1. **(A)** Diagram to show the axial packing arrangement of molecules in the collagen fibril. The fibrils have an axial periodicity of 67 nm and characteristic gap-overlap regions. The short nonhelical domains (telopeptides) at the ends of the molecules are shown. These telopeptides possess an axially contracted conformation relative to the main triple helical part of the collagen molecule. **(B)** Energy-filtered TEM image of a collagen fibril from bovine skin in vitreous ice. Defocus is at 1.5 µm to enhance phase contrast. The image has been aligned with the axial packing diagram **(A)**. The gap-overlap structure is apparent and the higher density telopeptide domains can be clearly distinguished. **(C)** Average axially projected structure of the collagen fibril in vitreous ice. This has been measured from images similar to **(B)** and is shown repeated over two *D*-periods.

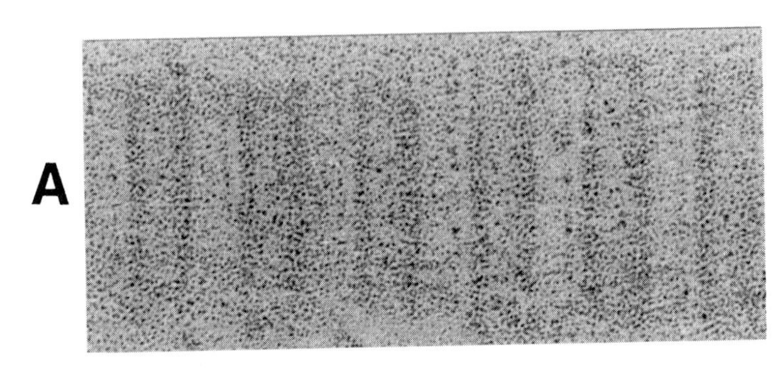

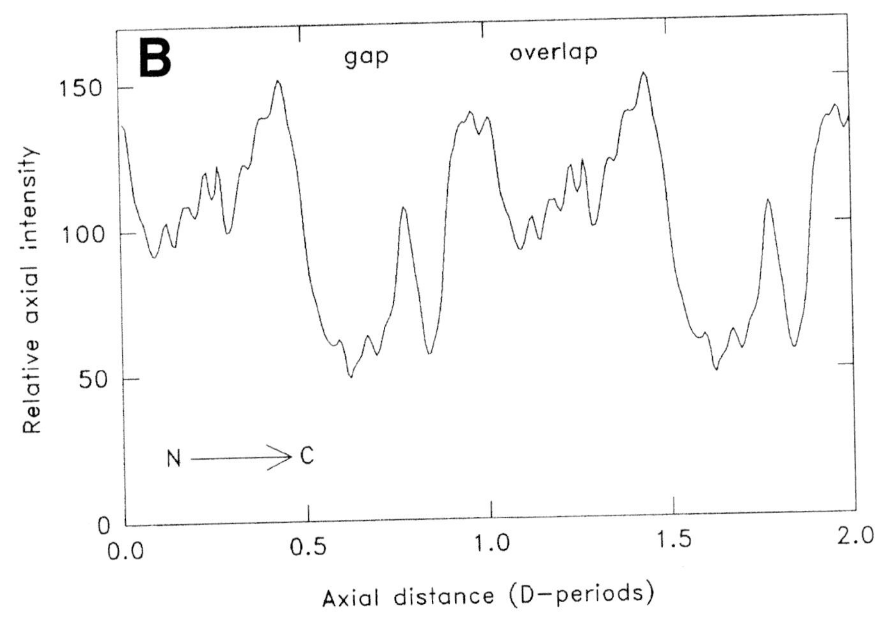

Fig. 2. **(A)** Electron image of reconstituted collagen fibril after freeze-etching using the mica ice wedge method and rotary shadowing with Pt/C at 11°. **(B)** Average axial projection from images similar to **(A)**, shown repeated over two *D*-periods. The intensity relates to the local thickness of platinum, revealing ridges on the fibril surface. There are three conspicuous ridges: two corresponding to the telopeptide domains and an additional ridge in the gap region. The underlying gap/overlap structure of the fibril is still apparent.

3. Acid-soluble type I collagen from bovine skin *(8)*.
4. Fine forceps (Agar Scientific).
5. Ultrapure water (Purite Stillplus HP).
6. 50-Grade filter paper (Whatman).
7. Aluminium spacer 15 mm × 15 mm × 0.25 mm.

8. Liquid nitrogen (BOC).
9. Cressington CFE-50C (Cressington Scientific Ltd., Watford).
10. Copper and nickel grids (400 and 600) (Agar Scientific).
11. Grid boxes (Agar Scientific).

3. Methods

3.1. Vitreous-Ice Procedures

3.1.1. Preparation of Holey Carbon Films

1. Prepare a solution of 0.2% formvar in chloroform.
2. To 90 mL of the formvar solution in a 100-mL Schott bottle, add glycerol and stir for 3 h. Increasing the glycerol from 0.01 mL and 5.0 mL, increases the hole diameters. Glycerol at 0.1 mL produced films with holes of ~2-μm diameter.
3. Place the emulsion in an ultrasonicator bath for 30 min. Ultrasonication produces a uniform droplet size and, consequently, carbon films with holes of a near-uniform diameter.
4. Clean a glass microscope slide with a solution of household dishwashing liquid in tap water, then rinse with ultrapure water and dry with a clean, dry cloth. Polish the slide with 3-in-1 oil to give the slide a thin, nonstick coating on its surface.
5. Dip the polished slide into the formvar-glycerol emulsion until 75% of its length is submerged, and hold for 10 s. Withdraw the slide at a rate of ~1.5 cm/s to thin the formvar film (now formed on the slide) in the chloroform vapor. Drain the slide by standing it vertically on filter paper.
6. When the slide is dry, mark around its surface with a pair of forceps about 1 mm from the edge. This aids detachment of the formvar film from the slide. Immerse the slide into a reservoir of ultrapure water, holding the slide at an angle of approximately 20° to the water surface. The formvar film should separate from the slide and float, intact, on the surface of the water.
7. Using watchmakers forceps, place transmission electron microscope grids (typically 18) on the formvar film. Collect the formvar-coated grids from the water by placing a strip of ordinary paper (weight ~80 g/m^2), just wider than the film, on top of the grids. Peel back the paper from the water with the grids and formvar film attached to it.
8. Air-dry the paper, with the formvar-coated grids attached, under a dustcover and then suspend it above a reservoir of acetone for 45-60 s; the acetone vapors should clear any pseudoholes in the formvar film.
9. Place the formvar-coated grids, still attached to the paper, in a carbon-coating unit. Carbon is evaporated onto the grids in the same way as that used to produce standard carbon films *(9)*, except the film used is thicker. This requires evaporation from a carbon spindle that is 2-mm long and 0.8 mm in diameter.
10. Detach the coated grids from their paper backing with watchmakers forceps and place them on 4 layers of 50-grade filter paper soaked in acetone (in a sealed glass container) for several h; this dissolves away the formvar leaving just a holey carbon film on the grids.

11. Place the grids in a conventional TEM to check the holey carbon films are suitable for use. Grids are then stored in grid boxes or on dry filter paper in Petri dishes prior to use.

3.1.2. Preparation of Samples in Vitreous Ice

1. Prepare the CEVS (*see* **Fig. 3**) *(7)* by filling the reservoirs with the same buffer used to prepare the samples. The buffer-soaked sponges provide a large surface area and thereby maintain the humidity of the chamber. Adjust the temperature of the chamber to 35°C and allow the system to reach equilibrium.
2. Grip holey carbon grids (*see* **Subheading 3.1.**) with the modified forceps fitted to the plunging arm of the CEVS.
3. Fill the freezing chamber with liquid nitrogen. Add the ethane gas slowly to the cryogen pot, where it first condenses, then starts to freeze. Condense more gas immediately prior to sample plunging to thaw the surface of the solid ethane.
4. Place 3 μL of the sample onto the holey carbon side of the grid and remove excess liquid by touching filter paper to both sides of the grid. This is aided by use of a binocular microscope. If chamber humidity is properly controlled, then the resulting thin film of sample will remain stable. The thickness of the film can be judged by eye.
5. When a suitable sample-film thickness (~100 nm) has been achieved, the grid is plunged into the liquid ethane. The grid is then transferred into a small container of liquid nitrogen. Traces of ethane will freeze on the surface of the grid protecting it from frosting during transfer to the microscope.

3.1.3. Electron Cryomicroscopy

1. Fill the microscope anticontaminator Dewar flask with liquid nitrogen at least 1 h before use.
2. Set the accelerating voltage prior to use to stabilize the H.T. voltage.
3. Place the cryospecimen holder under vacuum in the microscope column. Cool the sample to the working temperature (typically –180°C). The Dewar on the cryoholder should be pumped out periodically to maintain a vacuum adequate to permit the specified ultimate temperature to be achieved.
4. Transfer frozen grids, under liquid nitrogen, to a liquid nitrogen-cooled workstation close to the microscope. Remove the cryoholder from the microscope and move it to the workstation. Occasionally, splash the sample holder with liquid nitrogen to maintain the tip at low temperature.
5. Remove the grid-retaining ring with a cold tool and transfer the frozen grid to the holder with precooled forceps. Two pairs of forceps are often required because the grid can become charged with static electricity, making it difficult to handle. Ensure that the grid is sample-side up and refit the retaining ring with the cold tool. Most cryoholders have shields that can be brought into place to protect the grid from frosting as the holder is transferred to the microscope.
6. Pour liquid nitrogen over the tip of the holder immediately prior to transfer.

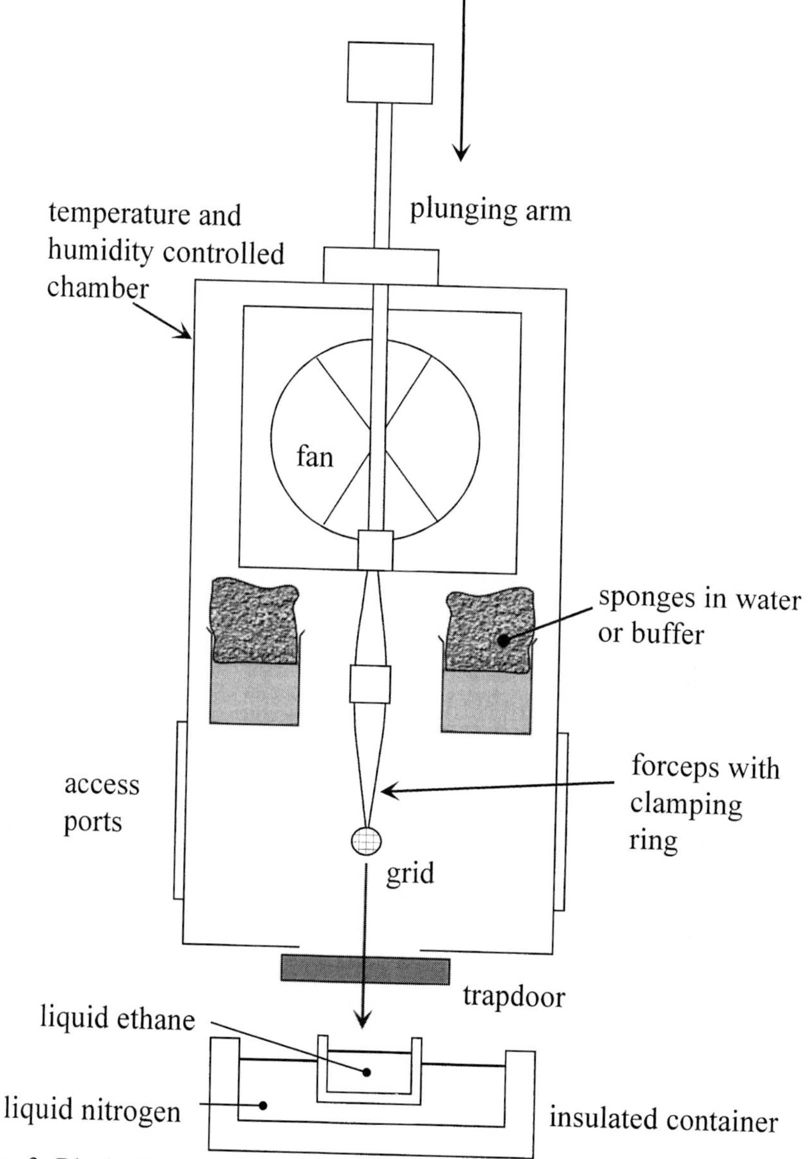

Fig. 3. Block diagram of the controlled environment vitrification system used to prepare vitreous-ice samples for transmission electron cryomicroscopy.

7. Transfer the holder quickly into the microscope specimen airlock to prevent warming up and to minimize frosting from atmospheric water vapor. Once the holder is in the high vacuum of the microscope column, leave it for 15 min to allow the temperature to stabilize and for the vacuum to remove any frosting

which may have occurred on the shields. It is often useful to switch the beam on and allow the resulting slight temperature increase to assist with frost removal.

3.1.4. Image Acquisition (See Fig. 4)

1. Frozen hydrated samples are extremely sensitive to beam damage, including mass loss, upon exposure to electrons.
2. Set the beam current at a minimum when first examining the specimen. Electron dose on the sample needs to be carefully controlled if damage is to be prevented.
3. Carry out searching, focusing, and basic alignment at low magnification, preferably with the assistance of a video rate image-intensifying camera.
4. Carry out higher magnification alignment and focusing on an area adjacent to the area of interest. Many microscopes have either beam-shift or image-shift controls to simplify this task. The only electrons that should strike the area of interest should be those used for image recording.
5. Ideally, keep the total accumulated electron dose in the area of interest below 10 electrons $Å^{-2}$. Frozen-hydrated samples exhibit poor contrast and under focus of the objective lens is often necessary to increase image contrast.
6. Only record structures lying over holes in the carbon film. Carbon-filmed areas are used only for recording images for image assessment purposes.
7. Notes on image analysis are in **Subheading 4.**, **Notes 1** and **2**. *See* **Note 3** on energy filtered TEM.

3.2. Mica Ice-Wedge Procedures

3.2.1. Formation of the Mica Ice Wedge

(*See* **Note 5**)

1. Cut a 75 mm × 25 mm × 0.15 mm sheet of mica into pieces of 15 mm × 10 mm × 0.015 mm with a rotary trimmer.
2. Prepare a sample of reconstituted type I collagen fibrils in phosphate buffer (I = 0.2, pH 7.4) by the "warm start" procedure *(8)*.
3. Disperse the resultant fibril gel into fibril fragments in Tris-buffered saline using ultrasonic disintegration.
4. Take one of the trimmed pieces of mica and cleave it. Onto the cleaved surface of one half of the mica, put 3 µL of the collagen sample and then place the other half of the mica on top so the cleaved faces are together. Leave this "sandwich" for 30–60 s to allow the fibrils to adsorb to the mica. (*See* **Note 5**.)
5. Submerge the "sandwich" in ultrapure water, separate the mica, and, holding each piece separately with watchmakers forceps, waft them in the water for 15–30 s. Reassemble the "sandwich," overlapping the longest edges by 2 mm, and withdraw it from the water.
6. Remove the excess water from the outer surfaces of the "sandwich" with filter paper and then place it on the stage of the Cressington CFE-50C freeze-fracture/etch device. Clamp the two halves of the mica together over the overlap and then insert a 15 mm × 15 mm × 0.25 mm aluminium spacer between the mica at the opposite unclamped end, so that it overlaps the lower piece of mica by 2 mm.

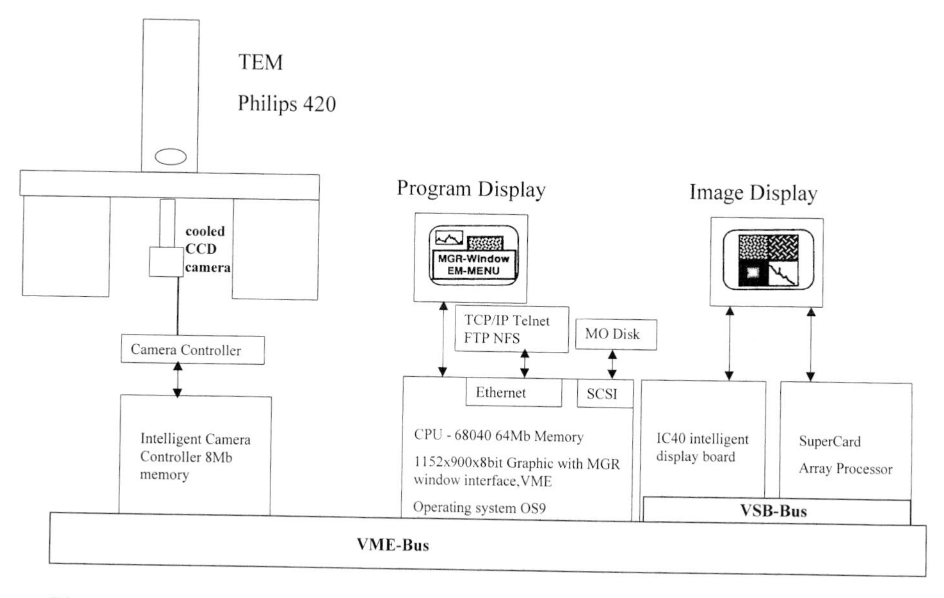

Fig. 4. Diagram to show the TVIPS system used for digital image capture by cooled CCD. The system allows efficient low-dose imaging and on-line image processing.

7. In the gap between the two halves of mica, pipet 8 μL of ultrapure water. This forms a "wedge" of water between the mica halves and keeps the adsorbed fibrils hydrated (**Fig. 5**). (*See* **Note 6**.)
8. Attach the stage to its transfer rod and then plunge it into nitrogen slush to freeze the sample. After 30 s remove the stage from the nitrogen slush, remove the spacer from the wedge and then unclamp the Mica Ice Wedge. Reattach it to the stage by clamping only the lower piece of mica (this allows the top piece of mica to be removed to facilitate the freeze fracture).
9. Via the airlock, transfer the stage with the mica ice wedge attached to it into the CFE-50C, using the transfer rod, for freeze fracturing/etching and rotary shadowing. The chamber of the CFE-50C should be at ~2×10^{-7} mbar and the stage holder precooled to around –189°C (**Fig. 6**).

3.2.2. Fracturing and Etching

1. The specimen stage is located on a revolving stage holder in the Cressington CFE-50C. The stage is cooled with liquid nitrogen to keep the temperature at ~–189°C.
2. Rotate the stage so that the thickest part of the mica ice wedge is facing the liquid nitrogen-cooled microtome of the Cressington. Lower the microtome blade so that it will just pass under the top piece of mica in the wedge.
3. Swing the microtome arm gently toward the sample so that the blade passes under the top piece of mica. This causes the mica to be lifted free and clear of the ice wedge, thereby producing the fracture that exposes the surface of the sample (**Fig. 7**).

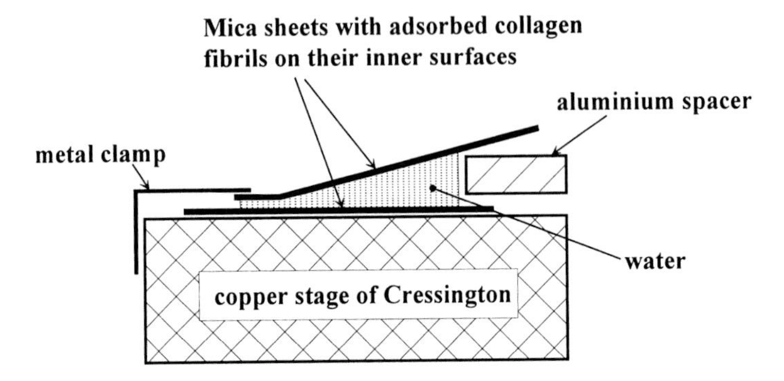

Fig. 5. Schematic representation of the first-stage of sample preparation of the ice wedge prior to freezing.

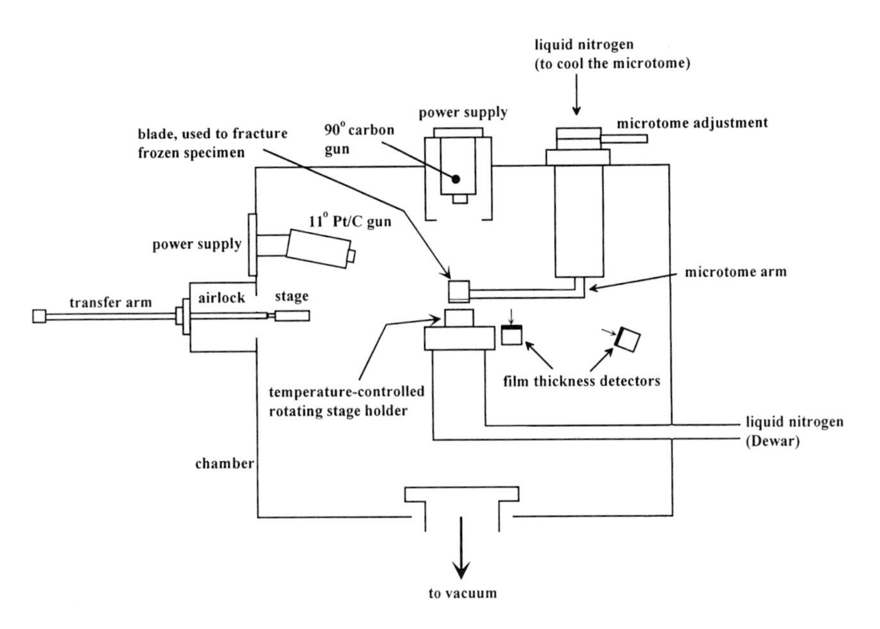

Fig. 6. Schematic representation of the Cressington CFE-50C instrument used here for freeze-fracture, etching, and low-temperature shadowing. The system allows fine temperature control (down to –204°C) of the specimen and well-controlled evaporation of the shadowing metal.

4. Raise the microtome slightly and then position it over the top of the sample to protect it from contamination during freeze-etching.
5. Set the thermal control of the stage heater to –100°C and turn on the heater. Once the stage temperature reaches –100°C etch the sample for 3 min and then turn off the heater, allowing the stage to return to –189°C.

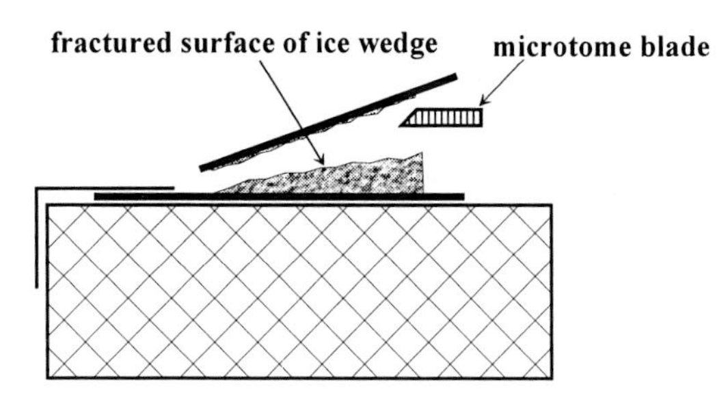

Fig. 7. Diagram to show the fracturing of the mica ice wedge. Fracturing is achieved by removal of the uppermost piece of mica using the microtome blade of the Cressington instrument.

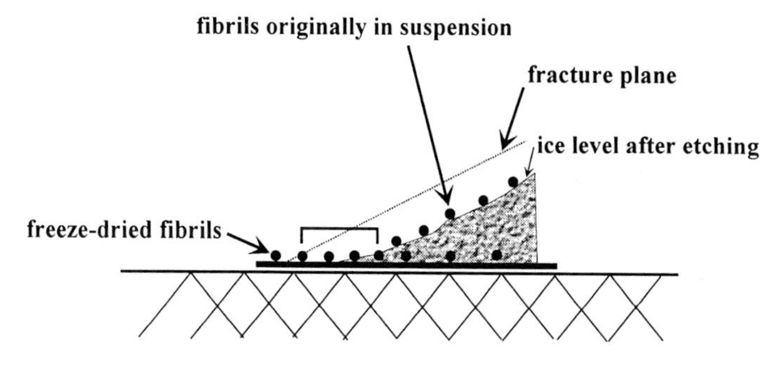

Fig. 8. Etching of the ice wedge by sublimation of water at –100°C generates molecules that are i) freeze-dried (those at the left-hand side of the mica, as shown); ii) molecules adherent to the mica and submerged in frozen water or buffer; and iii) molecules that are at the surface of the frozen water or buffer. Molecules adherent to the mica exhibit a progressive level of hydration (from left to right, as shown in the diagram). Bracket shows zone of progressive hydration.

6. Freeze-etching of the ice wedge causes the ice level to recede, revealing fibrils in a range of hydration states from containing bound water only, to being fully hydrated (**Fig. 8**).
7. Once the stage temperature has returned to –189°C shadowing and coating of the sample can take place.

3.2.3. Shadowing and Coating

1. Shadowing is achieved in the Cressington CFE-50C with a low-angle electron beam gun loaded with a platinum/carbon rod and set at 11° to the sample. This is then backed with carbon from an electron beam gun set at 90° to the sample.
2. The platinum/carbon gun is set to evaporate 1–2 Å of platinum in the plane of the mica substrate. The carbon gun is set to evaporate 5–8 nm of carbon for the film.

This is achieved by adjusting the current and voltage to the guns and the duration of shadowing. Film thickness is determined by quartz-crystal-film-thickness monitors built into the Cressington.

3. With the stage of the Cressington set at –189°C and the microtome still shielding the sample, degas the platinum and carbon guns. The guns are degassed at a preset current and voltage (lower than that used for evaporation), for a preset duration.
4. Start the stage rotating at approximately 60 rpm and move the microtome arm clear of the sample, and then start evaporating from the platinum/carbon gun. Check to be sure there is an appropriate thickness of platinum with the thickness monitor and then back it with carbon from the carbon gun (also to an appropriate thickness) to complete the replica.
5. When the replica is complete, stop the stage rotating and attach the transfer rod to it. Withdraw the stage from the vacuum chamber through the airlock. Detach the mica with the replica on it from the stage and allow it to reach room temperature under cover, in preparation for replica collection.

3.2.4. Collection of the Replica

1. Once the mica and attached replica are at room temperature, pick up the mica at the edge of one of its shorter sides with watchmakers forceps.
2. Slide the mica into a reservoir of ultrapure water at an angle of approximately 20°. The replica should detach from the mica and float on the surface of the water. (*See* **Note 7**.)
3. Looking at the replica, a thin band appears shiny (where the thinnest part of the wedge was originally). This is where freeze-drying has occurred and most of the water has sublimed off the mica. Next to the shiny band is a wider region of duller replica where there has been no freeze-drying of the sample (as the ice in the wedge was much thicker here) and an ice layer still persists on the mica surface. The region between these two bands is where the replica needs to be collected in order to get fibrils that have been exposed by etching, but retain much of their water.
4. Grip the edge of a transmission electron microscope grid with watchmakers forceps and submerge it in the reservoir of ultrapure water, away from the replica. Bring the grid underneath the appropriate part of the replica and raise it slowly, to attach the selected region.
5. Bring the grid clear of the water and place it carefully on grade-50 filter paper to remove any water from it, being careful not to damage the replica.
6. Once the replica has been collected on several grids, and they are dry, they can be stored in a grid box or on filter paper in a Petri dish before viewing in a conventional TEM.

3.2.5. Electron Microscopy

1. Examine the platinum/carbon replicas of the collagen fibrils in a transmission electron microscope operated at an accelerating voltage in the region of 100 kV.
2. Images can be collected on normal photographic film or recorded digitally on a CCD camera attached to the microscope.

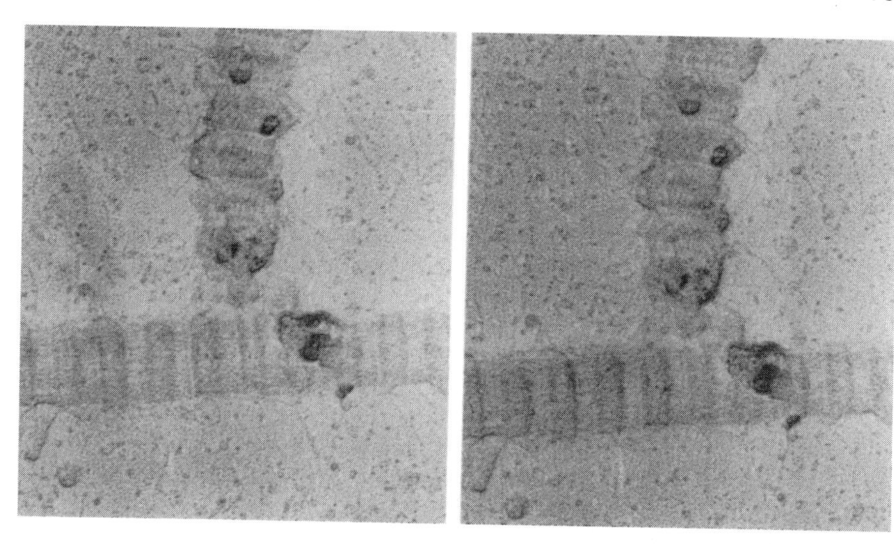

Fig. 9. Stereo pair images of reconstituted type I collagen fibrils prepared by the mica ice-wedge technique. The images were taken at + and – 6° tilt on a JEOL 1200 EX transmission instrument. The repeat period is ~67 nm.

3. Stereo pairs are a useful tool for visualizing the fibrils in 3-D. To achieve them take, two images of the same structure at + and – n° (5 to 15° is generally acceptable) of tilt on the goniometer stage of the microscope
4. It is important to keep the structures of interest in the specimen centralized in the field of view while tilting the stage. To achieve this, adjust the eucentric height of the stage so the area of interest does not move except on the tilt axis. (*See* **Notes 8** and **9**.).
5. Record two images at + and – the same degree of tilt (both at the same magnification and accelerating voltage) on photographic film and then print them onto photographic paper.
6. View the stereo pair prints on a stereo imager or use one of the other techniques for converting two images into one 3-D image (**Fig. 9**).

3.3. Image Analysis of Vitreous Ice and Replicated Samples

Axial projections were computed from fibril images using Semper 6.4 *(2)*. An average *D*-period was obtained from a set of individual *D*-periods after cross-correlation alignment.

4. Notes

Notes for vitreous-ice procedures.

1. For image analysis procedures, the image needs to be in a digital format. Images can either be captured using a slow scan-cooled CCD camera (which provides a

rapid and convenient method of obtaining digital images) or collected on high-resolution EM film, which is subsequently digitized with an appropriate scanning device. Both methods are efficient at electron detection. A typical CCD camera has 1K × 1K pixels compared with a possible 10K × 10K digitization of a sheet of film 8 cm × 8 cm. CCD cameras have, therefore, an inherently smaller field of view than photographic film at the same resolution. The CCD camera, however, has the advantage of immediate image assessment. The TVIPS system described in this chapter provides the opportunity for online processing and feedback to the microscope for automated data collection.

2. Fast Fourier transforms (FFT) of the image can be used to assess astigmatism, calculate defocus values (to permit contrast transfer function correction), and to perform cross-correlation functions for image alignment and averaging.
3. Image contrast and signal/noise ratio are low in images of frozen-hydrated biological specimens collected on conventional TEMs. Energy-filtered electron microscopy removes the inelastically scattered electrons from the image, thereby giving an enhancement of both contrast and signal/noise ***(10)***. Furthermore, energy filtered EM allows mass mapping of macromolecular assemblies in vitreous ice ***(11)***.

Notes for mica ice-wedge procedures.

4. This technique can be used to visualize many different macromolecules and macromolecular assemblies in the range of less than 10–100 nm in size.
5. Where molecules are damaged by washing with water and washing is required, the sample buffer minus any sample or extra additives may be used both for washing and the assembly of the wedge. It is normally only large assemblies, such as collagen fibrils, that can become masked after freeze-etching by buffer constituents. These might require washing.
6. Most samples containing smaller molecules are not left masked after etching and therefore do not require washing. In these cases the mica wedge is assembled (clamped at one end with the spacer at the other), 10 μL of the sample pipeted into the gap, the sample left for 30–60 s to allow the molecules to adsorb to the mica and then frozen/fractured/etched, and so on, in the same manner as washed samples.
7. If replicas fail to detach and float free, the mica may be placed in an atmosphere of acetic acid for several hours or days to try and free them. If this fails, it may be possible to dissolve away the mica with concentrated hydrofluoric acid, but this may cause disruption of the replicas and make collecting them on grids difficult.
8. By modifying the stereo pair idea, a series of images may be taken at *n* degree intervals from +60° to –60° of tilt and these might be used for 3-D reconstruction of the collagen fibrils.
9. For 3-D reconstruction of a tilt series or image analysis, the images need to be digitized, either by capturing them on the microscope with a CCD camera or by scanning negatives onto a computer with some form of scanner or digital camera.

Acknowledgments

We thank Dr. Rasmus Schröder for helping us to collect energy-filtered TEM images of collagen fibrils in vitreous ice, using a Zeiss 912 Ω instrument. This work was supported by the Wellcome Trust (019512). Karl E. Kadler is a recipient of a Senior Research Fellowship from the Wellcome Trust.

References

1. Chapman, J. A., Tzaphlidou, M., Meek, K. M., and Kadler, K. E. (1990) The collagen fibril—a model system for studying the staining and fixation of a protein. *Electron Microsc. Rev.* **3,** 143–182.
2. Holmes, D. F., Watson, R. B., Steinmann, B., and Kadler, K. E. (1993) Ehlers Danlos syndrome type VIIB. Morphology of type I collagen fibrils is determined by the conformation of the N-propeptide. *J. Biol. Chem.* **268,** 15,758–15,765.
3. Holmes, D. F., Chapman, J. A., Prockop, D. J., and Kadler, K. E. (1992) Growing tips of type I collagen fibrils formed in vitro are near-paraboloidal in shape, implying a reciprocal relationship between accretion and diameter. *Proc. Natl. Acad. Sci. USA* **89,** 9855–9859.
4. Dubochet, J., Adrian, M., Chang, J. J., Homo, J. C., Lepault, J., McDowall, A. W., and Schultz, P. (1988) Cryo-electron microscopy of vitrified specimens. *Quart. Rev. Biophys.* **21,** 129–228.
5. Raspanti, M., Alessandrini, A., Gobbi, P., and Ruggeri, A. (1996) Collagen fibril surface: TMAFM, FEG-SEM and freeze-etching observations. *Microsc. Res. Techn.* **35,** 87–93.
6. Heuser, J. (1989) Procedure for 3-D visualisation of molecules on mica via the quick-freeze, deep-etch technique. *J. Electron Microsc. Techn.* **13**, 244–263.
7. Bellare, J. R., Davis, H. T., Scriven, L. E., and Talmon, Y. (1988) Controlled environmental vitrification system: an improved sample preparation technique. *J. Electron Microsc. Techn.* **10,** 87–111.
8. Holmes, D. F., Capaldi, M. J., and Chapman, J. A. (1986) Reconstitution of collagen fibrils *in vitro*; the assembly process depends on the initiating procedure. *Int. J. Biol. Macromol.* **8,** 161–166.
9. Sherratt, M., Graham, H. K., Kielty, C. M., and Holmes, D. F. (1999) ECM macromolecules: rotary shadowing and scanning transmission electron microscopy. *Methods in Molecular Biology, Extracellular Matrix Protocols* (Streuli, C. and Grant, M., ed.), Humana Press, Totowa, NJ.
10. Schröder, R. R., Hofmann, W., and Menetret, J. F. (1990) Zero-loss energy filtering as improved imaging mode in cryoelectron microscopy of frozen hydrated specimens. *J. Struct. Biol.* **105,** 28–34.
11. Langmore, J. P. and Smith, M. F. (1992) Quantitative energy-filtered electron microscopy of biological molecules in ice. *Ultramicroscopy* **46,** 349–373.

10

Atomic Force Microscopy Measurements of Intermolecular Binding Strength

Gradimir N. Misevic

1. Introduction

Intermolecular binding forces are intrinsic property of cohesive structures and should be used as the main quantitative criteria for assessing and defining their functional contribution to the maintenance of the anatomical integrity of the adult and embryonal multicellular organism, to fertilization, to blood cell adhesiveness in normal and pathological conditions, to tumor cell adhesion, to parasite–host interactions, and to cellular associations in symbiotic organism. Recently, a novel technology of atomic force microscopy (AFM) measurements of binding strength between a single pair of cell adhesion molecules in various physiological solution was developed *(1)*. Binding studies, calorimetric, and spectroscopic analyses do not provide direct information about binding forces and are thus complementary kinetic methods to AFM measurements.

The AFM was initially built and is used mostly by physicists as a superb atomic imaging instrument for examination of solid-surface topography *(2)*. To be able to determine intermolecular binding forces with AFM complementary cell adhesion, molecules have to be covalently crosslinked to a silicon-nitride sensor tip and an atomically flat mica-silicon surface *(1)*. Because the AFM tip has a 10–20-nm diameter and most of the cell-adhesion molecules are over 20 nm in diameter, not more than one molecule could be attached to the tip assuring measurements between the single molecular pair. The crosslinking process consists of deposition of 10–30-nm gold on the two surfaces by evaporation in high vacuum, followed by formation of self-assembled monolayer of either 11-thio-undecanol or 11-thio-undecanoic acid. A very high density of hydroxyl and carboxyl groups are completely converted with carbonyldiimidazole in dry methanol to give highly reactive imidazole car-

From: *Methods in Molecular Biology, vol. 139: Extracellular Matrix Protocols*
Edited by: C. Streuli and M. Grant

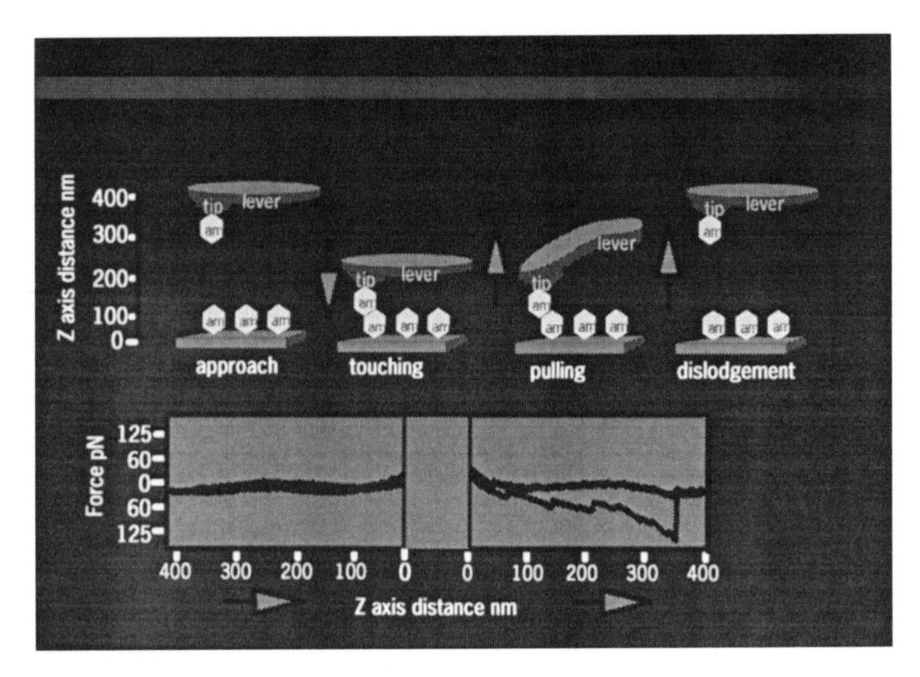

Fig. 1. Schematic representation of AFM measurements of intermolecular binding strength between glyconectin 1 proteoglycans in physiological solution. Yellow force distance curve represent measurements in artificial sea water with physiological 10 mM Ca^{2+} and red curve is obtained with 2 mM Ca^{2+}. Adhesion molecule abbreviated as **am** symbolizes glyconectin 1.

bamate and acylimidazoles, respectively. Amine-containing molecules (proteins, glycoproteins, proteoglycans, and glycans containing linkage amino acid) will be rapidly and quantitatively coupled to these groups to yield stable carbamate and amide linkages, respectively. Unreacted imidazole carbamate will hydrolyze in the presence of water to the hydroxyl groups and imidazole.

The cantilever tip with attached cell-adhesion molecule is carefully moved with subnanometer precision with AFM piezoelectric scanner toward the complementary receptor on the mica surface in the liquid cell containing physiological solution until contact between two molecules is made (**Fig. 1**). This approach is followed by retraction of the cantilever tip. During such a dislodgment, the cantilever bends until the pulling force becomes equivalent to the intermolecular binding strength between adhesion molecules crosslinked to the tip and receptor attached to the mica. When the power applied by AFM piezoelectric scanner tube exceeds ligand-receptor binding force, the lever jumps off the contact and straitens. The cantilever deflection and distance from the surface is permanently monitored by deflection of the laser from the cantilever to the position-sensitive photodetector. Stiffness of the cantilever is provided by the manufacturer and can be also directly determined by AFM. Thus, the registered cantilever hysteresis on the calibrated position-sensitive photodetector is a direct measure of the adhesion force and distance between molecules. Typically, such approach-retract cycles, sometimes also referred to as force-distance curves or force plots are repeated 50 times at five different points, with a speed of 0.01–1 Hz at room temperature, and in various physiological solutions.

Although the long-range electrostatic interaction between tip and mica solid surfaces coated with gold are shielded by selfassembled uncharged monolayer (*see* **Fig. 2**) to characterize and verify that the measured forces originate from an interaction between complementary cell adhesion molecules, measurements of both the force necessary to separate the ligand-functionalized sensor tip from the analogous receptor on the surface (final "jump-offs") and the percentage of interaction events should be investigated under different ionic conditions and using control measurements with non functinalized tip and/or surface ***(1)***. A further line of evidence that the atomic force microscopy-measured interactions originate from ligand–receptor binding can be provided by the use of specific monoclonal antibodies capable of blocking cell adhesion. Also, activators and other blockers of cell-adhesion molecules binding to receptors can be exchanged in the liquid cell during the course of experiment. More than 1000 measurements should be performed on the same preparation. The rupture force of a single covalent C–C bond is about 10 n*N*, whereas the strongest noncovalent binding forces measured with adhesion proteoglycan are about 25 times weaker (400 p*N*) ***(1)***. These findings explain why the adhesion structures shall remained intact throughout atomic force-microscopy experiments.

2. Materials

1. Atomic force microscope equipped with a liquid cell.
2. Cantilevers with silicon nitride tip 10–50 nm diameter with spring constant 0.01–0.1 N/m.
3. Mica.
4. Single and double stick Scotch tape, forceps.
5. Gold wire.
6. Turbo vacuum evaporator.
7. Cantilever holders consisting of teflon stub with slit for attaching cantilever. Teflon ring for fixing cantilevers into slit and mechanical protection (selfmade).
8. Flat bottom glass beakers (1–5 mL) with shift glass cover.
9. Plastic dishes filter paper and parafilm for humid chambers.
10. Analytical grade dry methanol.
11. 11-thio-undecanol, or 11-thio-undecanoic acid. Selfsynthesized. Keep dry at +4°C.

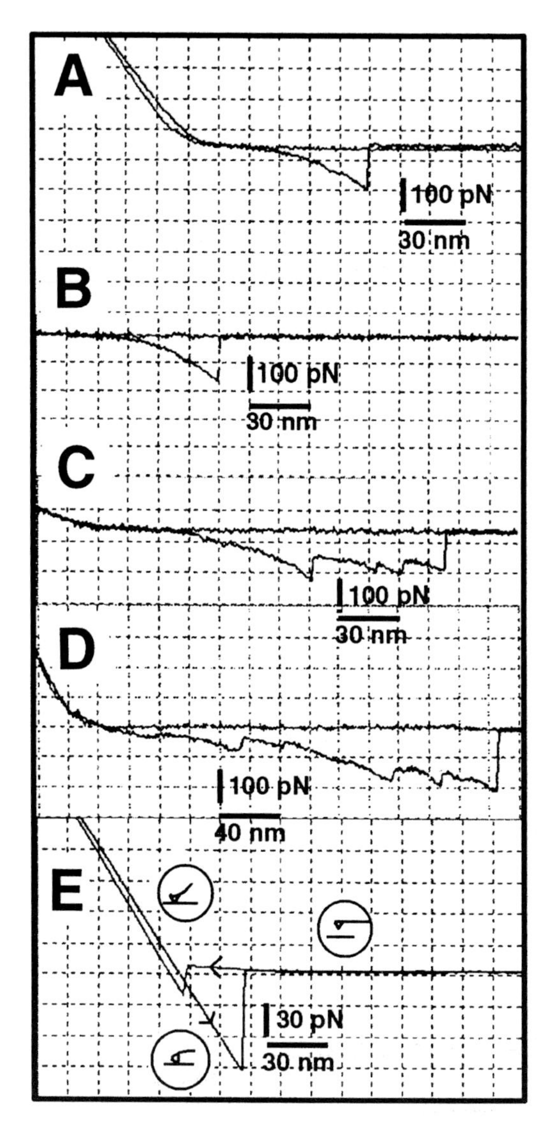

Fig. 2. Typical AFM approach and retract cycles for glyconectin 1 to glyconectin 1 interactions. The *x*-axis shows the vertical movement of the cantilever and the *y*-axis its bending and, thus, the force acting on it. **A, B, C** and **D** represent typical glyconectin 1 to glyconectin 1 interactions, whereas **E** is an example of the interaction between two gold surfaces not covered with selfassembled monolayers (11-thio-undodecanol).

12. Carbonyldiimidazole. Keep sealed in dry place at room temperature.
13. Crosslinking buffers. Choice depending of the nature of the molecule. Recommended is 0.1–0.5 *M* NaCl, 20 m*M* HEPES buffer pH 7.4. (*See* **Note 6.**)

3. Methods

3.1. Crosslinking

1. Attach cantilevers and freshly cleaved mica 1.5 × 1.5 cm with double-stick Scotch tape into a holder of a turbo vacuum evaporator equipped with a quartz crystal for monitoring thickness of gold (*see* **Note 1**). Place gold wire in an evaporation slot, turn vacuum on and adjust current of turbo vacuum evaporator. For a designated time, coat cantilevers and freshly cleaved mica to give thickness of 10–20 nm gold.
2. Immediately after gold coating, mount cantilevers in the teflon holders and place mica in 1–5 mL glass beaker with a shift glass cap. (*See* **Note 5**.)
3. Add 1–2 mL of freshly prepared 1 m*M* solution of 11-thio-undodecanol in analytical-grade methanol and incubate for 12 h at room temperature to form selfassembled monolayer.
4. Wash cantilevers and mica five times with analytical grade methanol.
5. Add 1–2 mL of 50 mg/mL of freshly prepared carbonyldiimidazole in analytical grade methanol and incubate for 20 min at room temperature.
6. Wash four times with 1 mL of methanol.
7. Dismount cantilever from teflon holder and place them on stretched parafilm in plastic dish humid chamber with tip facing upwards. Immediately add 10–20 µL of 1–20 µg of molecules to be crosslinked in 0.1–0.5 *M* NaCl 10 m*M* HEPES buffer pH 7.4. With forceps, take mica and do same as with cantilevers. Incubate 30 min to 1 h at room temperature in humid chamber.
8. Wash five times with buffer to be used for measurements.

3.2. AFM Measurements

1. Mount by double Scotch tape the back side of mica crosslinked with adhesion molecule to the AFM base. The upper part of mica with crosslinked molecules should always be covered wit 10–20 µL buffer.
2. Rapidly mount cantilevers with crosslinked molecules on a AFM equipped with a liquid cell filled with buffer to prevent drying.
3. Assure that no bubbles were in the liquid cell between cantilever tip and mica.
4. Use contact mode soft approach.
5. Perform about 50 force distance measurements of intermolecular binding strength at 0.01–1 Hz on 5–10 different locations (molecules on mica). Change solution to use blockers and/or activators of binding and repeat the same number of measurements. Return to original solution and repeat to test reversibility of blocking or activation.
6. At the end of the experiment measure the spring constant of the cantilever.

4. Notes

1. The author has used three following types of commercial AFMs equipped with a liquid cell: Digital Instruments Nanoscope III, TopoMetrix, and Park Instrument AutoProbe CP.

2. Some manufacturers glue cantilevers to a base. Test whether the glue is soluble in methanol before preforming crosslinking.
3. Cantilever with different spring constant, with and without coating for better laser reflection, as well as variety of sizes, shapes and material of tips are available. Principally, contact-mode silicon nitride cantilevers with spring constant 0.01–0.1 N/m, coated with gold on the top side for better reflection of the laser, with tips of 10–50-nm diameter should be used. For smaller molecules, smaller and sharper tips are recommended in order to match the size of molecule with a tip. This assures that only one molecule can be accommodated on the tip and thus measurements of adhesion forces between individual molecular pairs.
4. Different brands of turbo vacuum evaporator are available. Follow the manufacturer's instruction for proper coating of gold.
5. To keep gold more firmly attached to mica and silicon nitride, a 10–30-nm chromium layer can be deposited before gold coating ***(1)***.
6. Other crosslinking buffers such as 0.1 *M* bicarbonate pH 8.0 or 50–100 m*M* phosphate buffer pH 7.0–9.0 can be also used.
7. Different types of crosslinking procedures are possible with a variety of commercially available bifunctional crosslinkers for carboxyl amino and sulfhydryl groups ***(1,3)***. It is essential to form a selfassembled monolayer to shield long-range electrostatic forces between gold-coated surfaces (*see* **Fig. 2**).
8. Care should be taken when manipulating cantilevers. Tips should never be touched and not dried during crosslinking of adhesion molecules and mounting to AFM.
9. Each AFM manufacturer has slightly different computer software procedures for calibrating scanners, adjusting the photodetector, optimizing the signal of the reflected laser and for the contact mode approach, acquisition of force-distance curves, and direct determination of the spring constant of the cantilevers. Manufacturer instructions should be carefully read to avoid crashing the tip on the surface and to assure quantitative measurements.
10. Different brands and models of AFMs have different designs of liquid cells. Some are more complicated and require patient following of the procedure to avoid bubble formation. The simplest liquid cell operates in a droplet of solution. In this case, exchange is easy and bubble formation can be avoided, however, caution should be taken for evaporation after extensively long work, which can lead to changes of concentration. Thus, more frequent exchanges of solution are necessary. Prevent heating of the small volume of liquid by strong-illumination optical-microscope devices available with some AFM brands.

Acknowledgments

This work was supported by the private funds of Dr. G. N. Misevic, Conseil Régional Nord-Pas de Calais, CNRS, Marcel Mérieux Foundation, Swiss National Science Foundation, and Cancer League Basel.

References

1. Dammer, U., Popescu, O., Wagner, P., Anselmetti, D., Güntherodt, H.-J., and Misevic, G. N. (1995) Binding strength between cell adhesion proteoglycans measured by atomic force microscopy. *Science* **267,** 1173–1175.
2. Binnig, G., Quate, C. F., and Gerber, C. (1986) Atomic force microscopy. *Phys. Rev. Lett.* **56,** 930.
3. Hermanson, G. T., Mallia, K. A., and Smith, P. K. (1992) *Immobilised Affinity Techniques.* Academic, New York.

11

ECM Macromolecules: Rotary Shadowing and Scanning Transmission Electron Microscopy

Michael J. Sherratt, Helen K. Graham, Cay M. Kielty, and David F. Holmes

1. Introduction

The examination of ECM macromolecules by conventional transmission electron microscopy (TEM) or scanning transmission EM (STEM) can provide important information on macromolecular organization and interactions.

Conventional EM preparation techniques employ contrast enhancing heavy metal stains to visualize isolated macromolecules. Rotary shadowing (RS-TEM) minimizes staining artefacts by coating the surface of already dehydrated molecules. Heavy metal atoms are evaporated in a vacuum from a source set at an oblique angle relative to the substrate on which the specimen is mounted. The gaseous metal deposits on the specimen and substrate in layers of varying thickness, determined by the relief of the specimen and its substrate (**Fig. 1**). The image resolution of shadowed biological specimens is limited by specimen preparation techniques, which rely on dehydration, and by the granularity of the shadowed metal.

STEM is a well-established technique that is capable of providing quantitative mass distribution data on unstained and unshadowed macromolecular assemblies *(1)*. The techniques of mass measurement and mapping were originally developed on dedicated field-emission STEM instruments in which an annular dark-field detector (ADF) provided efficient collection of elastically scattered electrons from unstained biological samples *(2)*. The freedom from staining or shadowing allows the investigation of subtle changes in absolute mass and mass distribution within a system following experimental intervention. The assembly and growth mechanisms of collagen fibrils formed in vivo and in vitro have been investigated by STEM (**Fig. 2**) *(3)*. Fibrillin microfibrils

From: *Methods in Molecular Biology, vol. 139: Extracellular Matrix Protocols*
Edited by: C. Streuli and M. Grant © Humana Press Inc., Totowa, NJ

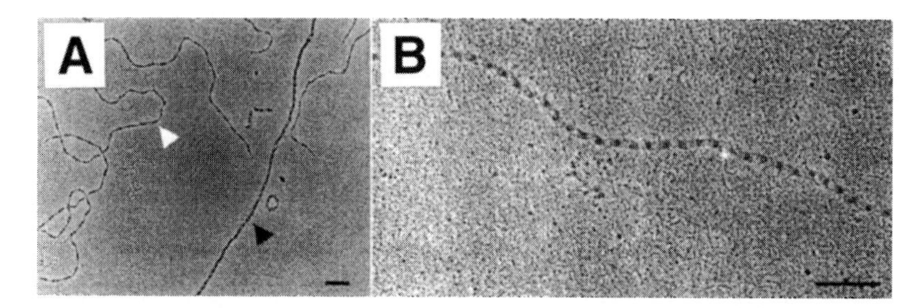

Fig. 1. Rotary shadowed fetal bovine dermal microfibrils. **(A)** Type VI collagen microfibrils (white arrow); fibrillin-containing microfibril (black arrow). **(B)** "Beads on a string" morphology of fibrillin-containing microfibrils. At higher magnifications the granularity of the background becomes apparent. Bar = 200 nm.

are complex multicomponent polymers which have proved amenable to STEM-mass mapping techniques. Microfibrils incubated in the presence of calcium chelating agents or with raised calcium concentrations (**Fig. 3**) demonstrate that their morphology is exquisitely sensitive to calcium *(4)*.

Multicomponent assemblies may be probed for constituent molecules using antibody binding. STEM analysis can detect the presence of antibodies by mass alone eliminating the need for an intrinsically inefficient secondary antibody-immungold complex binding step. The same techniques can be employed to quantify specific enzyme susceptibilities and binding affinities between native assemblies and other matrix components. Pathological and developmental processes can be investigated using STEM-mass mapping techniques to probe for the presence of distinct associated macromolecules. The number and orientation of molecules in a complex can also be determined using this technique.

The composition of multicomponent assemblies can also be probed using specific enzyme digests and STEM. Mass loss following enzyme digestions was shown following incubation of fibrillin microfibrils with chondroitinase ABC lyase *(5)*. These techniques are applicable to macromolecular assemblies ranging in mass from 10 kDa/nm to over 9000 kDa/nm (**Fig. 4**). The protocols in this chapter cover the isolation of fibrillar collagens and microfibrils, the production of carbon-coated grids, rotary shadowing of grids and mica, and the techniques of STEM and associated image analysis.

2. Materials

2.1. Extraction and Isolation of Collagen Fibrils and Microfibrils

All reagents are obtainable from BDH or Sigma except where stated. All solutions use double-distilled ultrapure water and may be stored at room temperature, unless otherwise stated.

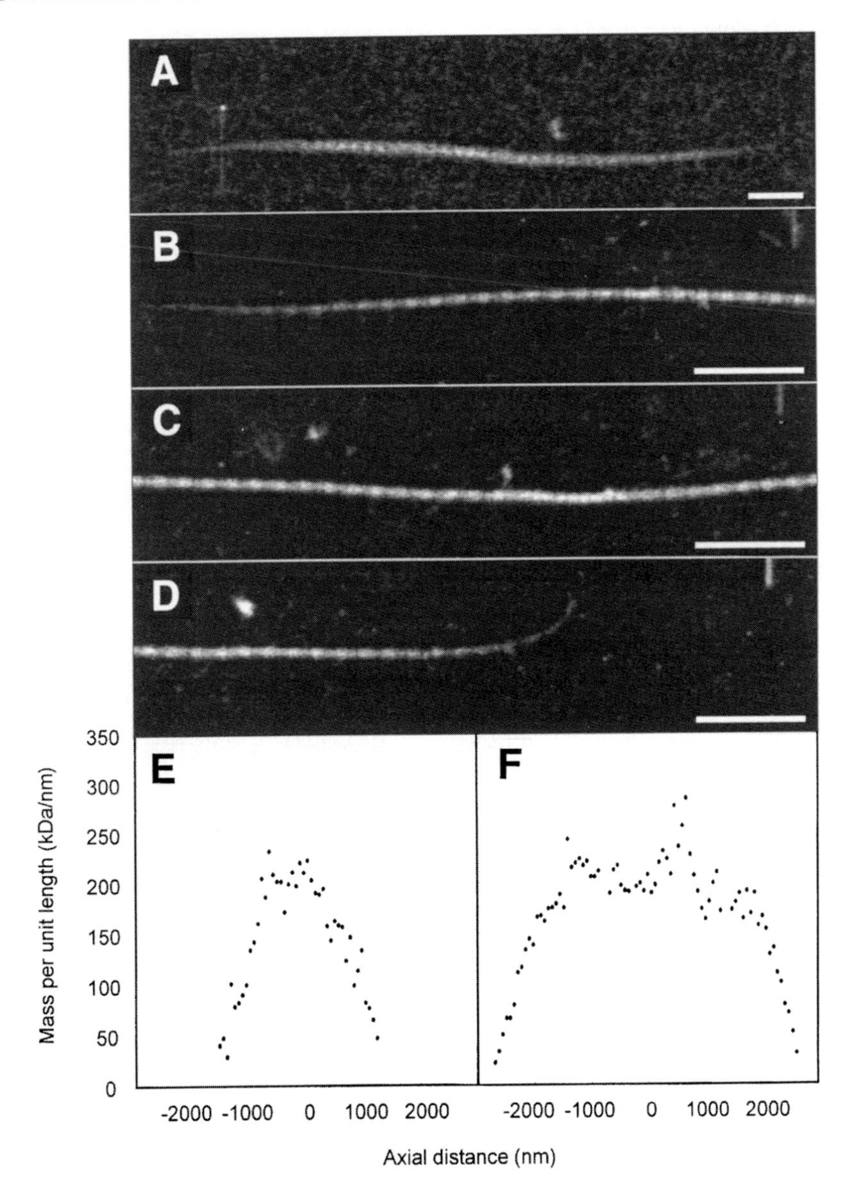

Fig. 2. STEM mass mapping of entire collagen fibrils. **(A)** dark field image of an entire fibril formed *in vitro* by reconstitution of acid extracted calf skin collagen. **(E)** axial mass distribution (AMD) of fibril (A), obtained by measurements of STEM digital images. Panels **(B–D)**; dark field STEM montage of an entire collagen fibril extracted from 14-d embryonic chick tendon. **(F)**; AMD of the entire collagen fibril **(B–D)** extracted from embryonic tendon. The tips show a linear AMD profile, typical of tissue fibrils. Bar = 300 nm.

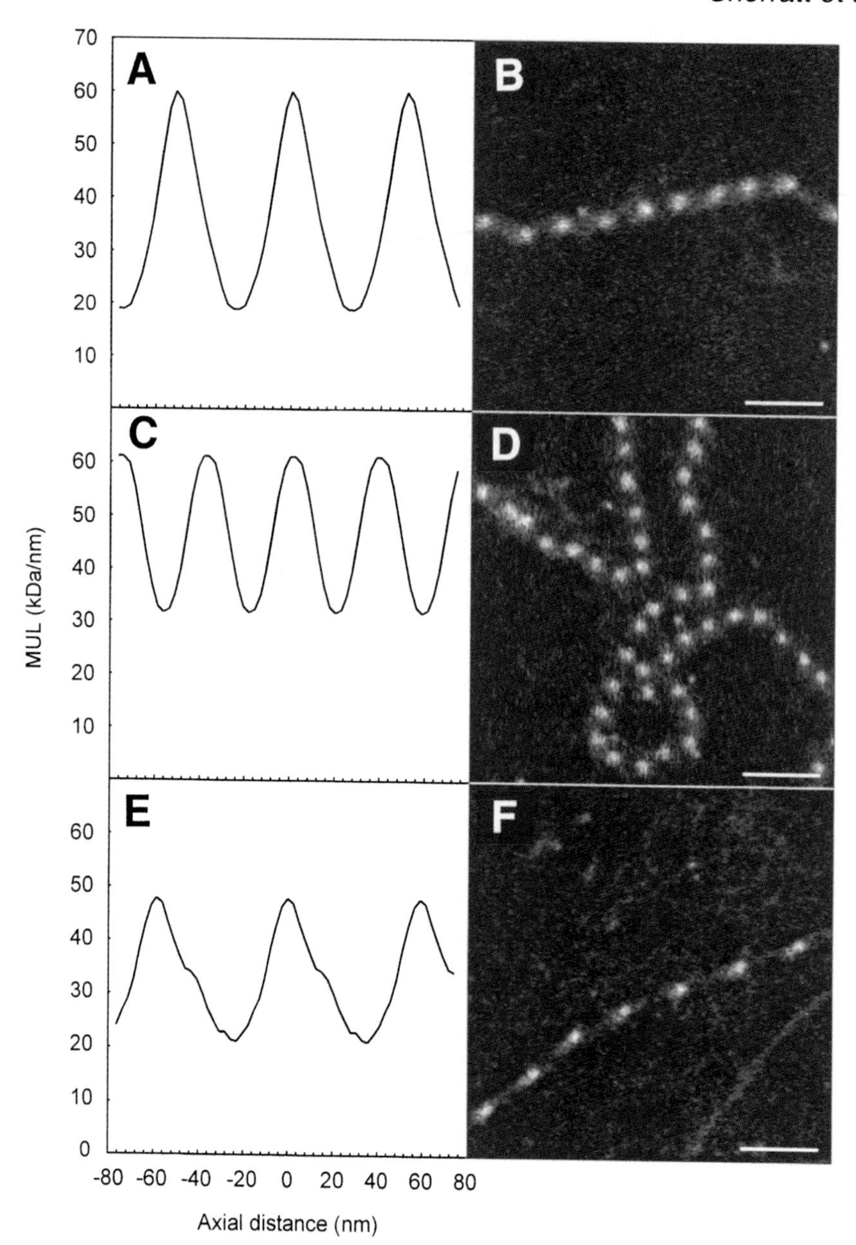

Fig. 3. Isolated fibrillin-containing microfibrils visualized by dark-field STEM **(B,D,F)**. Mean axial mass distributions of microfibrils **(A,C,E)**. Intact native bovine aorta microfibrils were visualized directly **(A,B)**, after incubation with 5 m*M* EDTA **(C,D)** and after incubation with 5 m*M* calcium chloride. STEM analysis revealed that Ca^{2+} removal or addition caused significant changes in microfibrillar mass distribution and periodicity.

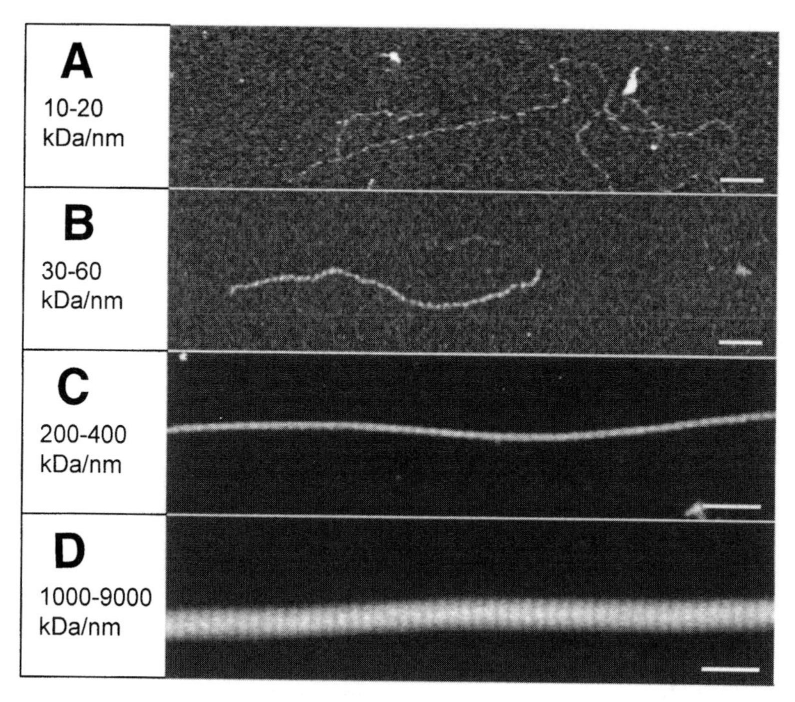

Fig. 4. STEM analysis of macromolecular assemblies within the mass range 10–9000 kDa/nm. **(A)** Type VI collagen microfibrils, **(B)** fibrillin-containing microfibrils. (**C** and **D**) Type I collagen fibrils of embryonic chick tendon and sea-urchin ligament (sea-urchin ligament micrograph courtesy of Prof. J. Trotter, Albuquerque, NM).

1. Fibrillar collagen extraction buffer: 50 m*M* Tris-HCl (pH 7.4), 50 m*M* EDTA, 100 m*M* sucrose, 150 m*M* sodium chloride.
2. Dounce homogenizer (Agar Scientific, Essex, U.K.).
3. Inactive collagenase buffer: 50 m*M* Tris-HCl (pH 7.4), 0.4 *M* sodium chloride, 10 m*M* calcium chloride.
4. Protease inhibitors: Prepare ×100 stock of 2 m*M* phenylmethanesulphonyl fluoride (PMSF) and 10 m*M* N-ethylmalemide (NEM) in methanol. Weigh out in a fumehood or wear a face mask. Stock solution may be stored at 4°C for 4–5 d.
5. Active collagenase buffer: to a 20-mL plastic universal tube add 10 mL collagenase buffer and 100 mL protease inhibitor stock. Weigh out 2 mg bacterial collagenase (type 1A) and add to mix to produce active collagenase buffer. The buffer may be stored at 4°C for 4–5 d.
6. Column buffer: 0.4 *M* sodium chloride, 50 m*M* Tris-HCl (pH 7.4).
7. Sepharose CL-2B column: equilibrate 30 mL Sepharose CL-2B with column buffer in a column of dimensions 1.5 × 25 cm at a flow rate of 0.2 mL/min overnight.
8. Peristaltic pump, fraction collector, ultraviolet (UV) detector, chart recorder.

2.2. Ultrastructural Investigations

Materials and preparative equipment for electron microscopy may be purchased from Agar Scientifc.

1. Carbon rods (6-mm diameter).
2. Braided carbon fiber.
3. Electron microscope grids (400 mesh copper or nickel).
4. Mica sheets (25 × 25 mm, 0.15-mm thick).
5. Tungsten wire (0.5-mm diameter).
6. Platinum wire (0.1-mm diameter).
7. Diffraction grating (2160 lines/mm).
8. Fine tweezers with clamping ring.
9. 0.2 *M* ammonium acetate (pH 6.0).
10. Liquid nitrogen.
11. Freeze-drying table (0.5-cm thick copper sheet supported by legs with a central handle).
12. High vacuum coating unit: large bell jar (30-cm diameter), rotating table with variable speed control (50–200 rpm), power supply providing 10 V/100 A.

3. Methods

3.1. Extraction and Isolation of Fibrillar Collagens

1. Dissect 0.5 g of the tissue sample (tendon, skin, cornea). Wash in fibrillar collagen extraction buffer (*see* **Note 1**) and cut into 1-mm^2 pieces.
2. Homogenize 0.5 g of the tissue sample in 0.5 mL fibrillar collagen extraction buffer using a hand-held Dounce homogenizer for 45 s (*see* **Note 2**).
3. Fibril-rich solutions are stable at room temperature for 2 wk.

3.2. Extraction and Isolation of Microfibrils

1. Dissect 1 g of tissue (skin, aorta, nuchal ligament, ciliary zonules) into 1 mm^3 pieces.
2. Incubate the tissue fragments with 1–2 mL active collagenase buffer on a rotary mixer until no visible fragments of tissue remain (*see* **Note 3**).
3. Following digestion, centrifuge the samples at 7840*g* for 10–30 min. The centrifugation step removes remaining aggregates (which could block the column).
4. Chromatograph the supernatant on the preequilibrated Sepharose CL-2B gel filtration column at a flow rate of 0.2 mL/min and a fraction size of 1 mL. High-M_r material in the excluded volume (V_o) includes fibrillin-containing and type VI collagen microfibrils (*see* **Note 4**).
5. Isolated microfibrils suspended in column buffer may be stored at 4°C for 7 d with no changes in gross morphology or mass distribution as determined by RS-TEM and STEM mass mapping techniques.

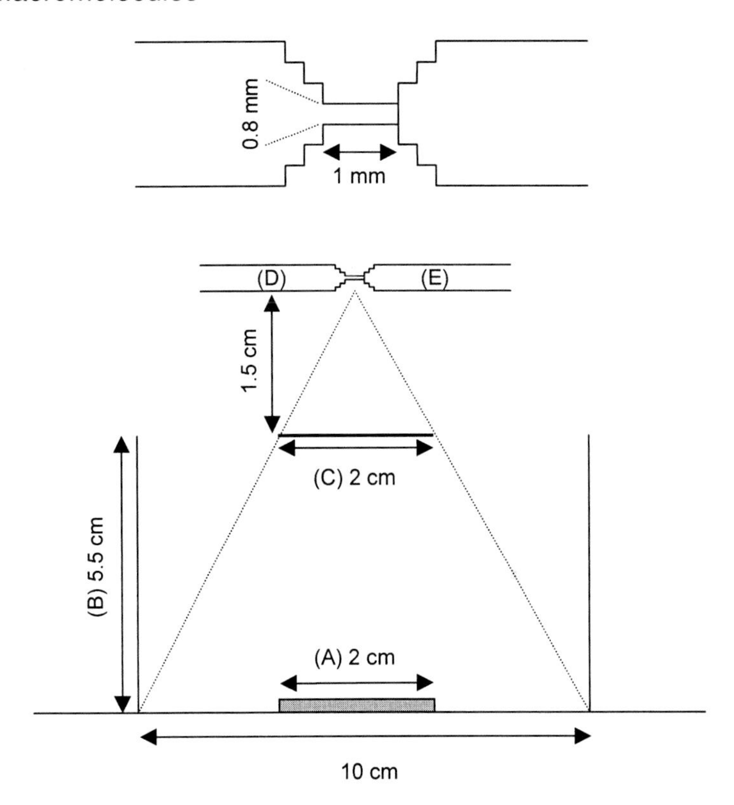

Fig. 5. Place the freshly cleaved mica sheet **(A)** within a clean ricochet cylinder **(B)** and below a mask **(C)** which shields the mica from direct carbon evaporation. Carbon rods (**D** and **E**), held in contact by a spring during the evaporation, are positioned 1.5 cm above the mask.

3.3. Preparation of Carbon-Coated EM Grids

1. Prepare thin carbon films by thermally evaporating a 1-mm long, 0.8-mm diameter spindle from a carbon rod using the geometry shown in **Fig. 5** (*see* **Note 5**). Using a clean ricochet cylinder *(6)*, evaporate the carbon onto a freshly cleaved sheet of mica (**Fig. 5**) (*see* **Note 6**). The film thickness should lie within the range 2.0–3.0 nm *(7)*. Film thickness may be determined using a film thickness monitor, assuming a carbon density of 1.9 g/cm^3 (*see* **Note 7**).
2. Incubate the carbon-coated mica in a humid atmosphere overnight to facilitate separation of the film.
3. Using watchmakers forceps, carefully space the EM grids 1–2 mm apart on the carbon film.
4. Using the forceps, slowly slide the mica at angle of 10–20° into a reservoir of ultrapure water. The carbon film should detach from the mica and remain intact on the surface of the water.

5. Break the carbon film into sections of approximately five grids per side. Cut strips of newspaper slightly larger than the raft of grids (*see* **Note 8**). Gently lay the paper onto the grids and carbon film. When the water has completely soaked through the newspaper, lift the paper and grids onto a piece of filter paper to dry. Store carbon films on dry filter paper in a Petri dish.

3.4. Preparation of Unstained Sample Grids

1. Prepare a range of sample dilutions (e.g., undiluted, one-third, one-sixth, and one-ninth) in fibrillar collagen extraction buffer or column buffer as appropriate.
2. Fold a circular filter paper into quarters. Using tweezers, with a clamping ring, grip the centre of the carbon-coated microscope grid. Pipet 5–6 μL of the sample onto the centre of the grid. Allow the sample to adsorb for 30 s (*see* **Note 9**).
3. Gently wick off excess sample against the folded side of the filter paper. Wash the grid with three successive drops of water then drain excess water from below the grid.
4. Allow the grid to air-dry for 5 min prior to examination in the microscope (*see* **Note 10**).

3.5. Rotary Shadowing

This protocol for rotary shadowing is based on the mica sandwich technique originally described by Mould *(8)* and subsequently adapted ***(9,10)***. Alternatively, unstained grids may be prepared as described in protocol **Subheading 3.4.**, examined by STEM and subsequently shadowed as described in **steps 8** onward.

1. Prepare a range of sample dilutions (e.g., undiluted, 1/3, 1/6 and 1/9) in 0.2 *M* ammonium acetate.
2. Cleave a 2 cm × 2 cm mica sheet (*see* **Note** 6).
3. Submerge a freeze-drying table in a liquid nitrogen bath to cool. Replenish the liquid nitrogen frequently maintaining a layer of nitrogen over the table.
4. Pipet 5 μL of the sample onto the inner surface of one mica sheet. Lay the second mica sheet onto the first (inner surface to inner surface) overlapping the edge by 1–2 mm.
5. After 10 min at room temperature, split the mica sandwich under 0.2 *M* ammonium acetate. Keeping the mica pieces immersed in the buffer, wash by gentle agitation for one min before reclosing the sandwich with a 1–2-mm overlap.
6. Plunge the closed mica sandwich into a liquid nitrogen bath. Using a pair of forceps in each hand, grip the opposing mica sheets and split open. Transfer the mica sheets onto the submerged freezing table, inner surfaces facing upwards.
7. Follow the manufacturers instructions to freeze-dry the mica sheets on the cold table. Freeze-dried mica sheets may be stored in a clean/dry environment indefinitely.
8. Cut a length of tungsten wire to a length appropriate to the dimensions of the holding jig. Cut 8 cm of platinum wire and wind in a spiral around the tungsten. Squeeze the wound platinum to cover a length of 1 cm. Place the tungsten/plati-

num wire in a jig at a distance of 10 cm from the rotating table. Adjust the wire/rotating table angle to 4° (*see* **Note 11**).
9. Attach mica sheets to the rotating table with double-sided adhesive tape. Carbon-film grids may be placed towards the centre of the rotating table without using adhesive tape.
10. Follow manufacturers instructions to evacuate the coating unit. Rotate the table to a speed of 60 rpm. Increase the voltage across the tungsten/platinum wire until the platinum begins *to bubble* (*see* **Note 12**).
11. Carbon coat the shadowed mica sheets (protocol **Subheading 3.3., step 1**, using carbon braided carbon fiber)
12. Leave the shadowed and carbon-coated mica over an atmosphere of 10% (v/v) acetic acid for 24 h.
13. Float the carbon films onto distilled water and pick up the carbon on 400 mesh copper or nickel grids.

3.6. Rotary Shadowing Transmission Electron Microscopy

Imaging of rotary shadowed macromolecules may be carried out in any conventional TEM.

3.7. Scanning Transmission Electron Microscopy

The digital STEM system employed in our laboratory is based on a JEOL 1200EX TEM with a JEOL ASID10 scanning unit and the standard bright field and annular dark-field detectors (*see* **Note 13**). A microcomputer interface (**Fig. 6**) permits digital scan control, signal acquisition, beam blanking, and digitization of the signal from the ADF detector *(11)*. Digital images are acquired using a spot size of approximately 3 nm. No correction for beam-induced mass loss is required where the specimen is exposed to an electron dose of less than 10^3 e nm^{-2}. Detector geometries are illustrated in **Fig. 7**.

Standard STEM images consist of 512 × 512 points with an acquisition time of 200 µs per point (*see* **Note 14**). Magnification calibration can be carried out with a carbon replica of a diffraction grating ruled at 2160 lines/mm Image analysis requires an interactive programming or macro-driven environment (*see* **Note 15**). Fibrillar collagens and microfibrils are filamentous structures, the mass characteristics of such assemblies are usually expressed as mass per unit length (in kDa/nm).

The following protocol demonstrates the steps in calibrating a STEM system using tobacco mosaic virus (TMV) particles which have a well-defined mass per unit length of 131 kDa/nm *(1)*. When calibrated, the system may be used to determine the absolute mass characteristics of a large range of ECM macromolecules.

1. Capture dark-field STEM images of TMV supported on carbon film (**Fig. 8**, panel A). Record the signal from the bright-field detector (BF) and the zero probe level

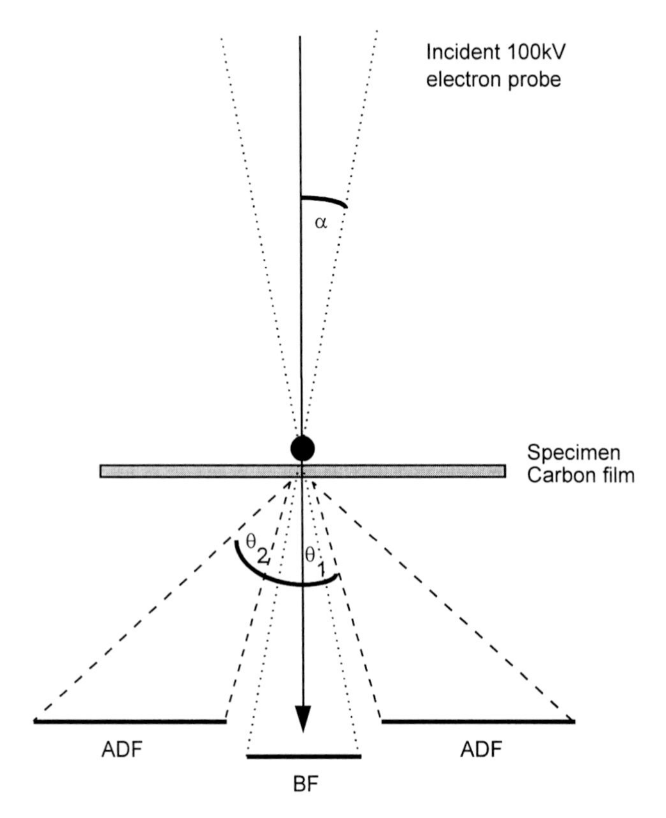

Fig. 6. Main components of a digital STEM system. Bright-field detector (BF), annular dark-field detector (ADF), analog-to-digital (A/D) and digital-to-analog (D/A) convertors.

(ZP). ZP defines the residual signal from bright- and dark-field detectors in the absence of an incident electron beam. **Note** the objective lens setting.

2. Extract a rectangular region centered on a TMV particle (**Fig. 8**, panel B).
3. Project (average) the extracted region along the long axis to form a 1D picture (**Fig. 8**, panel C).
4. Calculate the pixel intensity because of the filament (*pi*) (**Fig. 8**, panel C) from the filament mass (*fm*), left background (*lbg*), and right background (*rbg*).

$$pi = fm - lbg + \text{rbg} / 2 \quad (1)$$

5. Correct the pixel intensity for fluctuations in probe intensity and detector efficiency where $mlc = 1$, bf = bright-field panel and zp = zero probe level (**Fig. 8**, panel A). Correct for the loss of electrons due to the carbon film, for a thin (2–3 nm) film $cfi = 1.015$.

$$pi' = pi \times mlc \, / \, (bf - zp) \times cfi \quad (2)$$

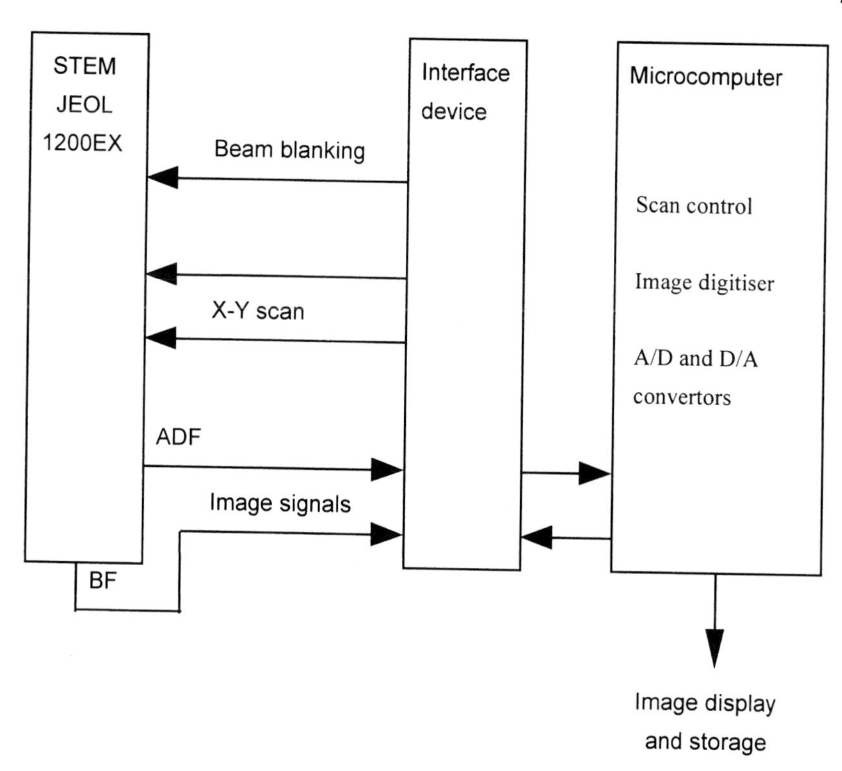

Fig. 7. Digital STEM system detector geometry. Standard values for 100 kV operation for the illumination angle (α) and the angular collection range (θ_1 to θ_2) of the ADF detector are: $\alpha = 12$, $\theta_1 = 25$ and $\theta_2 = 75$ (mrad).

6. Correct *pi'* for the image magnification *mag* (*see* **Note 16**).

$$pi'' = pi \times mlc \,/\, (bf - zp) \times cfi \qquad (3)$$

7. Determine the mean *pi''* for 20–30 TMV particles.
8. Calculate the mass per unit length correction factor (*mlc*) for TMV (assuming TMV = 131 kDa/nm).

$$mlc = 131 \,/\, pi' \qquad (4)$$

9. Capture successive STEM images of the sample of interest adjusting the specimen height to maintain an invariant objective lens setting.
10. Calculate the mass per unit length using **Eqs. 1–3**, substituting the calculated *mlc* value.

4. Notes

1. Fixing the tissue in 2% formaldehyde prior to homogenization may help to minimize loss of associated macromolecules during the extraction procedure.

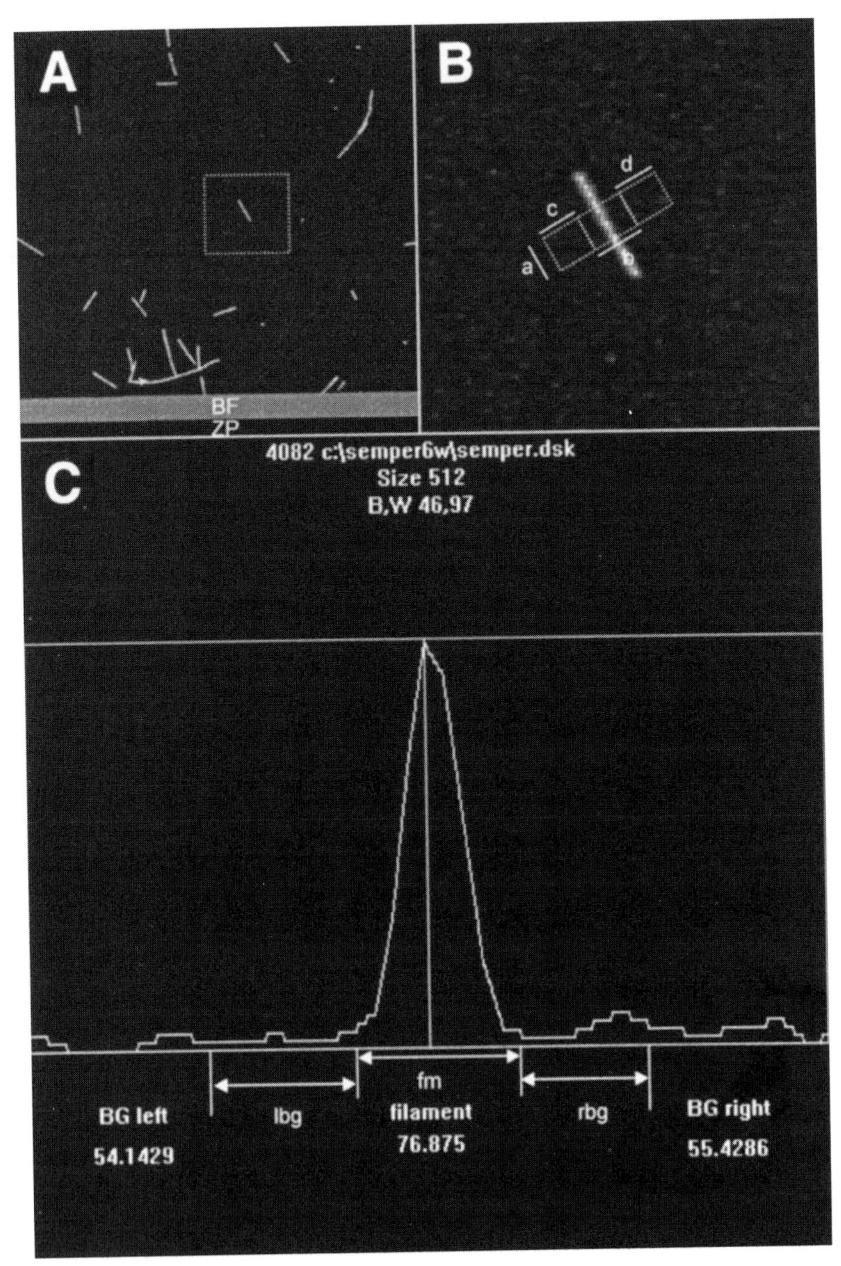

Fig. 8. **(A)** TMV particles supported on thin carbon film, ×20,000, bright field (BF) and zero probe (ZP) panels. **(B)** Extract a box of dimensions a by c+b+d where c and d define background carbon film and b defines the filament. **(C)** 1D extracted plot from **(B)**; mean pixels intensities can be calculated for the filament and left/right backgrounds.

2. Homogenization for 45 s using a Dounce homogenizer is usually sufficient to enable extraction of macromolecules from young embryonic or fetal tissue. Additional ultrasonic treatment and repeated Dounce homogenization may be required for older tissues. When ultrasonicating tissue samples, care must be taken to avoid heat denaturation.
3. The digestion rate varies with the tissue source. Skin from bovine fetuses will digest completely within 2–4 h at room temperature. Aorta and adult bovine skin require an overnight digestion.
4. Bacterial collagenase resistant type VI collagen microfibrils are copurified with fibrillin containing microfibrils by this technique. Dilute suspensions of microfibrils may be concentrated by centrifugation using microcon or centricon concentrators of 100 kDa cutoff (Amersham Life Sciences, Buckinghamshire, UK).
5. Using a lathe, cut a 6-mm diameter carbon rod to form a tip approximately 0.8 mm in diameter and 1 mm in length. Form the tip by progressively cutting concentric circles of decreasing diameter (**Fig. 5**).
6. Cut a 2-cm^2 sheet of mica and trim 1 mm from each edge. Using the side of a pair of forceps flatten one edge of the mica. Gently insert one blade of the forceps into the flattened side until the mica cleaves.
7. Where no monitor is available, a carbon-coating apparatus of the dimensions detailed in **Fig. 5** should produce films within the required film thickness range. Films should be self supporting, but not too thick. A thick carbon film significantly reduces the signal-to-noise ratio for low-mass samples such as microfibrils and collagen fibril tips.
8. An absorbent newspaper is ideal. Avoid areas of ink, particularly color printing.
9. The sample drop should spread over the surface of the grid. If problems are encountered, gently circle the pipette over the grid to spread the sample evenly. Adsorption times of greater than 30 s increase background contamination.
10. Samples may be freeze-dried as an alternative to air-drying.
11. The shadowing angle may be varied up to 10°. Larger angles improve image contrast but result in increased grain size.
12. Use welders goggles or glasses of a similar opacity.
13. Combined STEM/TEM instruments (capable of producing a 2–3-nm spot in STEM mode and equipped with efficient photomultiplier/scintillator ADF detectors) are available from JEOL Ltd. (Tokyo, Japan), Philips Electron Optics B.V. (Eindhoven, Netherlands), and Carl Zeiss Ltd. (Oberkochen, Germany).
14. Data acquisition systems (ES Vision) are available from Emispec Systems Inc. (Arizona, USA).
15. Our laboratory uses the Semper 6 image analysis language (Synoptics, Cambridge, UK). Semper 6 is no longer available commercially and has been replaced by the RAD tool (Visual Basic, Visual C^{++} and Delphi 3) Image Objects from the same company. Other commercial and public domain packages are also available.
16. Typical magnifications; TMV = ×20,000, microfibrils = ×80,000, fibrillar collagen = ×25,000/×50,000.

References

1. Muller, S. A., Goldie, K. N., Burki, R., Haring, R., and Engel, A. (1992) Factors influencing the precision of quantitative scanning transmission electron microscopy. *Ultramicroscopy* **46,** 317–334.
2. Engel, A. (1982) Mass determination by electron scattering. *Micron* **13,** 425–436.
3. Holmes, D. F., Watson, R. B., Chapman, J. A., and Kadler, K. E. (1996) Enzymic control of collagen fibril shape. *J. Mol. Biol.* **261,** 93–97.
4. Wess, T. J., Purslow, P. P., Sherratt, M. J., Ashworth, J., Shuttleworth, C. A., and Kielty, C. M. (1998) Calcium determines the supramolecular organisation of fibrillin-rich microfibrils. *J. Cell Biol.* **141,** 829–837.
5. Sherratt, M. J., Holmes, D. F., Shuttleworth, C. A., and Kielty, C. M. (1997) Scanning transmission electron microscopy mass analysis of fibrillin- containing microfibrils from foetal elastic tissues. *Int. J. Biochem. Cell Biol.* **29,** 1063–1070.
6. Baumeister, W. and Hahn, M. (1978) in *Principles and Techniques of Electron Microscopy* (Hayat, M. A., ed.), Van Nostrand-Reinhold, New York, pp. 1–112.
7. Holmes, D. F., Mould, A. P. M., and Chapman, J. A. (1991) Morphology of sheet-like assemblies of pN-collagen, pC-collagen and procollagen studied by scanning transmission electron microscopy mass measurements. *J. Mol. Biol.* **220**, 111–123.
8. Mould, A. P., Holmes, D. F., Kadler, K. E., and Chapman, J. A. (1985) Mica sandwich technique for preparing macromolecules for rotary shadowing. *J. Ultrastruct. Res.* **91,** 66–76.
9. Kielty, C. M., Cummings, C., Whittaker, S. P., Shuttleworth, C. A., and Grant, M. E. (1991) Isolation and ultrastructural analysis of microfibrillar structures from fetal bovine elastic tissues. Relative abundance and supramolecular architecture of type VI collagen assemblies and fibrillin. *J. Cell Sci.* **99,** 797–807.
10. Kielty, C. M. and Shuttleworth, C. A. (1993b) The role of calcium in the organization of fibrillin microfibrils. *FEBS Lett.* **336,** 323–326.
11. Holmes, D. F. (1995) Mass mapping of extracellular assemblies. *Biochem. Soc. Trans.* **23,** 750–725.

III

Molecular Biology of Extracellular Matrix Genes

12

Screening for Mutations in Cartilage ECM Genes

Michael D. Briggs

1. Introduction

1.1. Genetic Disorders of Cartilage

Genetic disorders of cartilage (chondrodysplasias) are a clinically and genetically heterogeneous group of diseases ranging in severity from relatively mild to severe and lethal forms *(1–2)*. There are over 100 unique well-characterized chondrodysplasia phenotypes and remarkable progress has been made in the last few years identifying the underlying genetic basis of many of these disorders *(3)*. In most cases, a molecular genetics approach was employed involving a combination of genetic linkage mapping, positional (candidate) cloning and DNA sequence analysis *(4–9)*. By its nature this approach requires extensive mutation screening in any potential candidate gene, first to determine if it is the disease gene and then subsequently to identify a range of disease causing mutations. In chondrodysplasia phenotypes this approach has been hampered by a difficulty in obtaining appropriate pathological tissue, such as cartilage, for the isolation of mRNA. This problem is compounded by the complex genomic structure of many genes that encode cartilage structural ECM molecules. For most cartilage diseases, a combination of these difficulties has necessitated screening for mutations in a large numbers of exons using a variety of techniques such as single-stranded conformational polymorphism (SSCP) *(10)*, conformational sensitive gel electrophoresis (CSGE) *(11)*, heteroduplex *(12)* and chemical cleavage mismatch (CCM) analysis *(13)*.

1.2. Multiple Epiphyseal Dysplasias and Pseudoachondroplasia

Pseudoachondroplasia (PSACH) and multiple epiphyseal dysplasia (MED) are clinically similar phenotypes manifesting in varying degrees of disproportionate short stature and early onset osteoarthritis *(2)*. Mutations in the gene-

From: *Methods in Molecular Biology, vol. 139: Extracellular Matrix Protocols*
Edited by: C. Streuli and M. Grant © Humana Press Inc., Totowa, NJ

encoding cartilage oligomeric matrix protein (COMP) have been shown to result in PSACH and some forms of MED ***(5–6,14–16)***. All of the COMP mutations reported to date have been found in exons, which encode the type III repeats, and C-terminal domains. MED can also result from mutations in the gene encoding the α2 chain of type IX collagen (*COL9A2*) ***(4,17)*** and there is evidence that mutations in the genes encoding the α1(IX) (*COL9A1*) and α3(IX) (*COL9A3*) chains of type IX collagen ***(19)*** may also result in MED phenotypes.

1.3. Ectopic (Illegitimate) Transcription

To complement mutation screening in genomic DNA by SSCP-analysis, we have used cultured patient skin fibroblast and transformed lymphoblastoid cell lines as a source of mRNA. Under normal culture conditions these cell lines can maintain a low basal transcription ("ectopic" or "illegitimate" transcription) of cartilage specific genes ***(20–22)***. Using such cell lines it is possible to amplify by the polymerase chain reaction (PCR), cDNA from the COMP, *COL9A1*, *COL9A2*, and *COL9A3* genes for direct mutation screening using DNA sequencing.

2. Materials

2.1. RNA and DNA Isolation

1. Trizol (Gibco-BRL, Gaithersburg, MD).
2. 99.7–100% v/v ethanol, chloroform (AnalaR–BDH, London, UK).
3. Isopropanol, diethyl pyrocarbonate (DEPC) (Sigma, St. Louis, MO).
4. QIAamp Blood Midi/Maxi Kit and QIAamp Tissue Kit (Qiagen, Chatsworth, CA).

2.2. Reverse Transcription of mRNA and Amplification of cDNA and Genomic DNA by the Polymerase Chain Reaction (PCR)

1. Superscript™ Preamplification system for first strand cDNA synthesis (Gibco-BRL).
2. *Taq* DNA polymerase.
3. 100 m*M* solutions of dCTP, dATP, dGTP, and dTTP (Boehringer–Mannheim, Mannheim, Germany).
4. Mineral oil (Sigma).
5. Oligonucleotide (PCR) primers.

2.3. Polyacrylamide and SSCP Gel Analysis

1. Protean II xi cell (20 cm × 20 cm) gel electrophoresis equipment (Bio-Rad).
2. DNA loading buffer: 30% glycerol (AnalaR-BDH), 0.25% bromophenol blue (Sigma), 0.25% xylene cyanol (Sigma) in 10 m*M* Tris-HCl, pH 7.5 (Gibco-BRL).
3. Denaturing DNA loading buffer: 95% Formamide (Sigma), 10 m*M* NaOH (AnalaR–BDH), 0.25% bromophenol blue (Sigma), 0.25% xylene cyanol (Sigma) in 10 m*M* Tris-HCl, pH 7.5 (Gibco-BRL).

4. 1 Kb and 100 bp DNA ladder (Gibco-BRL).
5. 6% polyacrylamide solution. For 500 mL: 100 mL of 30% w/v acrylamide solution (37.5:1 acrylamide:*bis*—Protogel™), 100 mL of 5X TBE solution, 300 mL of dH_2O.
6. 5X TBE solution. For 5 L: 270 g Tris, 138 g Boric Acid (AnalaR-BDH), 100 mL 0.5 *M* EDTA (Gibco-BRL).
7. SSCP—Silver stain solutions.
   a. Gel fixative solution: 10% ethanol, 0.5% acetic acid. 50 mL 100% v/v ethanol, 2.5 mL acetic acid glacial ≈100% (AnalaR-BDH) in 500 mL of dH_2O.
   b. Gel staining solution: 0.15% silver nitrate. 0.5 g silver nitrate (Sigma) in 300 of mL dH_2O.
   c. Developing solution: 2.25 g NaOH (AnalaR-BDH), 0.6 mL formaldehyde (Sigma) in 150 mL of dH_2O.
   d. Stop solution: 0.75% sodium carbonate. 0.75 g sodium carbonate (AnalaR-BDH) in 100 mL d H_2O.
8. Gel drying kit (Promega, Madison, WI).

2.4. Cloning of PCR Products

1. LB Broth. 20 g bacto-tryptone (Gibco-BRL), 10 g bacto-yeast extract (Gibco-BRL), 20 g NaCl (AnalaR-BDH) make up to 2 L with dH_2O, adjusted to pH 7.0 and sterilize by autoclaving.
2. For LB agar, add 15 g of agar (Gibco-BRL) to 1 L of LB broth and sterilize by autoclaving.
3. LB agar plates containing carbenicillin (Sigma) (50 mg/mL). Approx 1 h prior to transformation spread each plate with 50 mL of X-Gal (BIO-RAD) (40 mg/mL in dimethylformamide [Sigma]).
4. Original TA Cloning© Kit (Invitrogen, San Diego, CA).
5. QIAquick PCR Purification Kit and QIAprep Spin Miniprep Kit (Qiagen).

2.5. DNA Sequence Analysis of Cloned PCR Products

1. ABI PRISM™ BigDye™ Terminator Cycle Sequencing Ready Reaction Kit (Perkin–Elmer Applied Biosystems, Norwalk, CT).
2. M13 Forward (–20) primer (5'-g taa aac gac ggc gac–3') and M13 Reverse (5'-cag gaa aca gct atg ac–3') primer.
3. Sequencing reaction buffer: 450 m*M* Tris-HCl, pH 8.0 (Gibco-BRL), 10 m*M* $MgCl_2$ (AnalaR-BDH).

3. Methods

3.1. Isolation of genomic DNA and RNA

3.1.1 Isolation of RNA

The procedure described below is suitable for the isolation of total RNA from one 25 cm^2 flask of confluent cells (*see* **Note 1**).

1. Remove culture medium, add 2.5 mL of Trizol™ reagent and lay the culture flask flat so that all the cells are covered. Leave for 5 min then pipet up and down several times using a 5-mL sterile pipet and transfer sample into microcentrifuge tubes (1.25 mL into each tube). Add 250 μL of chloroform to each tube, vortex briefly then leave at room temperature for 5 min.
2. Centrifuge the samples for 10 min at 13,600 xg in a microcentrifuge. After centrifugation transfer the upper aqueous layer to a fresh microcentrifuge tube. Add 500 μL of isopropanol, mix by inverting the tube several times and incubate at room temperature for 10 min. Pellet the RNA by centrifugation at 13,600 xg for 10 min. Carefully remove the isopropanol from the tube without disturbing the pellet. Add 1 mL of 75% ethanol and respin. Carefully remove all traces of the 75% ethanol and allow the pellet to air-dry for 10 min.
3. Add 100 μL of DEPC-treated sterile H_2O to each RNA sample and incubate at 50°C if necessary to resuspend the RNA pellet.

3.1.2. Isolation of Genomic DNA

There are many commercial "Kits" available for the purification of genomic DNA from a variety of different sample sources, including blood, tissue and cells. We have found that the spin column-based kits manufactured by Qiagen are quick and easy to use and produce a good yield of high-quality genomic DNA.

3.2. Reverse Transcription of mRNA

1. Pipet 5 μg of RNA into an microcentrifuge tube, add 50 ng of random hexamers ($pd(N)_6$) or 500 ng of oligo-dt primer and DEPC-treated sterile H_2O to 12 μL. Incubate at 70°C for 10 min then place on ice for 2 min (*see* **Note 2**).
2. Add 2 μL of 10X RT buffer, 2 μL 25 m*M* $MgCl_2$, 1 μL of 10 m*M* dNTP mix and 2 μL of DTT, incubate the reaction for 5 min at room temperature when using $pd(N)_6$ or 42°C if using oligo-dt. Add 1 μL (200 U) of Superscript II RT enzyme and incubate at 42°C for 1 h. (When using $pd(N)_6$ preincubate the reaction at room temperature for 10 min prior to moving to 42°C.)
3. Following incubation, terminate the reaction by incubation at 70°C for 5 min, add 1 μL of RNase H, and incubate at 37°C for 30 min.

3.2.1. Amplification of Genomic DNA and cDNA by the PCR

1. For amplification by PCR, use 5 μL of cDNA (from first strand reaction, *see* **Note 2**) or 5 μL of genomic DNA (20–50 ng/μL) in a 100 μL reaction volume. For each pair of oligonucleotide primers, prepare a "master mix" containing PCR buffer, dNTPs, oligonucleotide primers and *Taq* DNA polymerase (**Tables 1–3**). Mix 95 μL aliquots of this to 0.5-mL microcentrifuge tubes containing either patient genomic DNA, cDNA, or appropriate control sample. Cover the reactions with two drops of light mineral oil and amplify by PCR using the following 40 cycle parameters:

Table 1
Components of a Typical Polymerase Chain Reaction

PCR reagents	Reagents for 100 μL reaction	"Master mix" for 5 × 100 μL reactions	Final concentration of PCR reagents
10X PCR buffer[a]	10 μL	50 μL	1X
100 m*M* Tris-HCl			10 m*M* Tris-HCl
15 m*M* $MgCl_2$			1.5 m*M* $MgCl_2$
500 m*M* KCl			50 m*M* KCl
pH 8.3 (@20°C)			
10 m*M* dNTP mix[b]	2 μL	10 μL	200 μ*M*
Forward primer (5 μ*M*) (**Tables 2** and **3**)	2 μL (10 pmols)	10 μL (100 pmols)	0.1 μ*M*
Reverse primer (5 μ*M*) (**Tables 2** and **3**)	2 μL (10 pmols)	10 μL (100 pmols)	0.1 μ*M*
Taq DNA polymerase	1 U	1 μL (5 U)	1 U
Sterile dH_2O	To 95 μL	394 μL	—

[a]Most *Taq* DNA polymerase enzymes are supplied with a 10X concentrated reaction buffer.

[b]10 m*M* dNTP mixes are available commercially. Alternatively dilute 100 m*M* dNTP stocks in 10 m*M* Tris-HCl, pH 7.5 to make a 10 m*M* dNTP mix containing all four dNTPs.

Table 2
Oligonucleotide Primers Used in the PCR-Amplification of COMP cDNA

COMP domain	Primer name	Primer location (nucleotide)	Primer sequences	Product size
Type III repeat domain	HT3-F HT3-R	824	5'-ggt cgc gac act gac cta gac–3'	789 bp
	HT3-R	1612	5'-ggt gag cgt gac ttc agc gtt–3'	
Carboxyl-terminal domain (exons 14–19)	HCt-F HCt-R	1598	5'-gaa gtc acg ctc acc gac–3'	702 bp
	HCt-R	2299	5'-cta ggc ttg ccg cag ctg atg g–3'	

Table 3
Oligonucleotide Primers Used in the PCR-Amplification of COMP Genomic DNA

COMP exon	Primer name	Primer location	Primer sequences[a]	COMP domain	PCR product size
10	I9F1	Intron 9	5'-tga gga gtg tga cct ttg cc–3' '	Type III repeat	279 bp
	I10R1	Intron 10	5'-agc cga atc ccg cct tcg gtg–3		
11	I10F1	Intron 10	5'-ctt ggg ctc tgg tcc cgt gg–3'	Type III repeat	181 bp
	I11R1	Intron 11	5'-gct tac cca gct gga gtc tg–3'		
12	I11F2	Intron 11	5'-att tcc tct gtc tga tta tgg–3'	Type III repeat	168 bp
	I12R1	Intron 12	5'-cca gag aca atg agc tct cca g–3'		
13	I12F2	Intron 12	5'-ggg tag cct ttg aca aaa cg–3'	Type III repeat	223 bp
	1492R	Exon 13	5'-gtt agg cac cag gcg gca g–3'		
14	I13F1	Intron 13	5'-tga ctt tag ccc acc gag gg–3'	Type III repeat/ C-terminal	281 bp
	I14R1	Intron 14	5'-ctc agc ata ggc ctc act gtg–3'		
15	I14F1	Intron 14	5'-cac agt gag gcc tat gct ga–3'	C-terminal	164 bp
	I15R2	Intron 15	5'-gtg gca gga tag cgc tgc tc–3'		
16	I15F1	Intron 15	5'-gcg ttc gga aag gcc act gc–3'	C-terminal	319 bp
	I16R1	Intron 16	5'-cta agt ggc tgt aaa ggg ttt–3'		
17	I16F1	Intron 16	5'-gcc cac cga ggt ctc tga cc–3'	C-terminal	275 bp
	I17R1	Intron 17	5'-ggc act ccc acc tgg gcc tg–3'		
18/19	2116F	Exon 18	5'-gcg att cta tga ggg ccc tga–3'	C-terminal	319 bp
	2342R	Exon 19	5'-gcg gtg agg gtg gct gtc at–3'		

[a]Refer to **ref. *10***.

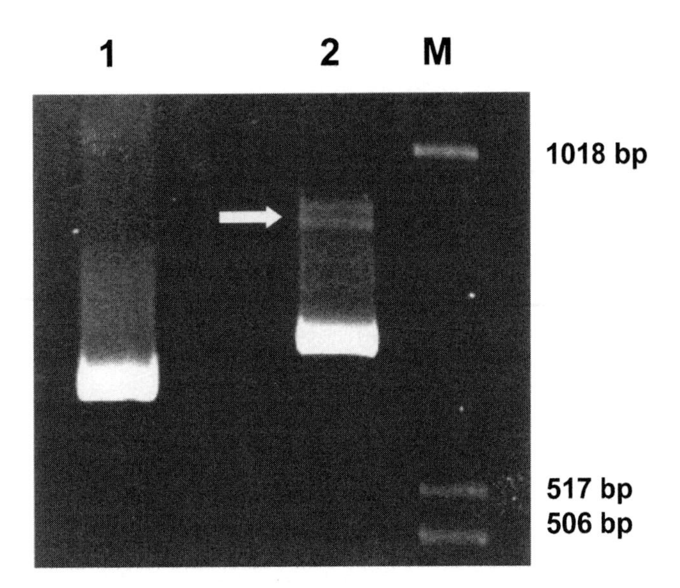

Fig. 1. Six percent PAGE-analysis of PCR amplified COMP cDNA. Total RNA was isolated from a skin fibroblast cell line established from a patient with pseudo-achondroplasia and reverse transcribed into cDNA with both oligo-dt and $pd(N)_6$ primers as described. **Lane 1,** COMP C-terminal domain cDNA PCR-amplified by primers HCt-F and HCt-R (702 bp). **Lane 2,** COMP Type III repeat domain cDNA PCR-amplified by primers HT3-F and HT3-R (789 bp). **M,** molecular weight marker. A white arrow indicates a putative heteroduplex. Sequence analysis of clones confirmed that this patient was heterozygous for a COMP gene mutation.

   a. 94°C for 1 min (template denaturation).
   b. 55°C for 1 min (primer annealing).
   c. 72°C for 1 min (template elongation).
2. After cycle is complete, remove 15 µL of PCR product, add 5 µL of DNA loading buffer and analyze on a 6% polyacrylamide gel in 1X TBE buffer (*see* **Note 3**). Stain the gel with ethidium bromide (10 µg/mL) and view DNA with an UV transilluminator (**Fig. 1** and **Fig. 2**).

3.2.2. SSCP Analysis of Genomic DNA Amplified by PCR

1. Remove 2 µL of PCR product to a fresh tube, add 10 µL of denaturing DNA loading buffer and heat samples to 95°C for 2 min. Place denatured DNA on ice for 2 min.
2. Load sample onto a 6% polyacrylamide gel precooled to 4°C and run at 300 V for 3–4 h (*see* **Note 4**).
3. After the run is complete, separate the plates gently, remove the gel, and place in 500 mL of gel-fixative solution for 20 min.
4. Wash the gel twice in dH_2O for 5 min.

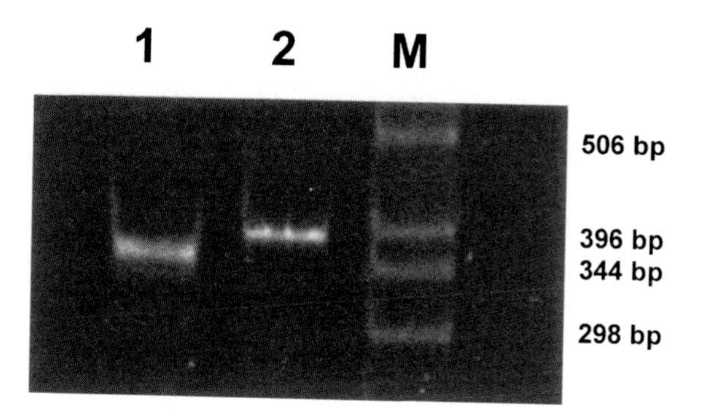

Fig. 2. Six percent PAGE analysis of PCR amplified *COL9A1* and *COL9A2* cDNA. Total RNA was isolated from a transformed lymhoblastoid cell line established from a patient with MED and reverse transcribed into cDNA with both oligo-dt and pd$(N)_6$ primers as described. **Lane 1**, *COL9A2* cDNA PCR-amplified by primers 9a2ex3-F (5'-aca atg ggc ccc ctg gaa aag c-3') and 9a2ex10-R (5'-cct tcc aga ccc tgg atg gtt-3') (348 bp). Lane 2, *COL9A1* cDNA PCR-amplifed by primers 9a2ex8-F (5'-cct gga gtt cca ggc atc gat-3') and 9a1ex16-R (5'gga tct cca tca tga aag cca a-3') (380 bp). **M**, molecular weight marker (2 Kb). These PCR-amplifed cDNA fragments encode equivalent regions of the COL3 domain of the α1(IX) and α2(IX) chains *(19)*.

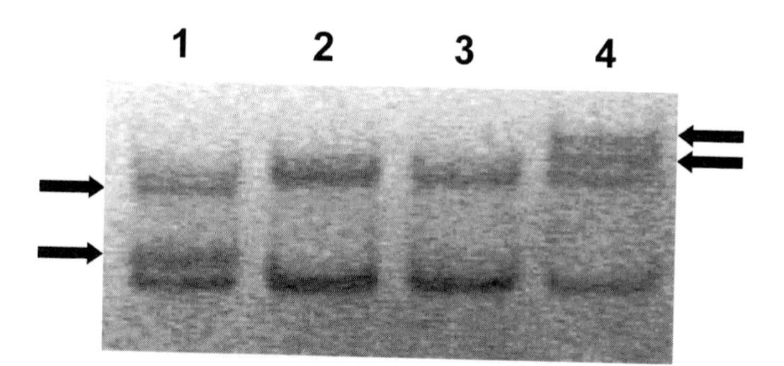

Fig. 3. Silver-stained 6% polyacrylamide gel showing SSCP analysis of PCR amplified genomic DNA from exon 13 of the COMP gene. **Lanes 1–4**, Four different patients with pseudoachondroplasia, the black arrows indicate the position of abnormally migrating single stranded DNA in patients 1 and 4. (For this particular DNA fragment, the single strands migrate at about 300-bp.) Sequence analysis of clones confirmed patients 1 and 4 were heterozygous for COMP gene mutations.

5. Stain gel for 15 min in 300 mL of gel-staining solution.
6. Wash the gel twice in dH_2O for 5 min.
7. Develop for 10–20 min in 150 mL developing solution.
8. Terminate developing by soaking gel in 150 mL of stop solution for 5 min.
9. Dry gel onto 3M Whatman paper or alternatively use a gel drying kit (**Fig. 3**).

3.3. Cloning of PCR Products

1. Prior to cloning the PCR product, remove excess primers using a spin column (QIAquick PCR Purification Kit) and elute the purified PCR product in 50 µL of Tris-HCl pH 7.5.
2. Set up a ligation reaction using 5 µL of purified PCR product, 1 µL of 10X ligase buffer, 2 µL of TA vector DNA and 2 µL of sterile dH_2O. Incubate the reaction overnight at 4°C.
3. Thaw one vial of competent cells (50 µL) for each ligation and place on ice. Add 5 µL of ligation reaction and 1 µL of β-mercaptoethanol. Allow the transformation to proceed on ice for 30 min. Heat shock the cells at 42°C for 30 s and place back on ice for 2 min. Add 250 µL of SOC medium and incubate at 37°C for 1 h in a shaking incubator.
4. Spread transformation onto LB agar plates containing 50 mg/mL Carbenicillin and X-Gal. Incubate inverted at 37°C for 16–18 h.
5. Using a sterile pipet tip pick 10 recombinant (white) colonies and with each one inoculate 5 mL LB broth containing 50 µg/mL Carbenicillin. Grow overnight at 37°C in a shaking incubator.
6. Isolate plasmid DNA from 1.5 mL of overnight culture using a QIAprep Spin Miniprep Kit. Elute purified plasmid DNA in 100 µL of 10 m*M* Tris-HCl pH 7.5.
7. Digest 10 mL of purified plasmid DNA with *Eco RI* (or other suitable restriction endonuclease) to confirm presence of DNA insert.

3.4. DNA Sequence Analysis of Cloned PCR Products

1. Quantify the amount of plasmid DNA using an OD_{260} spectrophotometer measurement and dilute to 50 ng/µL with 10 m*M* Tris-HCl pH 7.5.
2. Set up following sequencing reactions:

Plasmid DNA (50 ng/µL)	8 µL (400 ng)
Terminator ready mix	2 µL
Reaction buffer (450 m*M* Tris-HCl pH 8.0, 10 m*M* $MgCl_2$)	6 µL
Primer (0.8 pmols/mL) (M13 Reverse primer or M13 Forward (–20) primer—*see* **Note 5**)	4 µL (3.2 pmols)

3. Overlay the reaction with light mineral oil and use the following 25 cycle sequencing parameters:
   a. 96°C for 30 s (template denaturation).
   b. 50°C for 30 s (primer annealing).
   c. 60°C for 4 min (extension).
4. Remove sequencing reaction products from under the mineral oil and place in a clean microcentrifuge tube. Add 2 µL of 3M NaOAc pH 4.6 and 50 µL of ethanol (99%), vortex and incubate on ice for 10 min.
5. Spin samples at 13,600 xg in a microcentrifuge for 15 min then carefully remove the supernatant with a pipet.

6. Add 250 µL of 70% ethanol, vortex briefly, and respin. Remove all traces of ethanol with a pipet, if necessary respin sample to remove residual ethanol. Allow pellet to air-dry for 10 min.
7. Electrophoresis of samples on ABI Prism 377 DNA sequencer (*see* **Note 6**).

4. Notes

1. In many cases, transformed lymphoblastoid (LB) cells are cultured in suspension. If this is the case, remove the medium and pellet cells by brief centrifugation. Resuspend cell pellet directly in TRIZOL.
2. It is possible to use a gene-specific primer (GSP) in the first-strand cDNA synthesis, we have found, however, that this is more likely to result in spurious priming in the PCR. We usually set up two separate RT reactions, which contain oligo-dt and pd$(N)_6$, respectively, and then pool these to give 40 µL of cDNA for PCR. In some cases, a DNA product is not seen after the first round of PCR. There are several approaches that can be used to overcome this. The first is to take an aliquot of the PCR reaction and reamplify using the same or preferably nested PCR primers. Alternatively, it is possible to use more RNA in the RT reaction or more cDNA in the PCR. In the latter case, it must be remembered that there will be carry over of $MgCl_2$ and dNTPs, which will affect the specificity of the PCR. Reduce proportionally the amount of $MgCl_2$ and dNTPs added to the PCR to account for the carry over. A 10X PCR buffer can easily be made without $MgCl_2$, which is then added to the correct concentration from a 25-m*M* stock solution.
3. For DNA size determination, PAGE analysis of PCR products is far more accurate than agarose gel electrophoresis. It is also possible to detect heteroduplex DNA molecules on polyacrylamide gels (**Fig. 1**), which migrate slower compared to homoduplex DNA molecules. Heteroduplex molecules are usually formed when DNA is PCR amplified from a patient who is heterozygous for a mutation. In the final stages of the PCR, double-stranded DNA molecules are formed between normal and mutant single strands. This method of detection is particularly suited for small deletions, which result in very distinctive heteroduplex bands. It is possible to accentuate the formation of heteroduplexes by one of two ways. After the PCR has finished, heat the reaction to 96°C for 2 min then allow to cool slowly to room temperature. Alternatively, the PCR reaction volume can be reduced to 25 µL while using the same amount of DNA (100 ng), this results in a more concentrated PCR product which will favor heteroduplex formation, however, it may also result in spurious priming and nonspecific PCR products.
4. Run the gel so that the nondenatured double-stranded DNA is about 2–3 cm from the bottom of the gel (on a 6% PAGE, the xylene cyanol dye migrates at approx 300 bp). In this way, double-stranded heteroduplex molecules will be visible in addition to the individual single strands (**Fig. 3**). By running the gel in the cold room the voltage can be increased without the gel overheating and possibly disrupting the secondary structure of the individual single stranded DNA molecules.

5. Sequencing with the M13 forward and reverse primers usually allows the entire plasmid insert to be read (up to 500–600 bp per reaction). Sequencing with the PCR primers will result in the loss of the first 20–30 bp of the insert.
6. Most large university or research centers have a DNA-sequencing facility. If this is not readily available, there are numerous companies who provide this service.

Acknowledgments

This research was supported by the Arthritis Research Campaign

References

1. International Working Group on Constitutional Diseases of Bone (1992) International classification of osteochondrodysplasias. *Am. J. Med. Genet.* **44,** 223–229.
2. Rimoin, D. L. and Lachman, R. S. (1993) Genetic disorders of the osseous skeleton, in McKusick's Heritable Disorders of Connective Tissue, 5th ed. (Beighton, P., ed.), Mosby-Year Book, Inc., St. Louis, MO, pp. 557–689.
3. Francomano, C. A., McIntosh, I., and Wilkin, D. J. (1996) Bone dysplasias in man: molecular insights. *Curr. Opinion Genes Dev.* **6,** 301–308.
4. Muragaki, Y., Mariman, E. C. M., van Beersum, S. E. C., Perala, M., van Mourik, J. B. A., Warman, M. L., et al. (1996) A mutation in the gene encoding the alpha-2 chain of the fibril-associated collagen IX, COL9A**2,** causes multiple epiphyseal dysplasia (EDM2). *Nat. Genet.* **12,** 103–105.
5. Hecht, J. T., Nelson, L. D., Crowder, E., Wang, Y., Elder, F. F. B., Harrison, W. R., et al. (1995) Mutations in exon 17B of cartilage oligomeric matrix protein (COMP) cause pseudoachondroplasia. *Nat. Genet.* **10,** 325–329.
6. Briggs, M. D., Hoffman, S. M. G., King, L. M., Olsen, A. S., Mohrenweiser, H., Leroy, J. G., et al. (1995) Pseudoachondroplasia and multiple epiphyseal dysplasia due to mutations in the cartilage oligomeric matrix protein gene. *Nat. Genet.* **10,** 330–336.
7. Vikkula, M., Mariman, E. C. M., Lui, V. C. H., Zhidkova, N. I., Tiller, G. E., Goldring, M. B., van Beersum, S. E. C., de Waal, Malefijt, M. C., et al. (1995) Autosomal dominant and recessive osteochondrodysplasias associated with the COL11A2 locus. *Cell* **80,** 431–437
8. Hästbacka, J., de la Chapella, A., Mahtani, M. M., Clines, G., Reeve-Daly, M. P., Daly, M., et al. (1994) The diastrophic dysplasia gene encodes a novel sulfate transporter: positional cloning by fine-structure linkage disequilibrium mapping. *Cell* **78,** 1073–1087.
9. Warman, M. L., Abbott, M., Apte, S. S., Hefferon, T. W., McIntosh, I., Cohn, D. H., et al. (1993) A mutation in the human type X collagen gene in a family with Schmid metaphyseal chondrodysplasia. *Nat. Genet.* **5,** 79–82.
10. Orita, M., Iwahana, H., Kanazawa, H., Hayashi, K., and Sekiya, K. (1989) Detection of polymorphisms of human DNA by gel electrophoresis as single-stranded conformation polymorphisms. *Proc. Natl. Acad. Sci. USA* **86,** 2766–2770.
11. Ganguly, A., Rock, J. M., and Prockop, D. J. (1993) *Proc. Natl. Acad. Sci. USA* **90,** 10,325–10,329.

12. White, M. B., Carvalho, M., Derse, D., O'Brien, S. J., and Dean, M. (1992) Detecting single base substitutions as heteroduplex polymorphisms. *Genomics* **12,** 301–306.
13. Cotton, R. G. H., Rodrigues, N. R., and Campbell, R. D. (1988) Reactivity of cytosine and thymine in single-base-pair mismatches with hydroxylamine and osmiun tetroxide and its application to the study of mutations. *Proc. Natl. Acad. Sci. USA* **85,** 4397–4401.
14. Briggs, M. D., Mortier, G. R., Cole, W. G., King, L. M., Golik, S. S., Bonaventure, J., et al. (1998) Diverse mutations in the gene for cartilage oligomeric matrix protein (COMP) in the pseudoachondroplasia-multiple epiphyseal dysplasia disease spectrum. *Am. J. Hum. Genet.* **62,** 311–319.
15. Loughlin, J., Irven, C., Mustafa, Z., Briggs, M. D., Carr, A., Lynch, S., et al. (1998) Identification of five novel mutations in the cartilage oligomeric matrix protein gene in pseudoachondroplasia and multiple epiphyseal dysplasia. *Hum. Mut. Suppl.* S10–S17.
16. Ballo, R., Briggs, M. D., Cohn, D. H., Knowlton, R. G., Beighton, P. H., and Ramesar, R. S. (1997) Multiple epiphyseal dysplasia, Ribbing type: a novel point mutation in the COMP gene in a South African family. *Am. J. Med. Genet.* **68,** 396–400.
17. Briggs, M., Choi, H., Warman, M., Loughlin, J., Wordsworth, P., Sykes, B., et al. (1994) Genetic mapping of a locus for multiple epiphyseal dysplasia to a region of chromosome 1 containing a type IX collagen gene. *Am. J. Hum. Genet.* **55,** 678–684.
18. Deere, M., Blanton, S. H., Scott, C. I., Langer, L. O., Pauli, R. M., and Hecht, J. T. (1995) Genetic heterogeneity in multiple epiphyseal dysplasia. *Am. J. Hum. Genet.* **56,** 698–704.
19. Olsen, B. R. (1997) Collagen IX. *Intl. J. Biochem. Cell Biol.* **29,** 555–558.
20. Chan, D. and Cole, W. G. (1991) Low basal transcription of genes for tissue specific collagens by fibroblasts and lymphoblastoid cells. *J Biol. Chem.* **226,** 12,487–12,494.
21. Dharmavaram, R. M., Baldwin, C. T., Reginato, A. M., and Jimenez, S. A. (1993) Amplification of cDNAs for human cartilage-specific types II, IX and XI collagens from chondrocytes and Epstein-Barr Virus-transformed lymphocytes. *Matrix* **13,** 125–133.
22. Cooper, D., Berg, L., Kakkar, V., and Reiss, J. (1994) Ectopic (illegitimate) transcription—new possibilities for the analysis and diagnosis of human genetic disease. *Ann. Med.* **26,** 9–14.

13

Tissue-Specific KO of ECM Proteins

Emilio Hirsch, Mara Brancaccio, and Fiorella Altruda

1. Introduction

1.1. Gene Targeting

The analysis of phenotypes caused by null and mutant alleles is a very powerful means to understand gene function in vivo. Historically, this experimental approach has been widely and successfully used in invertebrate models. Now, thanks to the gene-targeting technology in ES cells, the genome of a mammalian organism such as the mouse can be artificially modified by precise alterations. The system exploits the ability of ES cells to be cultured and manipulated in vitro without losing their totipotency *(1,2)*. Mutations in specific genes can be achieved by in vitro selection of ES cell clones in which the locus of interest has been targeted by homologous recombination *(3,4)*. The peculiar property of being totipotent allows ES cells, once injected in the cavity of a blastocyst, to contribute to the formation of all cell types of a chimeric embryo. Whenever a chimeric mouse possesses ES-derived germ cells, the mutation can be propagated to its offspring. Heterozygous mice are then mated to generate the homozygous offspring needed for phenotypic analysis.

1.2. Tissue-Specific Gene Targeting

Recent advances in gene-targeting technology in ES cells now allow inducible mutations to be created in a tissue specific manner and at a precise developmental stage. Whereas the phenotype caused by germ-line mutations can be biased by epigenetic adaptations, induction of gene alteration in differentiated cells can result in clearer effects. Moreover, via this method, it is possible to study the consequences of ablating genes essential for cell survival *(5)*, identifying functions for distinct splice variants *(6)*, or tracking different gene functions at two or more developmental stages *(7)*.

From: *Methods in Molecular Biology, vol. 139: Extracellular Matrix Protocols*
Edited by: C. Streuli and M. Grant © Humana Press Inc., Totowa, NJ

The technique is based on the introduction of two or more short sequence tags called *loxP* sites at specific positions in the genome. The *P1* phage *Cre* recombinase can recognize two *loxP* sites oriented in the same direction, cut the intervening sequence, and rejoin the extremities ***(8)***. Inducible gene targeting can thus be achieved by mating a mouse in which important sites in the locus of interest have been flanked by *loxP* sites (a so-called floxed allele) with a transgenic mouse that expresses the *Cre* enzyme in a restricted pattern ***(9–11)***. Similarly, the floxed allele can be silenced by infection with a virus that transduces the *Cre* gene ***(12)***. In this way, a variable percentage of cells ranging from 10% to 100% ***(13)*** can be induced to undergo a controlled DNA rearrangement only when and where the recombinase is expressed.

1.3. Gene Targeting of ECM Proteins

Gene targeting has been widely used to study the function of ECM genes ***(14)*** and these experimentally induced mutations greatly extended the knowledge derived from the analysis of natural-occurring mutations ***(15)***. Interestingly, several of the knock-out mouse strains closely reproduce phenotypes of human hereditary disorders ***(14,16–18)***. In particular, mutations in ECM genes expressed in bone and cartilage provide genetic models that clearly link mutations to disease and greatly help to understand the molecular mechanisms underlying congenital connective tissue alterations ***(14)***.

Generation of mutant mouse strains has also resulted in the finding of new and unexpected functions of ECM genes. For example, the analysis of phenotypes of BM-40/SPARC and Thrombospondin 2-deficient mice drastically changed the view of their function ***(19,20)***. These two proteins turned out to be dispensable for morphogenesis but not for lens stability and lung homeostasis, respectively.

On the other hand, genetic ablation of ECM components can result in an unexpected lack of obvious phenotypes. Genes such as *tenascin-C*, although showing a complex and highly regulated expression pattern during development, are not essential for mouse development ***(21,22)***. These observations may suggest that some genes are of superfluous expression ***(23)***, but also that complex networks of compensation and functional redundancy are still to be discovered.

2. Materials

2.1. Isolation of the Gene

1. Cosmid library derived from 129/Sv mouse tissue.
2. Kanamycin or ampicillin (cosmid dependent).
3. LB: 10 g bactotryptone (Difco, Detroit, MI), 5 g bacto-yeast extract (Difco), 10 g NaCl. Fill to 1 L with deionized water, adjust pH to 7.0, and autoclave. LB can be

stored for a long time at room temperature (RT). Turbid, contaminated media must be discarded.

4. LB agar: Add 15 g agar (Sigma, St. Louis, MO) to LB, autoclave, and allow medium to cool to 50°C before adding antibiotics and pouring plates. Plates can be stored for 1 mo at 4°C. Dry the plates in 37°C incubator overnight before use.
5. Nylon membrane: for example, Hybond N+ (Amersham, Arlington Hts., IL).
6. Whatman 3MM papers.
7. Denaturation solution: 1.5 *M* NaCl, 0.5 *M* NaOH (can be stored at RT).
8. Neutralization solution: 1.5 *M* NaCl, 0.5 *M* Tris-HCl, pH 7.2 (can be stored at RT).
9. 1 *M* NaPi solution: 684 mL 1 *M* Na_2HPO_4, 316 mL 1 *M* NaH_2PO_4, pH 7.2 (can be stored at RT).
10. Church buffer: 500 mL 1 *M* NaPi, 330 mL 20% SDS, 1 mL 0.5 *M* EDTA, 10 µL sheared salmon sperm DNA, 10 g bovine serum albumin (BSA). Fill to 1 L with deionized water (can be stored at RT).
11. Random priming labeling kit for DNA. For example, Rediprime (Amersham).
12. ^{32}P-CTP (Amersham). The half-life of ^{32}P is approximately 14.3 d. Care should be taken when handling radioactive isotopes. Refer to local safety rules.
13. 20X SSC: 175.3 g NaCl, 88.2 g Na-citrate, adjust pH to 7.0, autoclave (can be stored at RT).
14. Wash solution 1: 2X SSC, 1% SDS (can be stored at RT).
15. Wash solution 2: 0.4X SSC, 1% SDS (can be stored at RT).

2.2. Generation of Constructs

2.2.1. Definition of the Probe

1. Restriction enzymes and buffers (store at –20C).
2. T7 sequencing kit (Pharmacia, Uppsala, Sweden).

2.2.2. Mutations Using the Selection Cassette and Cre/loxP System

1. Neomycin resistance cassettes: *pMC1 Neo* poly A vector (Stratagene, LaJolla, CA) containing a promoter, a neomycin resistance gene, and a polyadenylation signal; *pNeo* plasmid (Pharmacia) containing only the neomycin resistance gene; *pPNT* with a strong PGK promoter driving the neomycin resistance gene ***(24)***; *neo-tk* cassette flanked by *loxP* sites (Fässler, R., unpublished).
2. Restriction enzymes and buffers.
3. T4 DNA ligase and ligation buffer (Boehringer Mannheim, Mannheim, Germany).

2.3. Isolation of Feeder Cells

1. Mice expressing a neomycin (or hygromycin or puromycin) resistance gene.
2. C57BL6 female mice.
3. 70% Ethanol.
4. Sterile dissecting equipment.
5. 10X phosphate-buffered saline (10X PBS): 80.06 g NaCl, 2.01 g KCl, 14.42 g Na_2HPO_4, 2.04 g KH_2PO_4. Fill to 1 L with DDW and autoclave (store at RT).

6. Trypsin/EDTA solution: 10X stock (Gibco, Southersburg, MD, cat. no. 25090-028) (store at –20°C) diluted to 1X with 1X PBS (aliquot and store at –20°C, store aliquots in use at 4°C).
7. Feeder medium: DMEM high glucose with Glutamax-1 (Gibco, cat. no. 61965-026) supplemented with 10% fetal bovine serum (FBS) (Gibco, cat. no. 10270-106).
8. Freezing medium: 70% DMEM, 20% FBS, 10% DMSO.
9. ES medium: DMEM high glucose with Glutamax-1 (Gibco, cat. no. 61965-026) + Na-pyruvate (Gibco, cat. no. 11360-039) supplemented with 20% FBS (FBS needs to be tested for ES cell use, or can be bought from Gibco already tested, cat. no. 10270-106), 0.1 m*M* 2-mercaptoethanol, 5 mL 100X nonessential amino acids (NEA) (Gibco, cat. no. 11140-035), and 1000 U/mL leukemia inhibitory factor (LIF) (ESGRO from Gibco, cat. no. 13275-029).
10. Mycoplasma PCR Primer Set (Stratagene, cat. no. 302007).

2.3.2. Electroporation and Selection

1. Feeder medium (*see* **Subheading 2.3.**).
2. Feeder cells (*see* **Subheading 2.3.**).
3. Restriction enzymes and buffers.
4. Chloroform.
5. Phenol/chloroform: mix 1 vol of Phenol (equilibrated with 10 m*M* Tris-HCl, 1 m*M* EDTA, pH 8.0) with 1 vol of chloroform.
6. 3 *M* Na-acetate pH 5.2, adjust pH with glacial acetic acid (can be stored at RT).
7. 100% ethanol.
8. 70% ethanol.
9. Trypsin/EDTA (*see* **Subheading 2.3.**).
10. ES cells.
11. ES medium (*see* **Subheading 2.3.**).
12. 1X PBS (*see* **Subheading 2.3.**).
13. Burker's chamber.
14. Electroporation cuvets for 0.4 cm wide (Bio-Rad, Richmond, CA).
15. Gene Pulser™ (Bio-Rad).

2.3.3. Picking and Freezing of Resistant Clones

1. ES medium (*see* **Subheading 2.3.**).
2. G418 (Geneticin) (Gibco or Sigma).
3. Feeder cells (*see* **Subheading 2.3.**).
4. Feeder medium (*see* **Subheading 2.3.**).
5. 24-Well plates (Falcon, Los Angeles, CA).
6. 96-Well plates (Falcon).
7. Trypsin/EDTA (*see* **Subheading 2.3.**).
8. Stereomicroscope.
9. Cryovials (Sarstedt, Germany).
10. Cryovial rack (Sarstedt).
11. 1X PBS (*see* **Subheading 2.3.**).

12. Freezing medium (*see* **Subheading 2.3.**).
13. Dry ice.

2.3.4. Identification of Homologous Recombinants

1. Lysis buffer: 100 m*M* Tris-HCL, pH 8.5, 5 m*M* EDTA, 0.2% SDS, 200 m*M* NaCl, 100 μg/mL proteinase K (Sigma, USA, P0390). Keep proteinase K stock solution (10 mg/mL) at –20°C and always add freshly.
2. Isopropylic alcohol.
3. Restriction enzymes and buffers.
4. DNase-free bovine serum albumin (BSA) (New England Biolabs, Beverly, MA).
5. Agarose and ethidium bromide.
6. 10X TBE-buffer: 108 g Tris-base, 55 g boric acid, 9 mL 0.5 *M* EDTA pH 8.0, adjust to 1 L with deionized water (can be stored at RT).
7. Nylon membrane (*see* **Subheading 2.1.**).
8. Denaturation solution (*see* **Subheading 2.1.**).
9. SSC 20X (*see* **Subheading 2.1.**).
10. Hybridization plastic bags.
11. Church buffer (*see* **Subheading 2.1.**).
12. Random priming labeling kit for DNA *see* **Subheading 2.1.**).
13. ^{32}P-CTP (*see* **Subheading 2.1.**).
14. Wash solution 1 (*see* **Subheading 2.1.**).
15. Wash solution 2 (*see* **Subheading 2.1.**).
16. Autoradiography film.

2.3.5. Transient Cre Transfection

1. ES cells.
2. ES medium (*see* **Subheading 2.3.**).
3. 6-Wells microtiter dishes.
4. 9-cm tissue-culture dishes.
5. pIC-Cre plasmid *(5)*.
6. 3 *M* Na-acetate pH 5.2, adjust pH with glacial acetic acid (can be stored at RT).
7. 100% ethanol.
8. 70% ethanol.
9. Trypsin/EDTA solution (*see* **Subheading 2.3.**).
10. Gancyclovir (Syntex, cat. no. 00033-2903-48).

2.4. Generation of Mutant Mouse Lines

2.4.1. Generation of Vasectomized Males

1. Mice for vasectomy: 8- or more wk-old FVB males.
2. Avertin 100% stock: Dissolve 10 g 2,2,2-tribromoethyl alcohol (Fluka, Switzerland, cat. no. 90710) in 10 mL tert-amyl alcohol. For use, dilute the stock solution to 2.5% in PBS. Store both stock and use solutions at 4°C wrapped in aluminum foil to protect them from light.

3. 75% Ethanol.
4. Surgical equipment: fine dissection scissors; two pairs of watchmaker #5 forceps (sometimes manually sharpened); blunt, fine-curved forceps; serrefine clamp (1.5 in. or smaller); surgical silk or catgut suture with curved needle (e.g., size 10), 1-mL syringes with 26-gage hypodermic needle.

2.4.2. Preparation of Needles for Microinjection

1. Injection glass needles: with (Narishige #GD-1, Japan) and without (Narishige #G-1) internal filament.
2. Diamond glass cutter.
3. Needle puller (Narishige).
4. Microforge with 0.22-mm-thick platinum wire (Narishige, Japan).
5. Micropipet grinder (Narishige).
6. Teflon tube linked to a syringe.
7. 10% hydrofluoric acid (Sigma).
8. 100% ethanol.

2.4.3. Mouse Matings

9. C57B6 mice.
10. Vasectomized males.
11. CBA X C57B6 F1 females.

2.4.4. Isolation of Blastocysts

1. C57B6 females.
2. 70% ethanol.
3. Surgical equipment (*see* **Subheading 2.8.**).
4. Stereomicroscope.
5. Flush medium: High glucose DMEM, buffered with 20 m*M* HEPES pH 7.4.
6. 10-mL syringe.
7. 0.60 × 30 mm syringe needle.
7. Transfer pipet.
8. ES medium (*see* **Subheading 2.3.**).

2.4.5. Preparation of ES Cells for Microinjection

1. ES cells.
2. Feeder cells.
3. 6-cm tissue-culture dishes.
4. ES medium (*see* **Subheading 2.3.**).
5. 1X PBS (*see* **Subheading 2.3.**).
6. Trypsin/EDTA (*see* **Subheading 2.3.**).
7. 10-mL sterile tubes.

2.4.6. Microinjection of ES Cells

1. Microinjection setup: microscope with Hoffman or Nomarski optics (e.g., Olympus, Japan or equivalent). Left and right, water-driven micromanipulators

(Narishige, Japan). Two 10-mL syringes, each linked to a metal glass capillary holder (Narishige, Japan) via a silicon tube.
2. Injection chamber: lid of a 3-cm tissue-culture dish with a hole in the middle (about 1 cm in diameter).
3. Vaseline without any additives.
4. Siliconized coverslip: rinse the coverslips in chloroform 2% dimetildiclorosilane for 30 s and air-dry.
5. M2 medium: 94.66 m*M* NaCl, 4.78 m*M* KCl, 1.71 m*M* $CaCl_2$, 1.19 m*M* KH_2PO_4, 1.19 m*M* $MgSO_4$, 4.15 m*M* $NaHCO_3$, 20.85 m*M* HEPES, 23.28 m*M* sodium lactate, 0.33 m*M* sodium pyruvate, 5.56 m*M* glucose, BSA 4 g/L.
6. Petri dish (Falcon, Los Angeles, CA).
7. Ice.
8. Dimethyl polysiloxan (Sigma, cat. no. 9016-00-6).

2.4.7. Embryo Transfer

1. Microinjected blastocysts.
2. Pseudopregnant female mouse.
3. Avertin (*see* **Subheading 2.8.**).
4. Two stereomicroscopes.
5. Optic fibers illuminators.
6. Surgical equipment (*see* **Subheading 2.8.**).
7. Transfer glass pipet.

2.5. Mating of Chimeras

1. Adult chimeric males.
2. C57B6 females.
3. Adult 129 females.

2.6. Genotyping Offspring

2.6.1. Southern Blot Analysis of Tail DNA

1. 20–30-d-old mice.
2. Ear-clips (National Band and Tag Co.).
3. Rotary wheel.
4. Tail buffer-PK: 1 m*M* Tris-HCl, pH 7.5, 1 m*M* EDTA, 250 m*M* NaCl, 0.2% SDS and 0.1 mg/mL of freshly added Proteinase K (Sigma, cat. no. P0390).
5. Phenol/chloroform (*see* **Subheading 2.4.**).
6. Chloroform.
7. Isopropyl alcohol.
8. Sterile DDW.

2.6.2. DNA Preparation for PCR Analysis

1. Lysis buffer: PCR buffer X1 and 0.1 mg/mL proteinase K in DDW.
2. Thermomixer.

3. Sterile DDW.
4. *Taq* DNA polymerase (for example, Gibco, cat. no. 10342-012).
5. Agarose and ethidium bromide.

2.7. Generation and Analysis of Double KO ES Cells

1. ES cells.
2. ES medium (*see* **Subheading 2.3.**).
3. Feeder cells (*see* **Subheading 2.3.**).
4. 9-cm tissue-culture dishes (Falcon).
5. G418 (Geneticin).

2.7.1. Analysis of ES Contribution in Chimeric Mice

1. Surgical equipment (*see* **Subheading 2.8.**).
2. Extraction buffer: 50 m*M* Tris pH 8.0 0.1% Triton.
3. Disposable plastic pestles (Eppendorf, Germany).
4. Titan III cellulose acetate plate (HELENA Laboratories, Beaumont, TX, cat. no. 3033)
5. Super Z Applicator Kit (HELENA Laboratories, cat. no. 4093) containing Super Z applicator, Super Z aligning base, Super Z well plate.
6. Paper towels and Whatmann 3M filter paper.
7. Supre Heme buffer (HELENA Laboratories, cat. no. 5802).
8. Electrophoresis chamber.
9. Glass slides.
10. 1% agarose in 0.2 *M* Tris-HCL, pH 8.0.
11. Stock solutions for staining. Keep in 200-μL aliquots at indicated temperatures.
    - Magnesium acetate 0.25 *M*. Store at –20°C.
    - Fructose-6-Phosphate (Sigma, cat. no. F3627) 75 mg/mL. Store at –20°C.
    - Tetrazolium (MTT) (Sigma, cat. no. M2128) 10 mg/mL. Store at –20°C.
    - Phenazine methosulfate (PMS) (Sigma, cat. no. P9625) 1.8 mg/mL. Store at –20°C in the dark.
    - Nicotinamide adenine dinucleotide phosphate (NADP) (Sigma, cat. no. N0505). Store at –20°C.
    - Glucose-6-phosphate dehydrogenase (G6PDH) (Sigma, cat. no. G8878). Store at +4°C.
12. Stop solution: 5% acetic acid.

2.8. Analysis of Differentiation Abilities of Homozygous ES Cells in Embryo Bodies and Teratomas

2.8.1. Generation of ES Cells-Derived Embryoid Bodies

1. ES cells.
2. ES medium (*see* **Subheading 2.3.**).
3. Feeder cells (*see* **Subheading 2.3.**).
4. Feeder medium (*see* **Subheading 2.3.**).
5. 1X PBS (*see* **Subheading 2.3.**).

6. Trypsin (*see* **Subheading 2.3.**).
7. Burker's chamber.
8. 9-cm Petri dish.

2.8.2. Induction of ES Cells-Derived Teratomas

1. ES cells.
2. ES medium (*see* **Subheading 2.3.**).
3. Feeder cells (*see* **Subheading 2.3.**).
4. Feeder medium (*see* **Subheading 2.3.**).
5. 1X PBS (*see* **Subheading 2.3.**).
6. Trypsin (*see* **Subheading 2.3.**).
7. Sterile tubes (Falcon).
8. Burker's chamber.
9. 1-mL syringe.
10. Avertin (*see* **Subheading 2.8.**).

3. Methods

3.1. Isolation of the Gene

The first step in a gene knock-out experiment is to isolate a genomic clone spanning the locus of interest. To obtain high efficiency of homologous recombination, the gene must be isolated from a genomic library made with DNA from syngenic ES cells. Polymorphism present in DNA derived from other mouse strains has been shown to greatly reduce the targeting rate ***(25)***. Using R1 ES cell line ***(26)***, it is highly recommended that a cosmid library of 129sv mouse strain DNA is used.

1. Plate serial dilutions of an aliquot of the library. After overnight incubation at 37°C, count the colonies and calculate the concentration of cells per mL of the stock library.
2. Dilute the library stock to obtain about 2×10^5 cells in 1.5 mL of LB. Considering an average length of 45 Kb long inserts per cosmid, this number of cells should cover the whole mouse genome approximately 3 times.
3. Plate 100 µL of the dilution onto 15 LB-Amp Petri dishes. Grow cells overnight.
4. Cut round pieces of N+ nylon filters with a diameter shortly smaller than the Petri dish itself.
5. Overlay the colonies of one plate with a filter. Leave it in place for 1 min. In the meantime, to allow a subsequent correct alignment of filters to the Petri dish, make few punches with a syringe needle. Holes must not form a symmetric shape and should mark the bottom of the dish perpendicularly to the surface of the filter. Label marks with a black pen to ease their later detection.
6. Prepare three trays containing a large piece of Whatmann 3M paper. Soak the first with denaturing solution and the other two with neutralizing solution. Carefully peel off the nylon filter and overlay it upside down onto the first wet paper (avoid formation of air bubbles). Incubate for 5 min.

7. Transfer nylon filters to the paper soaked in neutralizing solution and incubate for 5 min. Repeat this step transferring filters to the second neutralizing solution wet 3M paper.
8. Repeat from **step 5** for each Petri dish.
9. Prehybridize all nylon filters in Church's buffer for at least 1 h at 65°C. Hybridize filters at the suitable temperature (*see* **Note 1**) overnight in fresh Church's buffer containing about 10^6 cpm/mL of radioactively labeled probe.
10. Wash with the required stringency (*see* **Note 1**) and expose filters. Mark the developed film with the position of the holes on the filters.
11. To isolate positive colonies, superimpose the marks on the film to the marks on the Petri dish. With a cut blue tip, excise a cylinder of agarose corresponding to the positive colony. Incubate the piece of agarose in 1 mL of LB-Amp to resuspend cells. Plate serial dilution of this aliquot to obtain single colonies. Repeat from **step 4** to **10** until positive clones can be isolated as single colonies.

3.2. Generation of Constructs

The next step, after isolation of a genomic clone is to generate, by general DNA cloning means *(27)*, a construct bearing the mutation that has to be introduced into ES cells DNA.

First, a detailed restriction map of the genomic clone must be obtained. To accomplish this task it is advisable to isolate subclones of the cosmid DNA and analyze their restriction patterns separately. Each subclone should be digested, following standard molecular biology protocols *(27)*, with as many as possible restriction enzymes. Contigs of subclones should be generated and DNA fragments that are supposed to contain interesting exons sequenced.

Once the structure of the isolated locus is determined, the clone must be ideally divided into two nonoverlapping regions. The first must contain the probe to isolate mutant clones by Southern blot and can be as short as few hundred bp (*see* **Note 2**). The second must comprise the sequences required to generate the construct. It has been shown that gene-targeting frequency strongly depends on the length of the ES DNA used *(28)*. Thus, a general strategy consists of finding a DNA fragment containing the region that has to be mutated (e.g., an exon or a promoter) flanked by at least 8 kb (*see* **Note 3**). Because ES cells are transfected with linear DNA, it is important to keep a single restriction site at the junction between either one of the arms and the vector.

3.2.1. Definition of the Probe

To initially establish a gene-targeting strategy, a probe that permits the identification of the mutants by Southern blot analysis must be isolated. Genomic DNA is interspersed by repetitive elements, which give high backgrounds in hybridization analysis. It is therefore essential to test several fragments of the cosmid clone to isolate a probe encoding a single genomic sequence.

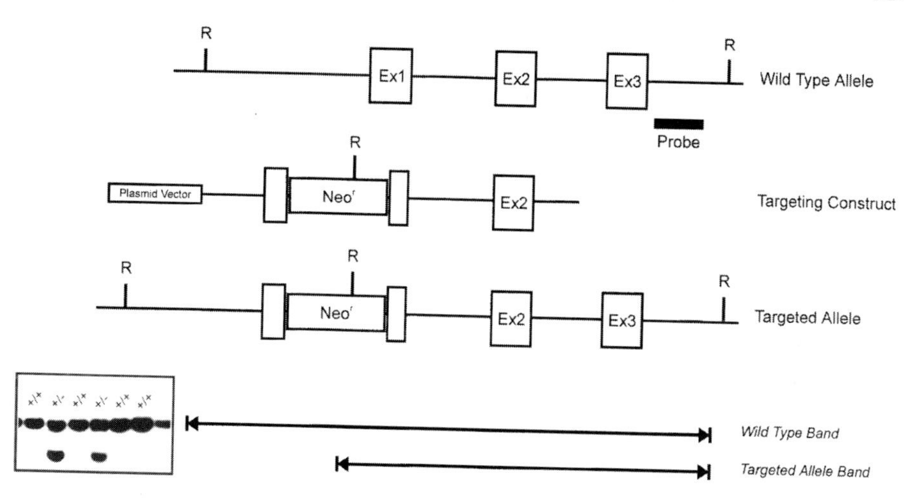

Fig. 1. Example of a gene-targeting strategy. The wild-type allele is substituted by a targeting construct that silences the gene. The null mutation is generated by insertion of the neomycin resistance gene into the sequences of *exon1*. R indicates the site recognized by a single restriction enzyme. Probe indicates the genomic fragment used for identification of homologous recombinants. As shown in the framed panel, in contrast to wild-type ES clones that show only the high molecular weight band (+/+), heterozygous cells show also the targeted allele-specific shorter band.

The probe must generate a diagnostic signal that distinguishes the targeted allele from the wild-type counterpart. This result can be obtained by finding a restriction enzyme that cuts outside the construct and generates a fragment that encompasses the probe. In the best situation this same enzyme cuts the targeted allele inside the resistance cassette and generates a fragment that is shorter than the wild-type (**Fig. 1**). Generation of shorter segments is recommended because the identification of a recombinant clone will not be confused by the presence of partially digested DNA.

If the search for a probe with the above features fails, it will be necessary to change the strategy outlined above and find alternatives such as extending the restriction map to new enzymes and considering other exons, deletions or resistance cassettes. Because cloning steps involved in construct preparation are often complex and time consuming, it is wise to start to build the construct only if a good probe has been found.

3.2. Mutations Using the Selection Cassette

In classical constructs, null alleles are made by inserting the neomycin resistance cassette in the middle of an exon. The choice of an exon near the 5' end of the gene enhances the chances of generating a nonfunctional product. Alternatively, the resistance cassette can be introduced in place of some genomic DNA. It has been reported that deletions of up to 10 kb are compatible with homologous recombination in ES cells *(29)*. The resistance cassette should be always inserted in an area that is not subjected by alternative splicing. It must be also kept in mind that the insertion of the cassette can cause aberrant splicing events that either generate out-of-frame mutations *(22)* or the skipping of the mutated exon. To avoid this latter case, a good strategy implies the insertion of the cassette in *exon 1*.

Several variations of the neomycin resistance cassette have been shown to function in targeting constructs. Consistent successful results have been obtained using the strongly active PGK (phsphoglyceratekinase) promoter driving the expression of the wild-type bacterial neomycin-phosphotransferase gene. Although weaker cassettes have been used, high-performance selectable genes are preferable: they allow the screening of more clones per single transfection experiment and are less susceptible to downregulation by the targeted locus.

3.2.3. *Mutations Using* loxP *Sites*

The use of the *Cre/loxP* system greatly enhances the ability to generate tissue specific knock-outs. This technique allows the creation of mutations that can be induced under controlled conditions *(9–11)*, the generation of homozygous mutant ES cells (*see* **Section 3.8.**), point mutations *(30)*, or large deletions *(31)*. These sophisticated gene-targeting experiments require the insertion of *loxP* sites in significant regions of the locus of interest. The *loxP* site is a 34-bp-long inverted repeat (5'-ATAACTTCGTATAGCATACATTATACGA AGTTAT-3') that is recognized by the *Cre* recombinase *(8)*. In the presence of two *loxP* sites oriented in the same direction, this enzyme deletes the genomic DNA between the inverted repeats reforming a single functional loxP site. If *loxP* sites are in opposite orientations, the *Cre* recombinase reverts the intervening sequence. Constructs for conditional knock-outs should usually contain at least two *loxP* sites that flank a DNA segment, which once deleted or flipped, leads to gene inactivation. For this purpose, *loxP* sites can be placed in noncoding regions so that they flank one or more exons (**Fig. 2**). In classic constructs three *loxP* sites are used: two are contained in a floxed selection cassette and a third is placed within one homology arm. This latter site must not be too far from the other two. It has been reported that integration by homologous recombination of this *loxP* site strongly decreases with the distance from the nonhomology region. The presence of floxed selection cassette allows to eliminate heterologous DNA from the targeted locus, leaving a presumptive functional allele. *Cre* mediated excision of the resistance cassette (*see* **Section 3.6.5.**) is a relatively efficient process; nonetheless, a *neo-tk* posi-

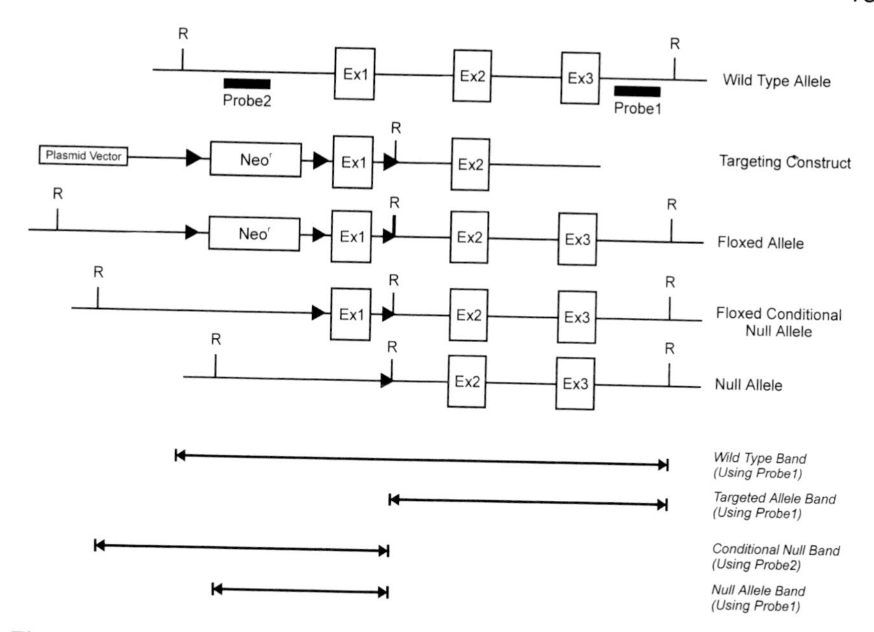

Fig. 2. Example of conditional gene targeting using the *Cre/loxP* technology. The wild-type allele is replaced by a targeting vector in which the resistance cassette and exon 1 are flanked by three *loxP* sites (filled triangles) showing the same orientation. R indicates a restriction enzyme site. Probe 1 serves to identify homologous recombinant clones after digestion with that enzyme. Transient expression of *Cre* allows generation of a *loxP* containing (active) and deleted (inactive) allele. Probe 2 identifies different Cre-mediated recombination events in R digested genomic DNA. Excision of the DNA segment comprised between the outer *loxP* sites leads to a null allele (type I deletion). Recombination involving *loxP* sites flanking the resistance cassette generate a conditional null allele (type II deletion). A third possible deletion involving *exon1* only is not shown. Mice homozygous for the floxed conditional null allele must be normal. Breeding with *Cre* expressing transgenic mice produces an offspring that should show the induction of null alleles.

tive/negative selection cassette can be used to ease the identification of deleted clones. Targeted clones are isolated in the presence of *G418*; successively, clones that lost the cassette are selected using gancyclovir. To identify the cells where *Cre* exerted its function, it is essential to find a restriction digest and a probe that can distinguish between the deleted and the original targeted alleles (**Fig. 2**).

3.3. Manipulation of ES Cells

ES cells to be used for knock-out experiments must be kept in an undifferentiated state. ES cells are small and round, with a large nucleus and few cytoplasm; they strongly adhere to each other and grow as aggregates. They are very sensitive to cell-culture conditions and, if not properly handled, tend to spontaneously differentiate in various cell types. Good ES cell colonies can be judged by microscopic analysis: they must be a multilayered aggregate that shows a clear cut, shiny boundary. On the contrary, differentiated ES cell colonies lose the glossy perimeter, tend to flatten, and/or to darken in the middle. Whenever such colonies are detected, it is advisable to discard the culture. The totipotency of the cells is correlated to the number of passages in culture and it is know to strongly decrease after 30 cycles of trypsinization/freezing. ES cell must be fed every day and, in certain critical concentrations, even twice a day. Colonies are, in fact, sensitive to density on the culture dish and as soon as they touch each other, cells start to differentiate. Thus, ES cell colonies must be split when they reach about 50% to 75% of confluency. It is important to passage colonies as a single cell suspension: if not properly dissociated they form large aggregates that are very prone to differentiation.

Several ES cell lines are currently available and most require a specific technique handling. Consistent results have been obtained using the R1 ES cell line ***(26)***. R1 ES cells grow on the top of a feeder layer of primary embryonic fibroblasts in a medium containing leukemia inhibitory factor (LIF). These cell-culture conditions assure very high levels of chimerism and an optimal rate of germ line transmission.

3.3.1. Isolation of Feeder Cells

Feeder cells provide a basal level of LIF production and a number of yet unidentified factors that sustain ES cells growth in the totipotent state. Feeder cells are derived from the carcass of a 14-d-old embryo. To allow selection of recombinant ES clones with antibiotics, embryos must derive from a transgenic mouse that expresses the proper resistance gene.

1. Mate a male homozygous for the resistance cassette transgene (*see* **Note 4**) with a C57BL6 female.
2. The following day, check for the presence of the vaginal plug, a whitish, solid sperm residue that indicates that the female mated during the night (*see* **Note 5**). Separate these females and keep them until needed.
3. After 14 d, sacrifice the pregnant females by cervical dislocation.
4. Thoroughly wet the animal in 70% ethanol and put it under a sterile hood above a paper towel.
5. Cut the skin with sterile scissors and carefully expose the abdomen. Grasp with fine forceps one end of the uterine horn and free with scissors the uterus together with embryos from mesometrium and cervix. Immediately transfer embryos (still inside the uterus) in a 10-cm bacterial culture Petri dish filled with sterile PBS.

6. With scissors separate, each implantation site from the other. With fine forceps, carefully free embryos from the uterine wall, yolk sac, and placenta. Repeatedly wash them in several Petri dishes filled with fresh sterile PBS, until bleeding stops.
7. With fine scissors, cut and discard the head. Open the abdomen with fine forceps and remove all internal organs.
8. Hold the carcass with forceps above a Falcon tube filled with trypsin (1 mL/embryo) and mince it with fine scissors to very small pieces that are let to fall into the trypsin solution.
9. Incubate the suspension at 37°C for 15 min. Break tissue pieces by pipeting up and down with a large gage pipet (e.g., 10-mL pipet). Incubate again at 37°C for 15 min. Thoroughly, dissociate cell clumps by pipeting up and down with a small gage pipet (e.g., a 2 mL or a Pasteur pipet).
10. Fill the falcon tube with feeder medium and let it stand for 5 min to let large aggregates sink. Transfer the supernatant to a fresh falcon tube and centrifuge it at 120*g* for 5 min. Discard the supernatant and dissolve the pellet with 2 mL/embryo of feeder medium. Seed 1 mL into a 15-cm cell-culture Petri dishes filled with 14 mL of feeder medium (2 Petri dishes/embryo).
11. Grow to confluency without changing medium in a 37°C, 5% CO_2 incubator. Wait for other 3 d. Wash plates with 15 mL PBS and incubate for 5 min inside the incubator with 2 mL of trypsin. Resuspend detached cells with 2 mL of feeder medium. Reseed the plate with 0.5 mL of cell suspension.
12. Pellet cells in a Falcon tube. Discard supernatant and irradiate the cell pellet with 6000 rad to inhibit cell division.
13. Resuspend the irradiated pellet with freezing medium (3 mL/Petri dish). Aliquot 1 mL of cell suspension/cryovial and freeze in a box kept in frozen carbon dioxide. Keep frozen stocks either at –80°C or in liquid nitrogen.
14. Add 14 mL of feeder medium to reseeded cells and repeat from **step 11** for a maximum of two times.
15. To check for sterility, thaw frozen aliquots at 37°C. Add cells to 5 mL of feeder medium in a 10-mL sterile tube. Centrifuge 5 min at 120*g*. Decant supernatant, resuspend pellet and seed it in a 10-cm cell-culture Petri dish. Culture cells for few days and then test supernatant for mycoplasma infection.

Frozen feeder fibroblasts can be thawed and plated in advance or just together with ES cells (in the presence of ES medium). To obtain a confluent layer of irradiated cells, thawing one frozen vial normally gives enough cells to cover the surface of a 10-cm diameter Petri dish. All other areas should be calculated using that rule of thumb.

3.3.2. Electroporation and Selection

Before transfection, ES cells must be expanded and kept growing minimizing passages. The construct is transfected into ES cells by electroporation. This method assures that in most cases only one copy of the exogenous DNA is inserted. In an average transfection experiment, 2×10^7 cells (the content of a full 10-cm Petri dish) are transfected with 30 μg of DNA.

1. Using feeder medium, plate irradiated fibroblasts (about 9 cryovials) onto nine 10-cm tissue-culture dishes. Change to ES medium when cells are adherent and spread.
2. Linearize 30 μg of the construct digesting with a single cutter enzyme (usually *Not*I) that cleaves at the boundary between one homology arm and the vector. Keep a final DNA concentration of at least 50 ng/μL. Incubate 1 h at the suitable temperature, using 30 or more enzyme units.
3. Extract the reaction mixture once with an equal volume of phenol/chloroform. Spin at maximum speed for 5 min. Collect the supernatant and extract it once with chloroform. Spin at maximum speed for 30 s. Save the supernatant.
4. Add to the supernatant 1/10 of the volume of 3 *M* Na acetate and 2 volumes of 100% ethanol. Mix thoroughly. A white DNA precipitate particle must appear. Using a yellow tip, transfer the precipitate in a sterile screw cap tube containing 1 mL of 70% ethanol/DDW. Keep in ice until needed.
5. Wash ES cells 2 times with PBS. Add 1 mL of warm trypsin. Incubate 5 min at 37°C. After adding 2 mL of ES Medium, dissociate colonies by pipeting up and down, avoiding the production of bubbles.
6. Dilute cells to 10 mL and count them using a Burker's chamber. Cells with large cytoplasm derive from the feeder layer and should thus be omitted from the count.
7. Spin 2×10^7 cells in a sterile 10-mL tube 5 min at 120*g*. Resuspend 5 mL of PBS. Repeat this step twice.
8. Spin again and resuspend in PBS to reach a final volume of 600 μL.
9. Spin the DNA precipitate for 30 s. Discard the supernatant under sterile conditions and let the pellet dry out for few min inside the hood. Resuspend DNA in 200 μL PBS.
10. Mix the cell suspension with DNA. Using a Pasteur pipet, transfer the mixture into a sterile cuvet for electroporation. Electroporate cells at 3 μF and 0.8 kV. The time constant should correspond to 0.1 ms. Electroporation leads to about 50% of cell death.
11. Quickly and carefully transfer electroporated cells into 8.5 mL of ES Medium. Mix and distribute 1 mL of transfected cells to each feeder plate.

3.3.3. Picking and Freezing of Resistant Clones

Twenty-four hours after electroporation, medium is changed to Selection Medium. In case a neomycin resistance cassette is used, cells can be selected by the addition of G418 antibiotic powder to ES Medium at a concentration of 400 μg/mL (*see* **Note 6**). In these conditions, resistant colonies appear within 5–6 d after transfection (*see* **Note 7**). In a typical experiment, enough recombinant clones can be detected in about 2–300 picked colonies. Not all colonies grow at the same rate, therefore picking and freezing steps can take 2–3 d each.

1. Thaw one cryovial of feeder cells at 37°C. Dilute cells in 10 mL of feeder medium. Spin 5 min at 120*g*. Decant supernatant and resuspend cells in 25 mL of feeder medium. Aliquot 1 mL of cell suspension into each well of a 24-well dish. Repeat this step for 10 or more (depending on the number of colonies that are to be picked) 24-well dishes. Let fibroblasts adhere and spread overnight.
2. Just before starting to pick the colonies, change medium of 24-well dishes to ES Medium.
3. With a 5-mL pipet, transfer 2 drops of trypsin into each well of a 96-well microtiter dish. Warm at 37°C.
4. Using a stereomicroscope under a laminar flow hood, gently scrape a colony with a P200 pipet equipped with a sterile yellow tip. Suck the cell aggregate in a maximum volume of 10 μL and transfer it in a 96-well filled with trypsin. Repeat this step for 12 colonies (*see* **Note 8**).
5. Incubate the 96-well dish at 37°C for 5 min. Open one 24-well dish with feeder and the 96-well dish under the hood.
6. With a fresh, sterile yellow tip collect about 100 μL of ES Medium from a feeder cells-containing well. Add it to a trypsin well. Dissociate the trypsinized colony by gentle pipeting. Carefully transfer the cell suspension to the same well from which ES Medium was taken. Make a few air bubbles to mark the well. Repeat this step for all trypsinized colonies (*see* **Note 8**).
7. Repeat **steps 4–6** until enough colonies have been picked.
8. The day after, change to fresh ES Medium all wells that received a colony.
9. Wait 3–4 d for ES cells to expand. As soon as the number and size of colonies are suitable for sibling, cells from each well can be frozen.
10. Mark cryovials with progressive numbers. Mark the same numbers on the bottom of each well to be passaged.
11. Set an empty cryobox into dry ice. Put labeled cryovials into the holding device.
12. Wash all marked wells of a 24-well dish with 1 mL PBS. Add 2 drops of trypsin into each well and incubate 5 min at 37°C.
13. Collect 900 μL of freezing medium with a P1000 equipped with a fresh, sterile blue tip. Thoroughly resuspend trypsinized cells of one well by gentle pipeting. Retrieve only about 600 mL of cell suspension and transfer it into the cryovial with the corresponding number. Place the cryovial into the box in dry ice. Repeat this step for all trypsinized wells.
14. Fill each well to maximum with feeder medium to dilute DMSO, which can eventually be toxic for cells.
15. Repeat steps 12–14 until all clones have been frozen.
16. The day after freezing, it is extremely important to change medium to feeder medium.

3.3.4. Identification of Homologous Recombinants

1. When medium inside a well turns to yellow, wait an additional day. Then discard the medium and add 500 μL of Lysis buffer. Keep at 37°C until all wells have been kept with Lysis buffer for at least one night.

2. Add 500 µL of isopropylic alcohol and let shake overnight at room temperature.
3. Label Eppendorf tubes correspondingly to lysed clones. Add 100 µL of sterile DDW to each tube.
4. Prepare a glass rod by flaming the tip of a Pasteur pipet. Collect with this instrument the white DNA precipitate that formed on the bottom of a 24-well dish well. Disperse the DNA in the water of the corresponding tube. Clean the glass rod in sterile DDW and dry it with a paper towel. Repeat this step for the precipitates of all different clones.
5. Let DNA dissolve overnight at 56°C. Store genomic DNA at 4°C.
6. Digest 15 µL of genomic DNA with the suitable restriction enzyme in 30 µL of a final reaction volume containing 0.3 µL BSA, 3 µL buffer, 30–40 U enzyme. Incubate overnight in an oven at 37°C.
7. Load digested DNA on a 0.8% agarose gel in 1X TBE, 1 µg/mL ethidium bromide. Separate for 6–8 h at 3 V/cm.
8. Photograph gel on an UV-table together with a ruler.
9. Blot the digested DNA onto N+ Nylon membrane in alkaline conditions.
10. Mark the position of the wells on the nylon membrane with a pencil. Neutralize the membrane twice in 2X SSC.
11. Place filters in a plastic bag containing the minimum amount of Church buffer to thoroughly wet them. Prehybridize for 1 h at 65°C. Radioactively label the probe by random priming following the instructions provided in the kit. Hybridize filters overnight at 65°C in Church buffer containing at least 1×10^6 cpm/mL of ^{32}P-labeled probe.
12. Wash filters at 65°C in 250–500 mL of hot Washing Solution 1 and 2 for 30 min each.
13. Expose with an autoradiography film until clear bands are visible.

3.3.5. *Transient* Cre *Transfection*

After homologous recombinant ES clones have been identified, the floxed selection cassette can be removed by transient *Cre* transfection. This procedure is particularly useful whenever inducible gene inactivation must be accomplished: removal of the selection cassette leaves only the short 34 bp *loxP* sequence in the locus. Modification of the genomic sequence is thus reduced to a minimum that leaves gene activity unaltered until a second *Cre* mediated recombination event takes place (*see* **Note 9**). Whenever a floxed neo-tk selection cassette is used, its deletion is essential because products of the tk gene can generate sterility in male chimeras ***(32)***.

1. Thaw ES cells in one well of a 6-well microtiter dish seeded with feeder cells. Expand the culture for few days to obtain one or more confluent 9-cm tissue-culture dishes.
2. Precipitate 20 µg of pIC-*Cre* plasmid with 1/10 the volume of 3 *M* Na acetate and 2 volumes of 100% ethanol. Leave the precipitate in 70% ethanol.

3. Electroporate 10^7 ES cells as described in **Subheading 3.3.2.** Plate about 5×10^6 ES cells/9-cm dish seeded with feeder cells.
4. The appearance of colonies of mixed genotype (floxed and deleted) is avoided by trypsinizing cell after the minimum amount of time required for *Cre* to exert its function: as soon as colonies appear (usually 2–3 d after transfection) trypsinize plates and reseed about 1/3 of the cells on the same dish. Add irradiated embryonic fibroblasts so that the feeder layer is always confluent. In case a *neo-tk* selection cassette is flanked by *loxP* sites, negative selection of deleted clones can be applied by adding 2 µ*M* Gancyclovir to the medium.
5. As soon as colonies appear again (3–4 d after **step 4**) proceed to pick and freeze as described in **Subheading 3.3.**
6. Identify the clones bearing the desired deletion by Southern blot as described in **Subheading 3.6.4.**

3.4. Generation of Mutant Mouse Lines

Chimeric mice are generated by microinjection of mutant ES cells into the cavity of a blastocyst *(33)*. Chimeras are then mated to generate an offspring that carries the ES cells genotype. To establish these techniques, an animal care facility must be available and treatment of mice must proceed in agreement to local laws regulating in vivo experimentation.

3.4.1. Generation of Vasectomized Males

1. Anesthetize a male mouse by intraperitoneal injection of 0.5 mL of diluted Avertin solution. Wash skin of lower abdomen (at the level of the top of the legs) with 75% ethanol and make a 1-cm cut with sharp scissors (1.5-cm large). Similarly cut the muscle of the body wall, avoiding the fat pad surrounding the genitals.
2. With bent blunt forceps, gently push the scrotum to move the right testicle into the abdominal cavity (until the white testicular fat pad appears at the edge of the incision). Expose the testicle by pulling the white fat pad. Note that around the testicle, the white coiled epididymis prolongs in a wider tube: the vas deferens.
3. Pierce with the tip of the forceps the thin membrane linking the vas deferens to the testicle and blood vessel. Apply two stitches around the freed tube, at a distance of about 5 mm from each other. With scissors, cut the vas deferens between the two stitches. By gently grasping the fat with forceps, reposition the testicle inside the abdomen. The same procedure is repeated for the left testicle.
4. Separately stitch 2 or 3 times muscle and skin, then put the mouse alone back into a fresh cage. Vasectomized mice can be used 15–30 d after surgery. Testing for residual breeding capacity is advisable.

3.4.2. Preparation of Needles for Microinjection

Two needles are needed: one is used to hold the blastocyst and the other to suck and inject cells (**Fig. 3**). Making such microinjection needles is a labori-

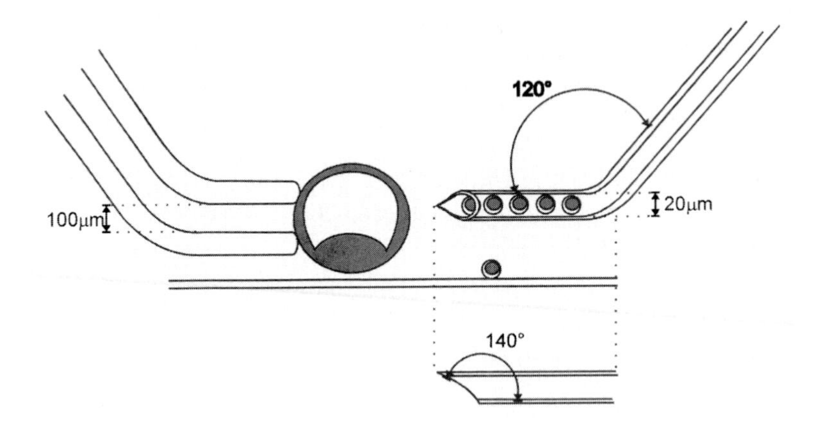

Fig. 3. Lateral view of the injection chamber. On the left: holding capillary used to block a blastocyst (dark grey). Edges of the glass walls have been rounded by flaming. The opening can vary from 50 to 100 μm. On the right: injection capillary holding some ES cells. The opening should be not smaller than the diameter of a cell (about 20 μm). A view of the tip from the above (as seen on the microscope) is shown below. Note the shape of the tip. Both holding and injecting capillaries should be bent (about 120 degrees) so that they remain parallel to the bottom of the injection chamber.

ous task. It is therefore advantageous to prepare sets of these tools in advance and store them until needed.

3.4.2.1. Preparation of Holding Capillaries

1. Heat the middle of a tube on a flame until the glass starts to melt. Quickly move out of the flame and pull the extremities of the tube by hand.
2. Select capillaries with a 2–3-cm long, 0.1-mm-wide tip.
3. With the help of a diamond tip, cut sharp and flush the end of the tip.
4. Adjusting the tip near the heating filament of the microforge, slowly melt the glass until the opening narrows to 1/4 of the original diameter.
5. Placing the capillary orthogonal to the heating filament bend the last 3–4 mm of the tip to form a 30° angle.

3.4.2.2. Preparation of Microinjection Capillaries

1. Using a puller, pull tubes so that they form a 1-cm long 1-μm wide tip.
2. Make a small ball of melted glass on the tip of the heating filament of the microforge.
3. Place the tip of the pulled capillary on the filament. Heat to a moderate temperature (when the ball gets a reddish glow) until the needle slightly fuses with the glass ball. Immediately switch the heating off. The capillary breaks leaving a sharp and flush opening (*see* **Note 10**).

4. Place capillaries with 20–30 µm-wide tips onto the grinding wheel at an angle of about 20–45 degrees. Leave until a beveled end is formed.
5. Connect the needle to a teflon tube linked to a syringe and wash it by sucking in and out at least 3 times with the following solution: 10% hydrofluoric acid, DDW 1, DDW 2, DDW 3, 100% ethanol.
6. Make a sharp fine tip at the beveled end. Heat the glass ball on the microforge filament to a moderate temperature. Gently push the needle against the ball and quickly withdraw it. The tip fuses with the glass ball and makes a sharp spike.
7. Place the capillary orthogonal to the heating filament and with the opening toward the operator. Bend the last 3–4 mm of the tip to form a 30 degree angle.

3.4.3. Mouse Matings

Blastocysts are obtained by natural mating of a relatively large number of C57B6 mice. The presence of the vaginal plug the morning following mating is considered 0.5 d postcoitum (dpc). Microinjected blastocysts are transferred to the uteri of a pseudopregnant female. To help embryos recover from the trauma of injection and to ensure a high rate of births, recipient females are one day delayed in respect to blastocysts.

1. Mate, at late afternoon, about 25 C576 males with 2 C57B6 females each (*see* **Note 11**).
2. The following morning check for plugs (*see* **Note 5**) and keep pregnant females in a separate cage.
3. At late afternoon of the same day, mate vasectomized males with 2 CBA X C57B6 F1 females each.
4. The following morning check for plugs (*see* **Note 5**) and keep pseudopregnant females in a separate cage.

3.4.4. Isolation of Blastocysts

1. To isolate blastocysts, sacrifice by cervical dislocation C57B6 females at 3.5 dpc.
2. Flush hair with 70% ethanol. Open the belly with scissors and expose the uterus below bowels.
3. Put the mouse under the stereomicroscope. Cut cervix (the single tube where uterine horns join) with sharp, fine, curved scissors. Grasp the cervix with forceps and gently lift the uterine horns. With fine curved scissors, cut mesometrium and vessels along the uterine horns, and also being sure to avoid hurting the muscular wall. Free the uterus by cutting at the utero–tubal junction. Place the uterus in a drop of flushing medium.
4. With forceps, hold a uterine horn near the utero–tubal junction. Using scissors with the other hand, open the tip by making a cut parallel to the horn. Repeat this procedure for the other horn. Place the uterus in a fresh drop of flushing medium.
5. Place the uterus in a dry, sterile watch glass under the stereomicroscope. With one hand, use forceps to hold the cervix. With the other hand, hold a 10-mL syringe filled with flushing medium. Insert the needle in the cervix towards one

horn. Flush with about 0.5-1 ml, tightening the muscular wall around the needle with forceps. If the uterine tube does not let liquid out, repeat the procedure in **step 4.**

6. Collect blastocysts under the stereomicroscope by sucking them into a transfer pipet. Transfer blastocysts in a watch glass containing ES Medium and keep them in the incubator until needed

3.4.5. Preparation of ES Cells

1. Plate ES cells and feeder on a 6-cm large tissue-culture dish 2–3 d before microinjection. Change ES Medium every day. ES cell colonies should reach an optimal density (75–90% confluency) without being passaged.
2. Wash cells 2 times with 5 mL PBS. Add 0.5 mL trypsin. Incubate 5 min at 37°C.
3. Add 3 mL ES Medium. Carefully dissociate colonies by pipeting up and down. Transfer the cell suspension to a 10-mL sterile tube. Spin 5 min at 1000 RPM.
4. Resuspend pellet in 5 mL ES Medium. Discard about 4 mL and spin the rest 5 min at 120*g*.
5. Discard the supernatant leaving just a drop medium. Resuspend cells by gently flickering the tube with fingers.

3.4.6. Microinjection of ES Cells

Blastocyst injection requires an inverted microscope equipped with phase contrast or Nomarski's optics. Two micromanipulators are set at the left and right sides of the microscope's stage. Each micromanipulator controls movements of a glass capillary. One needle (on the left) is used to hold the blastocyst and the other (on the right) to collect and inject cells. Each capillary is held through a hollow metal rod connected via a silicon tubing to an air filled syringe. Sucking and blowing of cells and embryos is controlled only by a gentle action on this device.

1. With vaseline grease, fix a siliconized coverslip to the injection chamber.
2. Put a large drop of M2 medium on the coverslip and, using the transfer pipet, add blastocysts at the upper left corner. In a similar way, add the suspension of ES cells in stripes along the drop.
3. Cool the stage of the microscope to about 10°C by placing on its top a glass Petri dish filled with ice. Put the injection chamber on the cooled stage.
4. Connect capillaries to their metal holder and syringe. Fill the pipets with M2 medium (*see* **Note 12**).
5. Adjust holding and injection pipets on micromanipulators. Put capillaries in focus. Pipets should be carefully turned until they show a straight horizontal orientation.
6. Overlay the M2 drop with dimethyl polysiloxan.
7. Collect about 150 healthy ES cells. Healthy ES cells are round and of medium size. They must have a smooth surface and a bright nucleus. Feeder cells can be easily avoided by their larger size and spiked shape (**Fig. 3**).

8. Using the holding pipet, gently suck a blastocyst. Moving the injection needle, turn the blastocyst until a thin depression between two cells is in focus. Hold the embryo in this position by delicate sucking (**Fig. 3**).
9. By a rapid horizontal movement, insert the injection needle into the blastocyst cavity. Inject about 15 cells. Gently withdraw the glass capillary and place the injected blastocyst on the lower right corner. After injection, blastocysts may collapse.
10. Repeat from **step 8** until cells are exhausted then repeat from **step 7** until all blastocysts were used.

3.4.7. Embryo Transfer

After microinjection, blastocysts should recover in the incubator for at least 1 h. Generally, embryos can be transferred as soon as they start to reexpand to form again their cavity. The number of transferred blastocysts for each pseudopregnant female can vary from 7 to 12.

1. Anesthetize a pseudopregnant female mouse by intraperitoneal injection of 0.5 mL of diluted Avertin solution. Put the mouse with the head facing 12 o'clock under the stereomicroscope equipped with optic fibers illuminators.
2. Disinfect the skin of the back and cut with scissors at about 1–2 cm above the hindleg. Wipe off hairs with a paper napkin. Detach the cut skin from the underlying muscle and displace it around to localize the underneath ovary (a reddish ball surrounded by a white fat pad). Expose the ovary by cutting the above body wall muscle, avoiding blood vessels (red) and nerves (white). Hold the fat with blunt-ended curved forceps and gently force the uterine horn out of the cavity.
3. On a second stereomicroscope, prepare the transfer pipet by filling it with a series of small air bubbles followed by ES Medium. Cautiously suck 3–6 embryos, minimizing the liquid in-between. Add a very small last air bubble at the tip. Store the transfer pipet undisturbed nearby the first stereomicroscope.
4. Using the fine forceps in one hand, hold the uterine horn near its apical end and gently stretch it outside the abdominal cavity. With the other hand, quickly grab the syringe needle and puncture the uterine wall avoiding the rupture of blood vessels. Insert the needle so that it reaches the lumen of the uterine horn without piercing again the musculature. Using this same hand, remove the needle and take the mouth pipet, possibly avoiding to leave the binoculars. Locate the hole again and insert the transfer pipet. Gently blow embryos until air bubbles are seen to move inside the uterus. Slowly withdraw the pipet (*see* **Note 13**). If liquid is expelled from the hole, quickly suck it into the pipet (*see* **Note 14**).
5. Put the ovary and uterus back inside the abdomen. Using surgical catgut thread, apply a few stitches to the muscle and to the skin.
6. Repeat the same procedure from **step 2** on the other uterine horn.
7. When finished, put the mouse back to the cage and keep it warm by covering it with straw.

3.5. Mating of Chimeras

To check for germ line transmission, adult chimeric males are mated to 2–3 adult C57B6 females. Transmission of the ES cell genotyope is evidenced by the birth of agouti pups. The agouti coat color, characteristic of the 129 mouse strain (from which ES cells derive), is dominant over the black coat color shown by C57B6 mice. The higher the chimerism, the better the chances for germ line transmission. If a chimera generates at least three litters of black pups, it is presumably not able to transmit the ES cell genotype. When chimeras show about 100% of ES cell contribution, it can happen to obtain agouti pups only. Transmission of the mutant allele should show a normal mendelian distribution among the agouti offspring.

Mice obtained by this mating strategy possess a mixed 129 and C57B6 genetic background. An easy way to obtain heterozygous animals of pure 129 inbred strain is to cross the chimera with an adult 129 female. In this case, the litter obtained shows only the agouti coat color. To identify 129 inbred animals, it is necessary to test the genotype for the presence of the mutant allele. Mice that bear the mutation virtually derive from the mating of two 129 inbred mice and being of pure 129 genetic background, they can be used to generate a 129 inbred mouse colony.

3.6. Genotyping of Offspring

Agouti pups must be genotyped to assess whether they are heterozygous for the mutation. For this purpose, genomic DNA must be extracted from tail biopsies and analyzed for the presence of the targeted allele. The high reliability of Southern blot techniques makes this it the system of choice for identification of heterozygotes in the very first litters. Faster but more prone to errors is PCR analysis, a procedure that is then suitable for genotyping of assessed mutant mouse strains.

3.6.1. Southern Blot Analysis of Tail DNA

1. Label 20–30-d-old mice by applying numbered ear clips. Mark number, sex, and mouse coat color in a notebook for further reference.
2. Cut the tail with sharp, strong scissors at about 1 cm from the tip and collect the tissue sample in a labeled sterile 1.5-mL tube. Dissolve tail tissue by incubation on a rotary wheel with 0.5 mL of tail buffer-PK overnight at 56°C.
3. Add 0.5 mL of phenol-chloroform. Incubate samples for 10 min on the rotary wheel at RT, then spin them at maximum speed on a bench centrifuge for 5 min.
4. Collect the DNA containing supernatant, avoiding precipitates at the interface, with a 1-mL disposable tip (cut to make the opening wider). Add the supernatant to a fresh tube containing 0.5 mL of chloroform. Incubate samples again on the rotary wheel for 10 min, then spin them for 5 min at maximum speed.

5. Collect the DNA containing supernatant as in **step 4** and add it to a fresh tube containing 0.5 mL of isopropyl alcohol. Invert the tube a few times and pellet the white DNA precipitate to the bottom by a short spin. Carefully eliminate the alcohol and air-dry the DNA for a couple of minutes.
6. Add to the DNA pellet 100 µL of sterile DDW and resuspend it overnight at 56°C. Store the DNA solution at 4°C.
7. Analyze the DNA samples as described in **Subheading 3.3.4., steps 6–13**. Using the probe defined for the screening of targeted ES clones, homozygous and heterozygous mice show one or two bands, respectively.

3.6.2. DNA Preparation for PCR Analysis

1. Design three oligonucleotides amplifying two short (<500 bp), but distinct diagnostic fragments each characteristic of a single allele. One oligonucleotide should be common and the other two should anneal to wild-type and mutant DNA sequences, respectively (**Fig. 4**). Best results are obtained with 20-mers having at least a 50% GC content.
2. Label mice as in **Subheading 3.6.1., step 1**. Collect no longer than 2–3 mm of the tip of the tail in a 1.5-mL tube containing 50 µL of lysis buffer.
3. Incubate samples at 60°C shaking in the Thermomixer for 2–3 h. Smash remnants of tail using a fresh yellow tip for each tube. Inactivate proteinase K by a 6 min incubation at 95°C. Spin at maximum speed for 5 min. Dilute 5 µL of the supernatant into 10 µL of sterile DDW.
4. Set up a PCR reaction following standard procedures *(27)* using 5 µL of the diluted sample in a final volume of 50 µL (*see* **Note 15**).
5. Amplify using the following cycle profile.
   - 10 cycles 95°C, 30 s
65°C–1°C each cycle, 30 s
72°C, 30 s
   - 25 cycles 94°C, 30 s
55°C, 30 s
72°C, 30 s
6. Check the reaction on 2% agarose gel. Amplification of one band with both combination of oligonucleotides indicates heterozygosity. Samples showing a band with either one of the two couples of oligos are to be considered wild-type or mutant homozygous depending on the length of the diagnostic fragment detected and on the presence of the wild-type or mutant allele specific oligonucleotide in the productive reaction.

3.7. Generation and Analysis of Double KO ES Cells

Knocking-out genes essential for embryonic development leads to lethality before birth and in many cases precludes the analysis of gene function in adult tissues. To overcome this limitation, it is necessary to generate *Cre/lox* conditional mutagenesis. A simpler alternative is to isolate ES cells homozygous for the mutation, which can be used to generate chimeras. The analysis of the

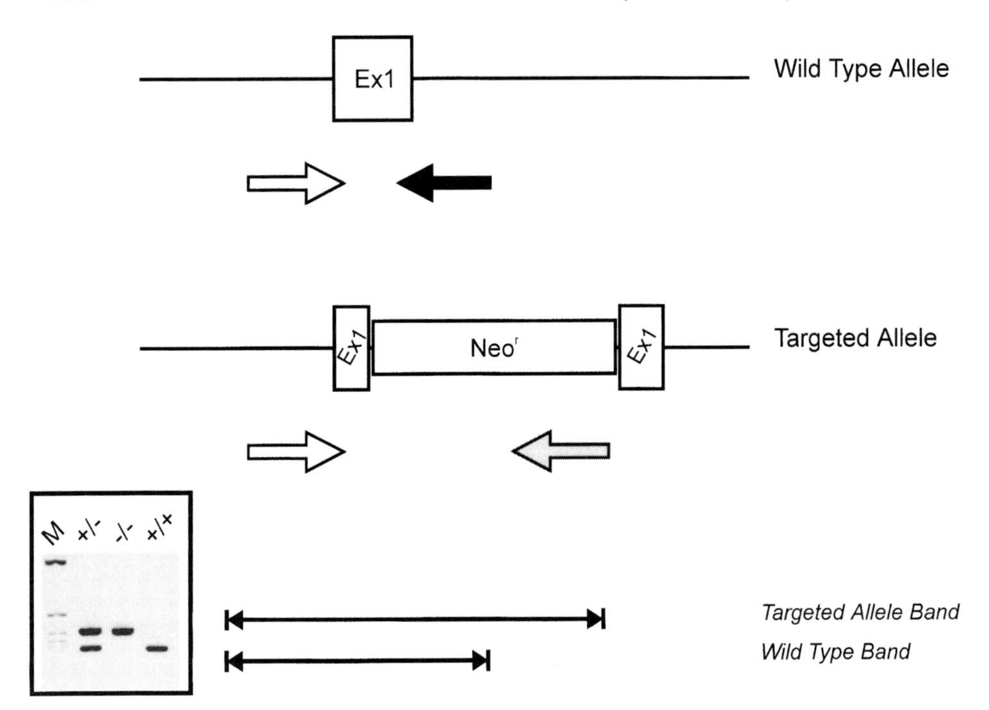

Fig. 4. Outline of the strategy to genotype litters by PCR. Arrows indicate the oligonucleotide from 5' to 3'. The white arrow indicates an oligonucleotide that can bind to both wild-type and mutant alleles. The black and hatched arrows correspond to the downstream oligonucleotides that anneal to the wild-type and targeted allele, respectively. Modulation of the distance from the common to the specific oligonucleotides allows to distinguish the two amplified fragments. As shown in the framed panel, amplification with the three oligonucleotides with homozygous and heterozygous DNA produces one or two distinct bands, respectively.

potential of these double mutant cells to contribute to the formation of different organs may give essential information on gene function in established tissues *(34)*.

Homozygous ES cells can be generated by electoporation of a second targeting construct that bears a different selection cassette (e.g., hygromycin in place of neomycin resistance gene). Alternatively, cells bearing a null allele generated by *Cre*-mediated excision of the selection marker gene (**Fig. 2**) can be transfected again with the same construct. In these two situations about 50% of the recombinant clones should show by Southern blot hybridization a homozygous-specific pattern of bands.

The simplest method, however, consists of growing heterozygous ES cells in a medium containing very high levels of selection drug *(35)* (*see* **Note 16**).

1. Thaw heterozygous ES cells carrying a neomycin resistance cassette on one allele. Grow and expand on feeder to obtain at least 105 cells.
2. Plate ES cells on 10 9-cm tissue-culture dishes with 10^4 cells each.
3. The next day, change medium with ES medium additioned with 4 mg/mL G418.
4. Change medium every day. Check for dying cells. If cells do not start to die in 4–5 d, it is likely that the technique will not succeed.
5. Wait for 6–7 d to start to see colonies. Treat resistant clones as described in **Subheading 3.6.3.**
6. Analyze the genotype as described in **Subheading 3.6.4.**

3.7.1. Analysis of ES Cell Contribution in Chimeric Mice

Genetic differences between ES and host embryo derived cells can be exploited to measure contribution of the injected cells in a chimeric tissue. Allelic variants of glucose phosphate isomerase (GPI) enzymes are often utilized for these measurements because they show distinct electrophoretic mobility on a cellulose acetate plate. The C57B6 derived GPI-B presents a higher electrophoretic mobility compared to the 129 derived GPI-A (*see* **Note 17**).

1. Dissect a small piece of tissue (10 mg or less) from the chimeric mouse.
2. Lyse the sample in 100–200 µL of extraction buffer using a small pestle. Lysates can be stored indefinitely at –20°C.
3. Mark the upper side of the plastic back of the plate. Soak the Titan III plate in Supre Heme buffer avoiding the formation of air bubbles inside the cellulose acetate.
4. Dilute the sample to the desired concentration and load 8 µL on the Super Z well plate.
5. Recover the Titan III plate and blot it dry between two paper towels. Fix the plate on the loading device with the cellulose on the top and so that samples will be loaded near the marked side.
6. Collect some sample by pressing the applicator inside the wells. Blot the applicator on tissue paper. Reload in the same way the applicator and finally press it against the cellulose acetate plate.
7. Fill buffer tanks of an electrophoresis device with Supre Heme buffer. Place a piece of filter paper in both buffer tanks so that they do not touch each other.
8. Overlay the plate onto the two pieces of filter paper so that the cellulose side faces the bottom of the chamber. In this way, the plate is electrically connected to the buffer tanks. Take care to place the marked side near the anode. Stabilize the extremities of the gel with 5–8 glass slides. Run (from anode to cathode) at 200 V, 4°C for 3 h.
9. Place the Titan III slab onto a glass plate. Melt the agarose and cool 10 mL at 55°C. Add 200 µL of each developing reaction component, mix well and pour it over the cellulose acetate plate. Leave the reaction in the dark for 2 to 15 min, depending on the concentration of samples.
10. Place the plate in stop solution as soon as the bands reach the desired intensity. Photograph immediately.

3.8. Analysis of Differentiation Abilities of Homozygous ES Cells in Embryo Bodies and Teratomas

In addition to the study of chimeric tissue formation, the differentiation abilities of ES cells can also be tested by in vitro differentiation assays and by the induction of ES-derived teratomas. The extraordinary potential of ES cells to differentiate in vitro can be utilized to study particular aspects of mutant phenotypes that cannot be easily approached in in vivo models. The case of embryonic lethal phenotypes provide the typical situation in which the study of the developmental abnormalities can be greatly extended with biochemical and cellular studies on differentiated homozygous cells. Analysis of the differentiation state at different time-points and comparison of mutant and wild-type cultures may give essential clues on the effects of the mutation ***(36)***.

Several protocols have been developed to differentiate cultured ES cells and most of them are specialized to generate a particular cell type. Nonetheless, ES cells can be aggregated in vitro to form so-called embryo bodies in which various tissues with distinct embryonic origin start to form. These cell aggregates can then be grown for several days and the formed tissues may be analyzed with the classical tools of biochemistry and cell biology.

An alternative method to generate ES cells-derived differentiated tissues, consists of injecting ES cells ectopically in syngeneic male mice (for example, under the skin). In these conditions, ES cells can form a teratoma, a noninfiltrating, benign tumors containing large numbers of highly differentiated cells often organized in epithelia, glands, vessels, and even nerves.

3.8.1. Generation of ES Cells-Derived Embryoid Bodies

1. Grow ES cells with standard methods to obtain at least 10^6 cells.
2. Wash cells two times with PBS and incubate with trypsin for 5 min at 37°C. Add ES medium and disperse aggregates by gently pipeting cells up and down. Add ES medium if needed and let feeder cells attach to the culture dish for 30 min in the incubator.
3. Gently wash the plate to collect nonadherent ES cells and save the supernatant in a sterile tube. Count the cells with a Burcker's chamber. Dilute cells so that 500–1000 are contained in 20 µL.
4. Pipet on the inside the lid of a sterile 9-cm Petri dish several 20-µL drops of cells. Fill the Petri dish with 5–10 mL of PBS. Gently turn the lid upside-down and close the Petri dish so that cells are held in suspension in the hanging drops.
5. Wait for 2 d until aggregates of ES form. Collect aggregates (embryoid bodies) by washing the lids with EF medium. Plate aggregates on bacterial culture Petri dishes in EF medium (supplement of serum to 20% may be required).
6. Embryoid bodies can be grown in suspension up to 30 d. Alternatively, they can be trypsinized and cells can be plated in tissue-culture dishes.

3.8.2. Induction of ES Cells-Derived Teratomas

1. Grow ES cells with standard methods to obtain at least 10^8 cells.
2. Wash cells two times with PBS and incubate with trypsin for 5 min at 37°C. Add ES medium and disperse aggregates by gently pipeting cells up and down. Add ES medium if needed and let feeder cells attach to the culture dish for 30 min in the incubator.
3. Gently wash the plate to collect nonadherent ES cells and save the supernatant in a sterile tube. Count the cells with a Burcker's chamber.
4. Centrifuge 10^7 cells 5 min at 120*g*. Resuspend the pellet in PBS.
5. Spin cells down and resuspend the pellet in 300 μL. Load the cell suspension in a 1-mL syringe.
6. Anesthetize a 129 male mouse and inject cells subcutaneously.
7. The tumor should be clearly visible in 15–20 d. When it has reached the desired dimension, excise it and treat it for histological analysis. As with embryoid bodies, teratomas can be collected in sterile conditions and trypsinized to culture differentiated cells.

4. Notes

1. Stringency varies depending on the homology of the probe with mouse DNA. In case the probe shows a homology from 80% to 100% hybridization can be carried out at 65°C and followed by two washes of 30 min at 65°C with 2X SSC, 1%SDS, and 0.4X SSC, 1%SDS.
2. Southern blot analysis of mutant ES cell clones is preferred over PCR analysis. Beside the fact that homologous recombination should always be confirmed by Southern blot, this method is more reliable than PCR because it essentially gives no false positive/negative signals.
3. Flanking arms can be asymmetrical, but better results have been obtained when the shorter arm is not smaller than 2 kb.
4. A viable homozygous knock-out mouse is often suitable for such purpose. The expression of a resistance gene (usually the *neo*^{r} gene) is guaranteed by the fact that the mutant ES cells from which the mouse line has been derived, had been selected in a similar way.
5. Plugs are unstable and easily lost during the day that follows mating. It is therefore important to check for plugs early morning and not later than 11 AM. To ease inspection, it is useful to use a blunt, sterile probe such as a flame-sealed tip of a Pasteur pipet.
6. In commercial preparations of G418, only a fraction of the total weight is the real active compound. The routine use of high concentration (400 μg/mL) of crude powder has proven to be adequate to avoid testing of different antibiotic batches.
7. Colonies should be picked as soon they start to be detectable by eye inspection. Choosing the picking time is a critical step: whereas waiting too long can result in differentiation of colonies, retrieval of small aggregates supplies too few cells for the subsequent expansion.

8. It is extremely important to put each colony in a separated well. Keeping track of the number of used yellow tips can help to avoid mistakes.
9. Obviously, it is essential to test that the floxed allele is fully functional before starting the conditional deletion experiments.
10. Keep the glass ball clean. Wipe debris away with a piece of cloth. By accumulating glass debris, the ball eventually changes its size and subsequently the temperature at which it can fuse with the needle.
11. Keep females well distributed in more than one cage. In this way, females will not show a synchronized oestrus, thus enhancing the chance of finding individuals able to mate.
12. Making a few small air bubbles in the thinner part of the injection needle can strongly increase the control over cell sucking and blowing.
13. Do not generate high pressure in the pipet. In this way transfer cannot be controlled and often results in the loss of the embryos. If the pipet becomes clogged, remove from the uterus and gently wash it in ES Medium, paying attention not to lose the blastocysts.
14. Immediately after embryos are blown into the uterine cavity, the muscular wall sometimes contracts and expels the transferred liquid together with injected blastocysts. It is, therefore, useful to suck into the transfer pipet the liquid that tends to overflow from the hole. In case embryos are expelled, they can be recovered in the pipet.
15. Initially test DNA of a proven heterozygous animal in two different reaction tubes each containing a separate combination of the two couples of oligonucleotides (common, wild-type, or common, mutant allele). To simplify the procedure, test whether all three oligonucleotides can work in one same reaction.
16. The success rate of this technique can vary depending on the nature of the targeted locus and must therefore be empirically tested.
17. The GPI enzymes are homodimers and because dimerization occurs inside the cell chimeric tissues show only two distinct bands. Three bands can be seen in particular situations such as in skeletal muscle extracts. Myofibers containing nuclei of distinct genotype can generate all three combination of subunits.

References

1. Evans, M. J. and Kaufman, M. H. (1981) Establishment in culture of pluripotent cells from mouse embryos. *Nature* **292,** 154–156.
2. Martin, G. (1981) Isolation of a pluripotent cell line from early mouse embryos cultured in medium conditioned by teratocarcinoma cells. *Proc. Natl. Acad. Sci. USA* **78,** 7634–7638.
3. Smithies, O., Gergg, R. G., Boggs, S. S., Koralewski, M. A., and Kuckerlapati, M. S. (1985) Insertion of DNA sequences into the human chromosomal β-globin locus by homologous recombination. *Nature* **317,** 230–234.
4. Thomas, K. R. and Capecchi, M. R. (1987) Site-directed mutagenesis by gene targeting in mouse embryo-derived stem cells. *Cell* **51,** 503–512.

5. Gu, H., Marth, J. D., Orban, P. C., Mossmann, H., and Rajewsky, K. (1994). Deletion of a DNA polymerase beta gene segment in T cells using cell type-specific gene targeting. *Science* **265,** 103–106.
6. Baudoin, C., Goumans, M. J., Mummery, C., and Sonnenberg, A. (1998) Knockout and knockin of the beta1 exon D define distinct roles for integrin splice variants in heart function and embryonic development. *Genes Dev.* **12,** 1202–1216.
7. Nagy, A., Moens, C., Ivanyi, E., Pawling, J., Gertsenstein, M., Hadjantonakis, A. K., et al. (1998) Dissecting the role of N-myc in development using a single targeting vector to generate a series of alleles. *Curr. Biol.* **8,** 661–664.
8. Sternberg, N. and Hoess, R. (1983). The molecular genetics of bacteriophage P1. *Annu. Rev. Genet.* **17,** 123–154.
9. Kilby, N. J., Snaith, M. R., and Murray, J. A. (1993) Site-specific recombinases: tools for genome engineering. *Trends Genet.* **9,** 413–421.
10. Kuhn, R., Schwenk, F., Aguet, M., and Rajewsky, K. (1995). Inducible gene targeting in mice. Science **269,** 1427–1429.
11. Rajewsky, K., Gu, H., Kuhn, R., Betz, U. A., Muller, W., Roes, J., and Schwenk, F. (1996) Conditional gene targeting. *J. Clin. Invest.* **98,** 600–603.
12. Akagi, K. Sandig, V., Vooijs, M., Van der Valk, M., Giovannini, M., Strauss, M., and Berns A. (1997) Cre-mediated somatic site-specific recombination in mice. *Nucleic Acids Res.* **25,** 1766–1773.
13. Betz, U. A., Vosshenrich, C. A., Rajewsky, K., and Muller, W. (1996) Bypass of lethality with mosaic mice generated by *Cre-lox*P-mediated recombination. *Curr. Biol.* **6,** 1307–1316.
14. Aszodi, A., Pfeifer, A., Wendel, M., Hiripi, L., and Fässler, R. (1998) Mouse models for extracellular matrix diseases. *J. Mol. Med.* **76**, 238–252.
15. Morrison-Graham, K., and Weston, J. A. (1989) Mouse mutants provide new insights into the role of extracellular matrix in cell migration and differentiation. *Trends Genet.* **5,** 116–121.
16. Fässler, R., Schnegelsberg, P. N., Dausman, J., Shinya, T., Muragaki, Y., McCarthy, M. T., et al. (1994) Mice lacking alpha 1 (IX) collagen develop noninflammatory degenerative joint disease. *Proc. Natl. Acad. Sci. USA* **91,** 5070–5074.
17. Mundlos, S. and Olsen, B. R. (1997) Heritable diseases of the skeleton. Part II: molecular insights into skeletal development-matrix components and their homeostasis. *FASEB J.* **11**, 227–233.
18. Bruckner-Tuderman, L. and Bruckner, P. (1998) Genetic diseases of the extracellular matrix: more than just connective tissue disorders. *J. Mol. Med.* **76**, 226–237.
19. Gilmour, D. T., Lyon, G. J., Carlton, M. B., Sanes, J. R., Cunningham, J. M., Anderson, J. R., et al. (1998) Mice deficient for the secreted glycoprotein SPARC/osteonectin/BM40 develop normally but show severe age-onset cataract formation and disruption of the lens. *EMBO J.* **17**, 1860–1870.
20. Lawler, J., Sunday, M., Thibert, V., Duquette, M., George, E. L., Rayburn, H., and Hynes R. O. (1998) Thrombospondin-1 is required for normal murine pulmonary homeostasis and its absence causes pneumonia. *J. Clin. Invest.* **101**, 982–992.

21. Saga, Y., Yagi, T., Ikawa, Y., Sakakura, T., and Aizawa, S. Mice develop normally without tenascin. *Genes Dev.* **6,** 1821–1831.
22. Forsberg, E., Hirsch, E., Frohlich, L., Meyer, M., Ekblom, P., Aszodi, A., et al. (1996) Skin wounds and severed nerves heal normally in mice lacking tenascin-C. *Proc. Natl. Acad. Sci. USA* **93**, 6594–6599.
23. Erickson, H. P. (1993) Gene knockouts of c-src, transforming growth factor beta 1, and tenascin suggest superfluous, nonfunctional expression of proteins. *J. Cell Biol.* **120**, 1079–1081.
24. Tybulewicz, V. L., Crawford, C. E., Jackson, P. K., Bronson, R. T., and Mulligan, R. C. (1991). Neonatal lethality and lymphopenia in mice with a homozygous disruption of the c-abl proto-oncogene. *Cell* **65,** 1153–1163.
25. te Riele, H., Maandag, E. R., and Berns, A. (1992) Highly efficient gene targeting in embryonic stem cells through homologous recombination with isogenic DNA constructs. *Proc. Natl. Acad. Sci. USA* **89,** 5128–5132.
26. Nagy, A., Rossant, J., Nagy, R., Abramow-Newerly, W., and Roder, J. C. (1993) Derivation of completely cell culture-derived mice from early-passage embryonic stem cells. *Proc. Natl. Acad. Sci. USA* **90,** 8424–8428.
27. Sambrook, J., Fritsch, E. F., and Maniatis, T. (1989) *Molecular Cloning. A Laboratory Manual.* Second Edition. Cold Spring Harbor Laboratory, Cold Spring Harbor, NY
28. Hasty, P., Rivera-Perez, J., and Bradley, A. (1991) The length of homology required for gene targeting in embryonic stem cells. *Mol. Cell Biol.* **11,** 5586–5591.
29. Zhang, H., Hasty, P., and Bradley, A. (1994). Targeting frequency for deletion vectors in embryonic stem cells. *Mol. Cell. Biol.* **14,** 2404–2410.
30. Reichardt, H. M., Kaestner, K. H., Tuckermann, J., Kretz, O., Wessely, O., Bock, R., et al. (1998). DNA binding of the glucocorticoid receptor is not essential for survival. *Cell* **93,** 531–541.
31. Li, Z. W., Stark, G., Gotz, J., Rulicke, T., Gschwind, M., Huber, G., et al. (1996) Generation of mice with a 200-kb amyloid precursor protein gene deletion by Cre recombinase-mediated site-specific recombination in embryonic stem cells. *Proc. Natl. Acad. Sci. USA* **93,** 6158–6162.
32. Braun, R. E., Lo, D., Pinkert, C. A., Widera, G., Flavell, R. A., Palmiter, R. D., and Brinster, R. L. (1990) Infertility in male transgenic mice: disruption of sperm development by HSV-tk expression in postmeiotic germ cells. *Biol. Reprod.* **43,** 684–963.
33. Hogan, B., Beddington, R., Costantini, F., and Lacy, E. (1994) *Manipulating the Mouse Embryo. A Laboratory Manual.* Second Edition. Cold Spring Harbor Laboratory, Cold Spring Harbor, NY.
34. Hirsch, E., Iglesias, A., Potocnik, A. J., Hartmann, U., and Fassler, R. (1996) Impaired migration but not differentiation of haematopoietic stem cells in the absence of beta1 integrins. *Nature* **380,** 171–175.
35. Mortenssen, R. M., Zubiaur, M., Neer, E. J., and Seidman, J. G. (1991) Embryonic stem cells lacking a functional inhibitory G-protein subunit (ai2) produced by gene targeting of both alleles. *Proc. Natl. Acad. Sci. USA* **88**, 7036–7040.
36. Sasaki, T., Forsberg, E., Bloch, W., Addicks, K., Fässler, R., and Timpl, R. (1998) Deficiency of beta 1 integrins in teratoma interferes with basement membrane assembly and laminin-1 expression. *Exp. Cell Res.* **238,** 70–81.

14

Homologous Gene Targeting to Study ECM Assembly

Francesco Ramirez, Friedrich Laub, and Hideaki Sumiyoshi

1. Introduction

Genetically engineered mutant strains of mice have become invaluable tools to decipher complex biological processes, such as ECM assembly, and to dissect phenotypic overlaps, like the genodermatoses. Homologous gene targeting in embryonic stem (ES) cells is increasingly favored over the transgenic approach because it allows the study of mutations in homozygosity and without altering the remainder of the genome *(1)*. Homologous gene targeting has been successfully employed to abrogate the expression or alter the structure of ECM-coding genes *(2,3)*. In both cases, the experimental protocol consists of the following three steps: a. construction of the targeting vector; b. homologous recombination in ES cells; and c. generation of mutant mouse strains.

The design of the targeting vector requires extensive knowledge of the distribution of restriction sites, the organization of the exons and introns, and the location of repetitive sequences within the genomic region in question. The ideal targeting vector consists of ≥5 kb of genomic sequences (homology arms) flanking the area to be targeted and where the promoter/gene-cassette for a positive selectable marker (neomycin/G418) has been placed. The neo-cassette contains diagnostic restriction endonuclease sites to verify the recombination by Southern blot analysis of ES-cell DNA. Traditionally, a promoter/gene-cassette for a negative selectable marker (HSV thymidine kinase/Gancyclovir) is placed immediately outside of the homology arm(s) to select against random insertion of the targeting construct into the genome *(1)*. Depending on the design, the neo-cassette can put the endogenous gene out-of-frame (insertion targeting; null mutation) or substitute one or more exons without altering its open-reading-frame (deletion targeting; structural mutation) *(1–3)*.

From: *Methods in Molecular Biology, vol. 139: Extracellular Matrix Protocols*
Edited by: C. Streuli and M. Grant © Humana Press Inc., Totowa, NJ

The following two requirements should be met before undertaking the construction of the vector. First, diagnostic restriction sites should be identified that distinguish the targeted from the normal allele and that digest DNA selected from ES cells fully and reproducibly. Second, hybridization probes should be selected that yield little or no background with the same DNA samples and that can "see" the recombination from the inside and the outside, and from upstream and downstream of the region covered by the homology arms of the vector.

Because the neo-cassette may occasionally affect proper expression of the targeted gene ***(4)***, one may wish to delete it after the ES clone selection. To this end, short recognition sequences (*loxP*) for enzyme-mediated recombination (*Cre*) can be placed 5' and 3' of the neo-cassette ***(5)***. Expression of *Cre* results in deletion of the DNA lying within the *loxP* sites arranged in a head-to-tail orientation. The *Cre* enzyme can be expressed transiently in expanded cultures of individual ES clones. It can also be expressed constitutively in siblings of crosses between targeted ES-derived mice and *Cre*-recombinase-producing transgenics. The less than optimal efficiency of *Cre*-mediated recombination may yield a heterogenous population of ES cells in tissue culture, and germ-line mosaicism in animals; both events can be readily monitored by careful Southern blot analyses.

ES cells are derived from a mouse strain whose coat color (agouti; 129/sv) is dominant over the one whose blastocysts they will eventually colonize (black; C57BL/6J). ES cells are maintained and grown on a feeder layer of mitotically inactivated and G418^r embryonic fibroblasts (EF) prepared from embryos homozygous for the neo-cassette ***(6)***. The double selection scheme (G418/ Gancyclovir) eliminates most, but not all, of the nonhomologous recombinants. Final validification rests on Southern analyses with probes 5' and 3' of, and within and outside the chromosomal region being targeted. The efficiency of homologous recombination varies greatly amongst different genes and within different regions of the same gene.

2. Materials

2.1. ES Cell Culture and Storage

1. Growth medium for ES and EF cells: Hepes-buffered 20 m*M*, pH 7.3 (Sigma, St. Louis, MO; #H-0887) Dulbecco's modified Eagles medium (DMEM, high glucose; Gibco, Gaithersburg, MD; #11995-040) supplemented with 15% heat-inactivated Fetal Calf Serum (FCS; HyClone, Logan, UT; # SH30070.03), 0.1 m*M* nonessential amino acids (Gibco; #11140-050), 0.1 *M* β-mercaptoethanol (Sigma; #M7522), 1X antibiotics (penicillin/streptomycin; Gibco; #15140-122) and 1000 U/mL of Leukemia Inhibitory Factor (LIF; Gibco; #13275-029). Medium is sterilized by filtering through a 0.22-μm pore size filter system (Costar; #430769) and stored at 4°C.

2. Freezing solution for ES- and EF-cell storage: 10% DMSO (Sigma; #D2650) in standard ES-cell medium; the freezing solution should be made fresh each time is used.
3. Trypsin: 0.25% trypsin/1 m*M* EDTA in Hepes-buffered saline (Gibco; #25200-056).
4. Gelatin (Fisher; #G8-500): 0.2% in ultrapure water, autoclaved and stored at room temperature.

2.2. ES-Cell Electroporation

1. Electroporation buffer: 20 m*M* Hepes pH 7.0, 137 m*M* NaCl, 5 m*M* KCl, 0.7 m*M* Na_2HPO_4, 6 m*M* glucose, 0.1 m*M* β-mercaptoethanol. Each stock solution should be autoclaved with the exception of glucose which should be sterilized through a 0.22-µm pore size filter. Once prepared, the electroporation buffer is again sterilized through a 0.22-µm pore size filter before being stored at 4°C.
2. Selection buffer: Same as the ES medium, except that G418 (Gibco; #11811-031) is added at a final concentration of 350 µg/mL. Make a 500X stock solution in H_2O and store it at –20°C. Because G418 activity can vary from batch to batch, it is best to determine the minimal concentration required to kill 100% of ES cells for each batch.
3. Electroporation set-up: Bio-Rad Gene Pulser™ and cuvettes (Bio-Rad, Richmond, CA; #165-2088).

2.3. ES-Clone Analysis

1. 96-Well V-bottom plates (Costar; #3894), 24-well plates (Costar, Cambridge, MA; #3524), 96-well flat-bottom plates (Costar;#3599).
2. Multichannel pipet (Costar; #4880).
3. Pipet tips from USA Scientific (#10110006) are especially well-suited for picking up ES colonies.
4. DNA lysis buffer: 100 m*M* Tris-HCl pH 8.5, 5 m*M* EDTA, 0.2% SDS, 200 m*M* NaCl, 100 µg/mL Proteinase K (Sigma; #P2308).

2.4. Blastocyst Microinjection

1. Microinjection set-up, *see* **Subheading 3.5., step 1.**
2. M2 medium (Sigma; #M5910).
3. M16 medium (Sigma; #M7292).

3. Methods

3.1. ES-Cell Cultures

1. Seeding ES cells: ES cells are always grown at 37°C in a CO_2 incubator with daily changes of ES medium and on a feeder layer of G418^r EF cells in gelatinized tissue culture plates or flasks *(6)*. Frozen vials of ES or EF cells are thawed as quickly as possible at 37°C and then sterilized with ethanol. To remove the cytotoxic DMSO, add the cell suspension to a sterile conical tube filled with ~10 times excess of ES medium. Pellet the cells at room temperature and resuspend

them in ES medium in order to seed ~2–5 × 10^6 cells/T_{25} flask. EF cells can be plated before or at the same time as the ES cells, and more EF cells can be plated together to increase cell density and optimize growth.

2. Culturing ES cells: Grow cells in ES medium and split them ~1:6 (numbers refer to the total plating area; 1 T_{25} into 2 T_{75} flasks) at ~70% confluency (i.e., 1–3 d depending on the growth characteristics of the ES cell line). Change the medium ~1 h before splitting the cells. Wash the cells twice with PBS before adding ~0.3 mL/T_{25} flask of trypsin. Make sure that the entire surface is covered and incubate for 3–5'. Neutralize trypsin with at least 2 vol of medium. Dissociate the cells by vigorous pipeting and transfer the suspension into a new flask. Although complete removal of trypsin is not critical, one should try to dilute it at least 20 times (*see* **Note 2**).
3. Storage of ES cells: Trypsinize and pellet the cells, and then resuspend in freezing solution at a density of ~10^7/mL. Divide cells into 0.5 mL aliquots in cryotubes and freeze. In contrast to thawing, ES cells should be frozen as slowly as possible. Accordingly, place the cryotubes overnight at –80°C first and then in a liquid nitrogen container for long-term storage.

3.2. Electroporation of ES Cells

1. Prepare DNA by double CsCl-ethidium bromide centrifugation being sure to remove all traces of salt and DNA-intercalating dye. Linearize the targeting vector at a restriction site placed at the junction between the plasmid and the insert (if only G418 selection is used), or at the junction between the plasmid and the 3' end of the TK-cassette (if double selection is used). DNA is ethanol precipitated, washed, and resuspended at ~1 µg/µL in TE buffer or deionized water *(7)*.
2. Trypsinize rapidly growing ES cells making sure they are completely dissociated. Dilute an aliquot of the suspension ~10-fold and count the cells with a standard counting chamber.
3. Wash the cells twice with 5 mL of electroporation buffer equilibrated to room temperature.
4. Transfer 50 µL of linearized DNA in a sterile electroporation cuvette and resuspend the cells in the electroporation buffer to a final concentration of ~ 2 × 10^7 cells/mL, making sure they are completely dissociated. Transfer 0.8 mL of the suspension to the electroporation cuvette and mix with the DNA by pipeting the solution several times without creating air bubbles. Put the cuvette in the electroporator, set a single pulse at 400 V-25 µF and leave it for 10' at room temperature.
5. Transfer the content of the electroporation cuvette to a tube containing 10 mL of ES medium, mix gently, but thoroughly, and plate cells on 100-mm plates which have been gelatin-coated and covered with the EF layer ~12 h in advance. Start selection 12–18 h later in G418-supplemented ES medium. If a positive–negative selection strategy is employed, gancyclovir should be also added at a final concentration of 1.0 µ*M*; in this case, a control plate with only G418 should be used to monitor the efficiency of gancyclovir selection. After 7–12 d of selection, pick singly or doubly resistant colonies.

3.3. Identification of Correctly Targeted ES Clones

1. Wash cells twice with sterile PBS and prepare the following items: 96-well V-bottom plates ("dissociation plate"), 24-well plates ("DNA plate"); 96-well flat-bottom plates ("stock plate"); trypsin, selection medium, gelatin, EF cells, multichannel pipetor, and sterile pipet tips. Coat the "DNA plate" and the "stock plate" with gelatin and add 2 mL/well of selection medium to the former and 150 µL/well of EF cells resuspended in the same medium to the latter. Add 20 µL/well of trypsin to the "dissociation plate" prepared just before picking the ES colonies.
2. Pull up colonies with a 10-µL micropipet set at 3.5 µL; transfer to the "dissociation plate" and resuspend the cells by multiple pipeting. Continue to pick clones (in groups of 8, 16, or 24 depending on your speed) and add 175 µL of selection medium after a few minutes, making sure all components are thoroughly mixed.
3. Transfer 30 µL/well from the "dissociation plate" into the "stock plate;" transfer remaining 150 µL/well into the middle of the wells in the "DNA plate."
4. Change the medium of the "stock plate" daily until the first clones reach ≥80% confluency (2–3 d). At that time, wash all wells with PBS, add 30 µL/well of trypsin and 150 µL of freezing solution a few minutes later; seal plates around the edges with parafilm and place them in a seal-a-meal plastic bag. Freeze and store cells at –80°C for up to six mo.
5. Wash the "DNA plate" with PBS once the cells reach confluence (~7 d). Add 0.5 mL/well of DNA lysis buffer and incubate overnight at 37°C *(8)*. Place the plate on a shaker for 30' at room temperature; add an equal volume of isopropanol and continue to shake for ~3 h until DNA is completely precipitated. Transfer DNA from the wells to Eppendorf tubes. Wash DNA with 80% ethanol, let air-dry, and resuspend the pellet in 5 µL of TE by overnight incubation at 55°C.
6. Digest 5 µL of the DNA sample (~10 µg) overnight at 37°C in 50 µL with 20 U/mg of DNA of the appropriate restriction enzyme and then perform Southern blot hybridization according to the standard protocol *(7)*.

3.4. Expansion of Positively Targeted Clones

Thaw the 96-well "stock plate" in a 37°C waterbath after placing it in another seal-a-meal bag. Transfer cells from the stock plate to a 24-well plate which has been covered with EF cells in ~2.0 mL of ES medium. Split the culture once the colonies reach a reasonable size. ES cells require a minimal density to grow and it is best to expand them in a step-wise manner (using 24, 12, and 6-well plates, etc.). If the density is too low, "split back" the cells into a smaller plating area; be careful because the dense layer of EF cells may cause some problems. Clones are usually grown enough to obtain several aliquots at ~10^7 cells/mL which are stored frozen until microinjection.

3.5. Microinjection of ES Clones into Blastocysts

1. Organize the set-up for ES microinjection into blastocysts; it consists of a fixed-stage inverted microscope with a cooled injection chamber and two microman-

ipulators and micrometer syringes. The whole set-up is mounted on a base-plate that stands on a vibration-damping table. Additional instruments include a mechanical puller, a microforge and a diamond-coated grinding wheel; they are used to prepare the injection and holding pipets. An extensive description of these instruments and how they should be used can be found in the laboratory manual by Hogan *et al.* ***(9)*** (*see* **Note 3**).

2. Recover blastocysts from pregnant C57BL/6J mice 3.5 d *postcoitum* by flushing the uterine horn with M2 medium. Select only mature blastocysts with large cavities and keep them in drops of M16 medium under mineral oil at 37°C in the CO_2 incubator. Use the other blastocysts as carriers when reimplanting those injected. Handling of the animals and surgical procedures are detailed in the aforementioned manual ***(9)***.
3. Seed positive ES clones and expand them in a T_{25} flask for 1–2 d and without selection. Add 0.5 mL of trypsin, incubate at 37°C for 4', and neutralize with ~2 mL of ES medium at ~1–2 × 10^6 cells/mL. Because cell clumps cannot be injected, it is imperative to obtain a single-cell suspension. Inject ~12–15 ES cells/blastocyst in ES medium without selecting agent(s). Store injected blastocysts in M16 medium drops under mineral oil at 37°C in a CO_2 incubator.
4. Reimplant 6–10 embryos (including two uninjected carrier blastocysts) into one uterine horn of an anesthetized Swiss-Webster pseudopregnant foster mother. Cage the animals individually 14 d after the reimplantation to avoid cannibalization of the offsprings. At ~2 wk after birth, assess the level of chimerism from the percentage of agouti color in the newborns' coat (*see* **Note 4**).
5. Cage 5–6-week-old male chimeras with C57Bl6/6J females to test germ-line transmission of the targeted allele. Extract DNA from 1.5 cm tails of 3-week old mice. Incubate tails in lysis buffer overnight at 55°C. Remove hair by centrifugation, add isopropanol, and precipate DNA for ~1 h. After ethanol precipitation and washing, dissolve DNA in 200 μL of TE overnight at 55°C and then use 10 μL of the sample for Southern blot analysis. Once germ-line transmission is demonstrated, cage the same male chimeras with 129/sv females to bring the mutation on pure 129/sv genetic background; be aware this cross will yield only agouti offspring. Alternatively, continue to back-cross the progeny of the [chimera × C57BL/6J] inter-cross for ~10 generations to bring the mutation on the C57Bl/6J inbred background. While these matings progress, study the phenotypic and morphologic consequences of the mutation in the randomly mixed 129/sv, C57Bl/6J background.

4. Notes

1. Even if not mentioned in the text, ES cells should always be placed in fresh ES medium 1 h before being trypsinized and washed with PBS. ES cells should be picked rather quickly because overtrypsinization reduces germ-line colonization. To this end, the pipet tips for the picking and processing of ES cells should have an even opening and fit tightly in the multichannel pipetor. ES colonies are picked under a standard microscope placed inside the laminar flow hood.

2. Unlike electroporation and blastocyst injection, routine culturing of ES cells does not need to achieve single-cell suspension after trypsinization.
3. The quality of the injection pipet is critical and particular attention should be paid to the type of glass that is used, the way the end of the pipet is bent and the dimension and sharpness of the tip.
4. Successful germ-line transmission is usually obtained with >60% chimerism. It is, therefore, best to inject many independent ES clones several times in order to identify those yielding the highest chimeras that will then be mated to test germ-line transmission.

References

1. Capecchi, M. R. (1989) The new mouse genetics: Altering the genome by gene targeting. *Trends Genet.* **5,** 70–76.
2. Rosati, R., Horan, G. S., Pinero, G. J., Garofalo, S., Keene, D. R., Horton, W. A., Vuorio, E., de Crombrugghe, B., and Behringer, R. R. (1994) Normal long bone growth and development in type X collagen-null mice. *Nature Genet.* **8,** 129–135.
3. Andrikopoulos, K., Liu, X., Keene, D. R., Jaenisch, R., and Ramirez, F. (1995) Targeted mutation in the col5a2 gene reveals a regulatory role for type V collagen during matrix assembly. *Nature Genet.* **9,** 31–36.
4. Pereira, L., Andrikopoulos, K., Tian, J., Lee, S. Y., Keene, D. R., Ono, R., Reinhardt, D. P., Sakai, L. Y., Biery, N. J., Bunton, T., Dietz, H. C., and Ramirez, F. (1997) Targeting of the gene encoding fibrillin–1 recapitulates the vascular aspect of Marfan syndrome. *Nature Genet.* **17,** 218–222.
5. Gu, H., Marth, J. D., Orban, P. C., Mossmann, H., and Rajewsky, K. (1994) Deletion of a DNA polymerase beta gene segment in T cells using cell type-specific gene targeting. *Science* **265,** 103–106.
6. Wurst, W. and Joyner, A. L. (1993) Production of targeted embryonic stem cell clones, in Gene Targeting (Joyner, A. L., ed.), IRL, Oxford, UK, pp. 33–61.
7. Sambrook, E., Fritsch, E. F., and Maniatis, T. (1989) *Molecular Cloning: A Laboratory Manual.* Cold Spring Harbor, Cold Spring Harbor, NY.
8. Laird, P. W., Zijderveld, A., Linders, K., Rudnicki, M. A., Jaenisch, R., and Berns, A. (1991) Simplified mammalian DNA isolation procedure. *Nucleic Acids Res.* **19,** 4293.
9. Hogan, B., Constantini, F., and Lacy, E. (1986) *Manipulating the Mouse Embryo: A Laboratory Manual.* Cold Spring Harbor, Cold Spring Harbor, NY.

15

Enhancer Analysis of the α*1(II)* and α*2(XI)* Collagen Genes in Transfected Chondrocytes and Transgenic Mice

Noriyuki Tsumaki, Ying Liu, Yoshihiko Yamada, and Paul Krebsbach

1. Introduction

Extracellular matrix (ECM) plays a critical role in normal development, tissue repair, and is altered in several disease states. During these processes, expression of ECM genes is regulated temporally and spatially. Identification of regulatory elements that direct tissue- and stage-specific expression is important for understanding of the role of ECM genes in cellular differentiation and function. A significant number of ECM genes have been cloned and their regulatory elements have been identified. Reporter gene systems are widely used to study gene regulation and function ***(1–3)***. In this chapter, we describe methods for the identification of promoters and enhancers of cartilage collagen genes for the α*1(II)* and α*2(XI)* chains using transgenic mice and transient transfection in cell cultures ***(4–9)***.

Transient transfection of plasmid DNA into cultured cells is a well established method for studying gene regulation. Traditionally, the CAT (chloramphenicol acetyltransferase) gene has been used as a reporter gene to analyze activity of promoter and enhancer elements in transfection assays ***(1,2)***. More recently, the *lacZ* (β-galactosidase) and luciferase genes have often been used because of their high sensitivity and nonradioactive methods of analyzing these gene products. In addition, several reagents and equipment (e.g., Helios Gen Gun and Biolistic PDS-1000/He System from Bio-Rad, Richmond, CA) for DNA transfection have been developed to improve transfection efficiency ***(10,11)***. Although the transfection in cell cultures is rapid and convenient, it is difficult to evaluate comprehensive cell-type specificity of promoter and

From: *Methods in Molecular Biology, vol. 139: Extracellular Matrix Protocols*
Edited by: C. Streuli and M. Grant © Humana Press Inc., Totowa, NJ

enhancer activity. Further, some cell types (e.g., ameloblasts and odontoblasts) are difficult to culture in vitro.

The generation of transgenic mice provides a powerful molecular tool for identification of tissue- and developmental stage-specific enhancers of genes. This method allows the analysis of promoter and enhancer activity in all tissues at various developmental stages. A drawback of this approach is that it is a time-consuming and expensive process. However, the problem may be overcome by analyzing generation 0 (G0) embryos instead of establishing transgenic mouse lines.

2. Materials

2.1. CAT Reporter Gene Assay in Transfected Chick Chondrocytes

1. Ham's F-12 medium.
2. Fetal bovine serum (FBS).
3. Collagenase B (0.4% w/v): prepare by dissolving 100 mg collagenase B (Boehringer Mannheim, Mannheim, Germany) in 25 mL Hank's balanced salt solution (HBSS). Sterilize by passing solution through a 22-µm filter apparatus.
4. Sterile instruments for dissection of sterna (small scissors, forceps, and scalpel).
5. Sterilized Nintex membranes (Tetko, New York, NY) and Swinex filters (Millipore, Bedford, MA).
6. Sterile stirring bar and magnetic stir plate.
7. HBBSS: 0.14 g $CaCl_2$ $2H_2O$, 0.40 g KCl, 0.06 g KH_2PO_4, 0.098 g $MgSO_4$ (anhyd.), 8.0 g NaCl, 0.048 g Na_2HPO_4, 1.00 g D-glucose made up to 900 ml; adjust pH with 2 *M* NaOH to 7.4, adjust final volume to 1 L. Sterilize by autoclaving (*see* **Note 1**).
8. Complete growth medium: Ham's F-12 medium, 10% FBS, 100 U per mL penicillin, and 50 µg/mL streptomycin.
9. pDAS1BB5 containing the CAT reporter gene under the control of the promoter and intron enhancer of the *α1(II)* collagen gene ***(5)***.
10. For LipofectAMINE™ (Life Technologies, Bethesda, MO) transfections, prepare the following solutions in sterile 15 mL tubes:
    Solution A: Dilute 5 µg DNA in 300 µL of serum-free medium for each transfection.
    Solution B: Dilute 10 µL of LipofectAMINE reagent (Life Technology) in to 300 µL of serum-free medium for each transfection.

2.2. CAT Assay

1. 5 m*M* chloramphenicol: dissolve 16 mg of chloramphenicol in 10 mL of ethanol, store at –20°C.
2. CAT scraping buffer: 0.04 *M* Tris-HCl pH 7.4, 1 m*M* EDTA, and 0.15 *M* NaCl.
3. CAT extraction buffer: 0.25 *M* Tris-HCl pH 7.8.
4. [^{3}H] Acetyl CoA (200 mCi/mmol, New England Nuclear, Boston, MA).
5. Water-immiscible scintillation fluid (Econofluor, New England Nuclear).

2.3. Luciferase Reporter Gene Assay in Transfected RCS Cells Using FuGene

1. FuGene 6 transfection reagent kit (Boehringer Mannheim).
2. Dulbecco's modified Eagle's medium high glucose without pyruvate, (Life Technologies, Inc).
3. RSC cells ***(4)***.
4. Penicillin (50 U/mL), streptomycin (50 µg/mL), 10% heat-inactivated FBS (Hyclone, Laboratories, Inc., Logan, UT) for RCS cells and 10% heat-inactivated calf serum (BALB/3T3 cells) (Hyclone, Laboratories, Inc.).
5. Dual-Luciferase™ Reporter Assay System (Promega, Madison, WI).
6. The *742Luc/Int* reporter gene construct derived from the *pGL3*-Basic (Promega) containing the firefly luciferase reporter gene. Control pRL family vectors (e.g., *pRLSV40*, *pRL-CMV*, Promega) for renilla luciferase as an internal control.
7. Luminometer (e.g., Model MLX from DYNEX, Chantilly, VA).

2.3. Transgenic Mice

1. The most commonly used reporter gene in transgenic mice is the bacterial *lacZ* (β-galactosidase) gene. The basic reporter gene vector is commercially available (e.g., *pNASSβ*, Clontech, Palo Alto, AC). The *742LacZInt* construct contains the *lacZ* reporter gene under the control of the promoter and enhancer of the *α2(XI)* collagen gene ***(6,8)***. (*See* **Note 2**.)
2. LB broth with ampicillin (100 µg/mL), Plasmid preparation kit (e.g., Maxi Kit, Qiagen, Chatsworth, CA), QIAquick Gel Extraction Kit (Qiagen).
3. Fertilized ova isolated from female donors that are produced through timed mating with fertile males. Recipient females are made by timed mating with vasectomized males.
4. PBS (pH 7.4)
5. 4% paraformaldehyde, Dissolve 4 g paraformaldehyde in 100 mL PBS (pH 7.4) at 80 °C under stirring in a chemical hood. Prepare fresh.
6. Washing solution: 0.1 *M* phosphate buffer (pH 7.4), 2 m*M* $MgCl_2$, 0.01% sodium deoxycholate, 0.02% Nonidet P-40.
7. Staining solution: 0.1 *M* phosphate buffer (pH 7.4), 2 m*M* $MgCl_2$, 0.01% sodium deoxycholate, 0.02% Nonidet P-40, 5 m*M* potassium ferricyanide, 5 m*M* potassium ferrocyanide, 20 m*M* Tris (pH 7.4).
8. X-gal stock, 25 mg/mL in dimethylformamide (store in the glass tube at –20°C protected from light). Just before use, dilute X-gal stock to give a final concentration of 1 mg/mL.

3. Methods

3.1. CAT Reporter Gene Assay in Transfected Chick Chondrocytes

Although there are a few established cell lines that exhibit a chondrogenic phenotype, primary chondrocytes derived from the sterna of embryonic chickens offer a reproducible cell population for analysis of cartilage matrix genes. The transfection of plasmid DNA into primary chick chondrocytes has been successful using a number of experimental approaches. The liposome-mediated transfection procedures have yielded high efficiency plasmid DNA transfer, low toxicity and reproducible results *(12)*. Several other techniques such as electroporation, and modifications such as cationic liposome-mediated delivery systems have been used to facilitate the entry of nucleic acids into other cell types *(13,14)*. The method used may need to be optimized for the particular cell type of interest.

1. Obtain timed fertilized chicken eggs from a local vender and incubate at 37°C until embryonic d 15.
2. On d 15, sacrifice chick embryos and dissect individual sterna using a dissecting microscope (if necessary) using sterile techniques.
3. Rinse sterna in sterile HBSS.
4. Transfer sterna to a sterile 50-mL flask containing a magnetic stir bar.
5. Add 10 mL of collagenase B, incubate at 37°C for 15 min with gentle stirring.
6. Rinse sterna two times in sterile HBSS.
7. Add 15 mL of collagenase B to sterna and incubate for approximately 2 h while stirring.
8. Inactivate collagenase with F-12 medium containing 10% FBS.
9. Filter digested cells through a sterile Nintex membrane.
10. Centrifuge cells at 1000*g* for 10 min and resuspend in complete growth medium.
11. Count cells and plate 5×10^5 cells in 60-mm tissue-culture plates. Incubate primary chick chondrocytes in 5 mL complete growth medium at 37°C and 5%CO_2.
12. Transfect the primary chondrocytes on the following day. (*See* **Notes 3** and **4**.)
13. Combine solutions A and B. Mix gently and incubate for 15 min at room temperature to allow DNA/lipid complexes to form.
14. Rinse chondrocytes 2X with sterile serum free medium. Add 3.0 mL of serum-free medium to each tube containing the DNA/lipid complex and mix gently.
15. Remove serum-free medium from cells and add the DNA/lipid complex (3.3 mL) to the cells.
16. Incubate the cells with the transfection complexes for 6 h. Replace the medium with fresh complete growth medium and incubate at 37°C and 5% CO_2 for 24 to 72 h, depending on cell type and promoter activity.

3.2. CAT Assay

The fluordiffusion protocol measures the activity of the bacterial enzyme chloramphenicol acetyl transferase (CAT) in extracts from cultured chondrocytes transfected with the plasmid. The assay is rapid and allows for the quantitation of multiple samples in a single assay. Using this assay, linearity of the enzyme activity with time can easily be assessed. This assay exploits the differential solubility of acetylated and nonacetylated chloramphenicol. CAT catalyzes the acetylation of the substrate chloramphenicol. Whereas chloramphenicol is soluble in aqueous solutions, acetylated forms of chloramphenicol are lipid soluble and diffuse into the scintillation cocktail. Accumulation of radioactivity in the water-immiscible phase is measured by liquid scintillation counting. The samples are incubated at room temperature (or at 37°C) and counted hourly. CAT activity is represented as the linear regression slope of the data plotted as cpm of product vs time of incubation and is expresses as cpm/h/μg protein after normalizing to the protein content of the sample assayed. (*See* **Notes 5** and **6**.)

1. Rinse cells 2X in ice-cold PBS.
2. Scrape cells with a rubber policeman in 1 mL of CAT scraping buffer and transfer to 1.5-mL tube.
3. Recover cells by centrifugation at 12,000*g* for 15 s.
4. Resuspend cell pellet in 100 μL of CAT extraction buffer.
5. Extract proteins with three consecutive freeze/thaw cycles (freezing on dry ice and thawing at 37°C for 1 min).
6. Heat cell extracts at 65°C for 15 min to inactivate potential deacetylase activity.
7. Centrifuge extract for 5 min at 12,000*g* at 4°C and place supernatant into a clean 1.5-mL tube.
8. Store extract at –80°C until assayed.
9. Label scintillation vials for each sample and include a blank and a positive control.
10. Pipet up to 50 μL of the sample into a scintillation vial; adjust volume up to 50 μL with CAT extraction buffer. For the blank, add 50 μL CAT extraction buffer. For positive control, add 49 μL CAT extraction buffer and 1 μL of purified CAT (0.1 U/μL).
11. Prepare enough CAT reaction mix for total samples to be assayed (200 μL for each sample). *See* **Table 1**.
12. Add 200 μL of the CAT reaction mix to each sample.
13. Gently overlay sample with 5 mL of scintillation cocktail. Do not mix.
14. Incubate at room temperature (or 37°C).
15. Count each h for 4 h. The reaction is usually linear for at least 4 h.
16. For each sample, plot cpm versus time and calculate the linear regression line. CAT activity (cpm/h) is normalized to protein or DNA content of the sample (see **Note 5**).

3.3. Luciferase Reporter Gene Assay in Transfected RCS Cells Using FuGene

The luciferase assay is more sensitive than the CAT assay. The use of two luciferase reporter enzymes, *firefly* and *Renilla* luciferase, make it possible to cotransfect the "experimental" reporter construct with the "control" reporter construct as an internal control. Because the two enzymes have different substrate specificity and kinetics, their activity can be measured in a single test

Table 1
Preparation of CAT Reaction Mixture

	Number of Samples to be			
Solution	15	30	45	60
0.25 M Tris-HCl	0.375 mL	0.75 mL	1.125 mL	1.5 mL
5 mM Chloramphenicol	0.75 mL	1.5 mL	2.25 mL	3.0 mL
H_2O	1.875 mL	3.75 mL	5.625 mL	7.5 mL
[^{3}H] Acetyl CoA	3 µL	6 µL	9 µL	12 µL
Total Volume	3 mL	6 mL	9 mL	12 ml

tube, thus the dual luciferase system provides speed and reliability for the assay. RCS cells, a cell line derived from rat chondrosarcoma, produce a number of proteins specific to chondrocytes and have been used for study of cartilage gene regulation *(4)*. Unlike LipofectAMINE, FuGene is a nonliposomal based reagent and can be used for transfection in the presence of serum. FuGene has become popular because it produces high levels of transfection of many cell types with minimum optimization and less cytotoxicity.

1. Plate RCS cells the d before the transfection to achieve about 1 – 3 × 10^5 cells in a six-well dish. It is not necessary to wash cells prior to transfection.
2. Prepare FuGene 6 reagent/DNA complex by the manufacturer's protocol (3–6 µL FuGene 6 reagent per 1–2 µg of DNA). DNA solution contains the two reporter constructs *742Luc/Int* and *pRL-SV40* at 10:1 ratios, 1 µg DNA per a six-well plate. (*See* **Notes 7** and **8**.)
3. Dropwise, add mixture of FuGene 6 reagent/DNA to the cells. Swirl the wells to ensure even dispersal.
4. After incubation of the cells for 48 h, wash the wells and add 200 µL lysis buffer per well (PLB, Promega Dual-Luciferase Kit). Scrape the cells and transfer the cell suspension to a 1.5-mL microcentrifuge tube. Lysis the cells by freezing and thawing in dry ice and 37°C water bath twice. Remove debris by centrifuging the lysate for 2 min.
5. The cleared lysate (50 mL) is used to measure activity for *firefly* and *Renilla* luciferase sequentially by a luminometer.

3.4. Transgenic Mice

The purity of DNA is important for the successful production of transgenic mice. Plasmids can be purified by banding on CsCl two times. However, plasmid preparation kits from Qiagen also work well if the samples are not overloaded in the column purification step. The kits are more convenient than the CsCl method.

1. Usually, an overnight bacterial culture in 100 mL LB broth with ampicillin yields enough DNA for microinjection to create transgenic mice. Prepare DNA with Maxi (Qiagen) following the manufacturer's protocol.
2. The DNA to be microinjected should be free from the vector sequence. Digest the plasmid with *EcoRI* and *PstI*. The transgene containing the promoter, enhancer, and reporter gene are then purified by electrophoresis on an agarose gel. After ethidium bromide staining, the transgene fragment is eluted from the gel slice using a Gel Extraction Kit (Qiagen). The DNA fragment should be resuspended at concentrations of about 5 ng/μL in TE (10 m*M* Tris HCl pH 7.5, 0.25 m*M* EDTA).
3. Microinject about 5 pL of the purified DNA into the pronuclei of fertilized mouse ova ***(15,16)***. Sacrifice the pregnant female mice at the desirable stages of gestation to obtain embryos. Embryos older than 13.5 dpc, may require the dissection of tissues to allow full penetration of the fixing and staining solutions.
4. Place the embryos or tissues into six-well tissue-culture dishes on ice. Wash them with cold PBS once, and fix with 4 % paraformaldehyde for 20 min on ice with a gentle rocking motion to increase penetration of the fixative.
5. Aspirate the solutions, paying attention not to damage the embryos. Wash with PBS for 10 min on ice while rocking. Wash the samples two more times with PBS.
6. Add the X-gal staining solution. Put the cover on the dish and seal it with Saran wrap. Incubate the samples for 1 to 24 h at 30°C in the dark. The levels of the staining depends on the expression level of the β-galactosidase gene. (*See* **Note 9**.)
7. After staining with X-gal, wash the samples briefly with PBS. Fix the samples again with 4% paraformaldehyde to prevent the leakage of X-gal stain.
8. The X-gal stained tissues can be sectioned for histological analysis. Dehydrate them with alcohol and embed them in paraffin wax. Counterstaining is useful for identification of cell types expressing the β-galactosidase gene. (*See* **Note 10**.)
9. The presence of the transgene in embryos can be tested with DNA isolated from placentas ***(8)***. PCR amplification of genomic DNA with specific primers can be used for rapid detection of the transgene *(17)*. Southern blot analysis can be performed to determined a copy number and integration sites of the transgene. (*See* **Note 11**.)

4. Notes

1. PBS can be substituted for HBSS.
2. β-gal is a convenient reporter gene to identify the specific cells that express the transgene. However, if the experimental goal is to quantitate small differences between different constructs, then CAT or luciferase reporter systems are required.
3. It is essential to keep the cell density consistent for each transfection experiment. The cells should be between 50–70% confluent at the time of transfection, although this may need to be optimized for each cell type.
4. The most important aspect of successful transfection is the careful optimization of transfection conditions for each cell type. Several factors may influence the

expression transiently of transfected DNA constructs; the amount and purity of DNA, the ratio of DNA to transfection reagent, the cell density at time of transfection, and the transfection time may require optimization for each construct or cell type.
5. If CAT activity is to be normalized to DNA content, rather than protein content, centrifuge the sample prior to heat inactivating. Save the cell pellet for DNA content analysis and use supernatant for CAT analysis.
6. Cotransfection with a control plasmid utilizing a different reporter gene is required to determine the transfection efficiency between experimental groups.
7. It is important to perform preliminary cotransfection experiments to optimize both the amount of vector DNA and the ratio of coreporter vectors added to the transfection mix.
8. The *pRL* family of coreporter vectors for cotransfection assays are useful to optimize transfection efficiency. However, promoter activity of the control coreporter gene may affect activity of the promoter and enhancer of the test gene.
9. Background staining of the endogenous β-galactosidase activity can be decreased by staining tissues at pH 7.4. Inclusion of an intron cassette at the 5' leader sequence and poly-A signal at the 3' end of the construct often increases reporter gene activity probably because these sequences stabilize mRNA for the transgene.
10. Activity of the endogenous β-galactosidase gene may increase in certain tissues (e.g., bone and kidney) in late developmental stages and cause increased background staining. Green fluorescent protein (GFP) may also be used as a reporter gene.
11. If the level of reporter gene expression is low in G0 embryos, homozygous transgenic mice can be established and analyzed. Higher levels of expression of the reporter gene can be obtained in these mice because of double dosages of the transgene in the homozygous mouse.

References

1. Gorman C. M., Moffat L. F., and Howard B. H. (1982) Recombinant genomes which express chloramphenicol acetyltransferase in mammalian cells. *Mol. Cell. Biol.* **2,** 1044–1051.
2. Neumann, J. R., Morency, C. A., and Russian, K. O. 1996 A novel assay for chloramphenicol acetyltransferase gene expresssion. *BioTechniques* **5,** 444–447.
3. Norstedt. G., Enberg, B., Francis, S., Hansson, A., Hulthen, A., Lobie, P. E., et al. (1994) Cell transfection as a tool to study growth hormone action. *Proc. Soc. Exp. Biol. Med.* **206,** 181–184.
4. Mukhopadhyay, K., Lefebvre, V., Zhou, G., Garofalo, S., Kimura, J. H., and de Crombrugghe, B. (1995) Use of a new rat chondrosarcoma cell line to delineate a 119-base pair chondrocyte-specific enhancer element and to define active promoter segments in the mouse pro-α1(II) collagen gene. *J. Biol. Chem.* **270,** 17,711–17,719.
5. Krebsbach, P. H., Nakata, K., Bernier, S. M., Hatano, O., Miyashita, T., Rhodes, C. S., and Yamada, Y. (1996) Identification of a minimum enhancer

sequence for the type II collagen gene reveals several core sequence motifs in common with the link protein gene. *J. Biol. Chem.* **271,** 4298–4303.

6. Tsumaki, N., Kimura, T., Matsui, Y., Nakata, K., and Ochi, T. (1996) Separable cis-regulatory elements that contribute to tissue- and site-specific α2(XI) collagen gene expression in the embryonic mouse cartilage. *J. Biol. Chem.* **134,** 1573–1582.
7. Bridgewater, L. C., Lefebvre, V., and de Crombrugghe, B. (1998) Chondrocytes-specific enahncer elements in the Col11α2 gene resemble the Col2a1 tissue-specific enhancer. *J. Biol. Chem.* **273,** 14,998–15,006.
8. Tsumaki, N., Kimura, T., Tanaka, K., Kimura, J. H., Ochi, H., and Yamada, Y. (1998) Modular arrangement of cartilage- and neural tissue-specific *cis*-elements in the mouse α2(XI) collagen promoter. *J. Biol. Chem.*, in press.
9. Leung, K. K. H., Ng, L. J., Ho, K. Y., Tam, P. P. L., and Cheach K. S. E. (1998) Different cis-regulatory DNA elements mediate developmental stage- and tissue-specficif expression of the human COL2A1 gene in transgenic mice. *J. Cell Biol.* **141,** 1291–1300.
10. Johnston, S. A. and Tang, D. C. (1997) Gene gun transfection of animal cells and genetic immunization. *Meth. Cell. Biol.* **43**, 353–365.
11. Bernasconi, A. G., Rebuffat, A. G., Lovati, E., Frey, B. M., Frey, F. J., and Galli, I. (1997) Cortisol increases transfection efficiency of cells. *FEBS Lett.* **419,** 103–106.
12. Tseng, W. C., Haselton, F. R., and Giorgio, T. D. (1997) Transfection by cationic liposomes using simultaneous single cell measurements of plasmid delivery and transgene expression. *J. Biol. Chem.* **272,** 25641–25647.
13. Neville, C., Rosenthal, N., and Hauschka, S. (1997) DNA transfection of cultured muscle cells. *Meth. Cell. Biol.* **52,** 405–422.
14. Osborne, B. A., Smith, S. W., Liu, Z. G., McLaughlin, K. A., and Schwartz, L. M. (1995) Transient transfection assays to examine the requirement of putative cell death genes. *Meth. Cell. Biol.* **46,** 99–106.
15. Hogan, B., Beddington, R., Constantini, F., and Lacy, E. (1994) *Manipulating the Mouse Embryo: A Laboratory Manual, 2nd Ed.* Cold Spring Harbor Laboratory, Cold Spring Harbor, NY.
16. Bonnerot, C. and Nicolas, J.-F. (1993) Application of lacZ gene fusions to postimplantaion development, in *Guide to Techniques in Mouse Development* (Wassarman, P. M. and Depamphilis, M. L., ed.), Academic, San Diego, CA, pp. 451–469.
17. Hanley, T. and Merlie, J. P. (1991) Transgene detection in unpurified mouse tail DNA by polymerase chain reaction. *BioTechniques* **10,** 56.

16

Retroviral Delivery of ECM Genes to Cells

Kyonggeun Yoon and Vitali Alexeev

1. Introduction

The introduction of recombinant DNA has become a common tool for studying functional and structural properties of a wide variety of proteins. Functional analysis of protein can be studied by suppression of gene expression, thus introducing a plasmid which expresses an antisense RNA in mammalian cells. Several extracellular matrix proteins require an assembly of subunits to form functional heteromultimers with distinct features. Thus, expression of a specific mutant protein can interfere with the assembly of multimeric protein, resulting in a dominant-negative phenotype. In these studies, an efficient delivery of DNA into appropriate target cells represents a critical step. Although many procedures of transfection of the plasmid DNA into mammalian cells are available, viral infection represents a far superior mode of delivery because retroviral vectors were shown to transduce genes of interest into tissue-culture cells with success rates approaching 100%. The retrovirus inserts the viral genome into the chromosome of the infected cell permanently, usually without any measurable effect on the viability of the infected cells. The result is an efficient gene-transfer system in which most recipient cells will incorporate and express the transduced gene.

The retrovirus genome consists of an RNA molecule of 8500 nucleotides packaged into each viral particle *(1)*. The retrovirus enters the cells via interaction between the viral envelope protein and the appropriate viral receptor protein on the target cell. Once inside the cell, the enzyme reverse transcriptase brought in with the capsid makes a DNA copy of the viral RNA molecule to form a DNA-RNA hybrid duplex. After degradation of RNA, the reverse transcriptase completes a second DNA strand synthesis, generating a double-stranded DNA copy of the RNA genome and two long terminal repeats (LTRs)

From: *Methods in Molecular Biology, vol. 139: Extracellular Matrix Protocols*
Edited by: C. Streuli and M. Grant © Humana Press Inc., Totowa, NJ

at both 5' and 3' ends. This double-stranded DNA is integrated into the host chromosome, catalyzed by the viral integrase. Once DNA is integrated, new viral RNA synthesis is carried out by the host-cell RNA polymerase, producing a large number of viral RNA molecule. These RNAs are translated to produce the capsid, envelope, and reverse transcriptase proteins. Initiation of viral assembly begins by binding of capsid proteins to the RNA packaging signal (ψ sequence) of the viral RNA. Mature viral particles bud out from the host cells containing two copies of RNA genome, capsid, envelope, and reverse transcriptase proteins.

The single most important advance in the development of retroviruses as gene-transfer vectors was achieved by complementation of two systems, packaging cells and retroviral vectors *(2)*. A specialized cell line (termed "packaging cells") was generated in order to permit the production of high titers of replication-defective recombinant virus, free of wild-type virus (**Fig. 1**). For this purpose, the structural genes necessary for particle formation and replication, *gag*, *pol*, and *env*, are integrated into cell lines without the RNA packaging signal ψ sequence. Thus, viral RNA in packaging cells provide proteins necessary for particle formation but cannot be packaged into a viral particle. The other component is a retroviral vector in which most structural genes have been deleted for insertion of the gene of interest. The retroviral vector contains the ψ sequence for RNA packaging and two LTRs that provide transcription, polyadenylation of viral RNA, and integration of double-stranded viral genome (**Fig. 2**). When the retroviral vector is transfected into the packaging cells, the viral RNA containing the gene of interest is transcribed and packaged into viral particle by gag, pol, and env proteins provided by the packaging cells. This recombinant virus is utilized to infect the target cells, delivering a gene of interest at high efficiency. In principle, because no genetic information for virus production is transferred from the packaging cells, transduced cells are unable to perpetuate an infection and spread virus to other cells.

Many different packaging cells are available currently: *NIH3T3* based packaging cells, (*PA317* and *PT67*), and 293 based packaging cells, (*BOSC23* and Φ*NX (2–5)*. The viral env protein expressed by the packaging cell line determines the cellular host range of the packaged virus and allows infection of different cell types through recognition of specific cellular receptors. The *PT67* packaging cells, a *NIH3T3* based line expressing the *10A1* viral envelope, can enter cells via two different surface molecules, the amphotropic retrovirus and the gibbon ape leukemia virus receptors *(3)*. Thus, they exhibit a broader host range than other packaging cells. Among many packaging cells, we found Φ*NX* cells to be most convenient for a small-scale generation of a high-titer recombinant virus. These cells can be obtained from ATCC (Rockville, MD) upon permission from Dr. Nolan (http://www/leland.stanford.edu/group/nolan/

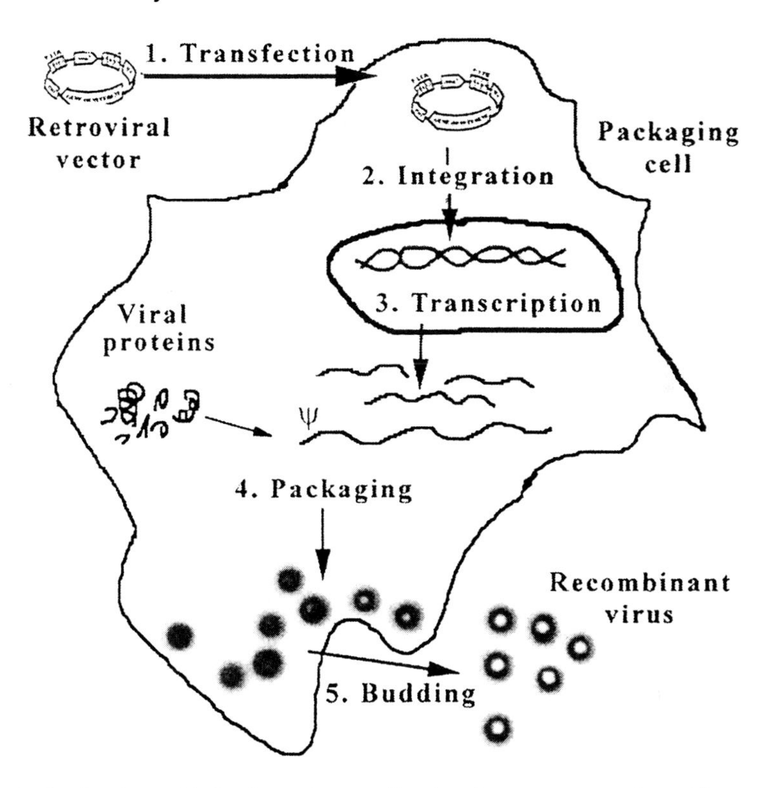

Fig. 1. Packaging of infectious, but replication incompetent, virus. The retroviral vector is transfected into packaging cells that provides viral proteins, gag, pol, and env, necessary for particle formation, which have been deleted in the recombinant viral vector. The full-length viral transcript containing the gene of interest is packaged into viral particle upon binding of capsid protein to the ψ sequence. The virus released from packaging cells is infectious but lacks the viral genes, thus preventing retroviral production from subsequently infected cells.

NL-Homepage. html) at Stanford University. Previously, a production of high-titer recombinant virus required laborious and lengthy processes: transfection of retroviral vector into the packaging cells, selection of transfected cells by antibiotic resistance, and cloning of a high-titer producer cells. The Φ*NX* packaging cells are based on the *293T* cell line, a human embryonic kidney line transformed with adenovirus *E1a* and carrying a temperature sensitive simian virus large T antigen ***(5)***. The unique feature of this cell line is that a high frequency of transfection (greater than 50%) can be achieved by either calcium phosphate or lipid-based transfection. Owing to an efficient transfection, high-titer recombinant virus can be generated by transient transfection of retroviral vector into the Φ*NX* packaging cells. The advantages over previous, stably

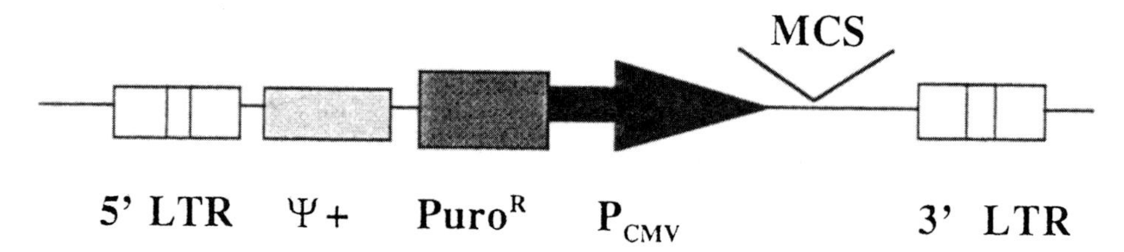

Fig. 2. Diagram of the recombinant retroviral vector. Retroviral vectors consist of the 5' LTR containing viral promoter and enhancer, an extended ψ sequence for efficient RNA packaging, a drug-resistant gene, a multiple cloning site (MCS) where a cDNA can be inserted, and the 3' LTR containing polyadenylation site. Upon transfection into packaging cells, a retroviral vector can transiently express (or integrate and stably express) a transcript containing the ψ sequence, the puromycine resistance gene and the inserted gene. In this vector the 5' LTR controls the drug resistance gene ($puro^R$), whereas the CMV early promoter controls expression of inserted gene.

integrated systems are that virus could be produced in days rather than months. The Φ*NX* packaging cells contain integrated *gag-pol* and *env* coding sequences driven by two different promoters. Separate introduction and integration of the structural genes into the packaging cells minimize the chances of producing replication-competent virus because of recombination events. In addition, the CD8 surface marker cDNA sequence was placed downstream of the reading frame of the *gag-pol* construct. Thus, monitoring of CD8 expression reflects directly an intracellular *gag-pol* expression and the stability of the packaging cell population's ability to produce *gap-pol* production. Two cell lines, Φ*NX-eco* and Φ*NX*-ampho are used to infect rodent host range and all host range target cells, respectively.

Most retroviral vectors are derived from Moloney murine leukemia virus and consist of the 5' LTR containing viral promoter and enhancer, an extended ψ sequence for efficient RNA packaging, drug-resistant gene for selection and a cDNA of a gene of interest, and the 3' LTR containing polyadenylation site (**Fig. 2**). In addition, it contains the β-lactamase gene and a plasmid origin of replication for bacterial propagation. Two promoters of different strength are incorporated to allow cloning and expression of the gene of interest and antibiotic resistant gene, neo^R or $puro^R$, for selection of transfected cells. The cytomegalovirus (CMV) early promoter is typically a stronger promoter than the viral LTR, but both promoters exhibit cell line-specific variation . A self-inactivating (SIN) retroviral vector was originally developed as a safer alternative to be used in human gene therapy *(6)* and was recently shown to sustain a prolonged gene expression in vivo *(7)*. Deletion of promoter and enhancer sequences at the 3' LTR in SIN vector results in inactivation of transcription by viral promoter since the 5' LTR is ultimately replaced by the deleted 3' LTR during viral replication. Thus, gene expression is instead driven by an internal promoter. Absence of enhancer and promoter sequences in both LTRs of the integrated provirus minimizes the possibility of activating cellular proto-oncogenes and provides a safer vector for gene therapy ***(6)***. Prolonged expression of SIN vector was attributed to the lack of methylation of the LTR and to the absence of a heterochromatin-induced inactivation of transcription, which occurs frequently in the integrated viral sequence in mammalian cells ***(7,8)***.

An important feature of the gene delivery system is the ability to regulate the expression of a delivered gene. Efficient delivery of regulatable genes by retroviral vectors makes it possible to analyze the population of transduced cells, circumventing a lengthy process and clonal variation, often observed in transfection of plasmids in mammalian cells. Several inducible gene-expression systems such as those controlled by heat shock, steroids, or metallothionein suffer from either high basal levels of gene expression or pleiotropic effects on host cell genes ***(9)***. Tetracycline (Tc) -inducible gene expression is restricted to the regulation of the gene of interest only because DNA response elements were derived from *Escherichia coli*, thus preventing pleiotropic effects on host cell genes.

The tetracycline-resistance operon of the Tn10 transposon are negatively regulated by the Tet repressor (*TetR*), which blocks transcription by binding to the tet operator sequences (*tetO*) in the absence of Tc ***(10,11)***. There are two components in the Tc-inducible system. The first one is a hybrid regulatory protein base on *TetR* under the control of the *SV40* promoter. To convert *TetR* from a repressor into a Tc-controlled transactivator, amino acids 1-207 were fused to the C-terminal 127 amino acids of the VP16 protein of herpes simplex virus ***(10,11)***. The hybrid protein binds to the tetracycline-responsive element (TRE) and thereby activates transcription in the absense of Tc (Tet-off system). The second component is the response element which expresses the gene of interest, cloned in the multiple cloning site, under the control of TRE. The TRE consists of seven copies of the 42-bp *tetO* sequence and is located upstream of the minimal immediate early promoter of CMV. These two components can reside in the same retroviral vector or can be divided into two retroviral vectors ***(11,12)***. Regulation of gene expression can be turned off by Tc (Tet-off system) as described above or turned on by Tc (Tet-on system). The four amino acid changes in *TetR* results in a "reverse" Tet repressor (*rTetR*), which binds the TRE sequence in the presence of Tc ***(11,13)***. Therefore, when fused to the VP16 activation domain, *rTetR* activates transcription in the presence of Tc (Tet-on).

Selection of the retroviral vectors will depend on the system where gene expression will be studied. In general, constitutive vectors are likely to yield a high-titer virus in comparison to a self-inactivating or a Tc-regulatable vector. Once the retroviral vector is chosen, viral transduction can proceed by transfection of the retroviral vector into packaging cells, virus production, and infection of target cells (*see* **Subheading 3.**).

2. Materials

2.1. Retroviral Vectors

Retroviral vectors listed below are available from Clontech (Palo Alto, CA).

1. Constitutive retroviral vector: pLXSN (#K1060-B), pLNCX (#K1060-C).
2. Tc-regulatable vector: *RevTet*-On vector (#K1659-1), *RevTet*-Off vector (#K1640-1).
3. Self-inactivating vector: pSIR (#6063-1).

2.2. Packaging Cells

1. 293-based packaging cells: Φ*NX*-eco (ATTC), Φ*NX*-ampho (ATCC).
2. NIH3T3-based packaging cells: *PA317* (ATCC #f-13677, 9078-CRL), *PT67* (Clontech #K1060-D).

2.3. Tissue Culture, Transfection, and Selection

1. Tissue culture. All standard tissue culture reagents can be obtained from Gibco-BRL (Bethesda, MD). Dulbecco modified Eagle medium (DMEM), fetal bovine serum (FBS), glutamine, streptomycin, penicillin, trypsin, PBS.
2. Transfection of packaging cells: CalPhos Maximizer™ Transfection kit (Clontech #K2050-1), ProFection mammalian Transfection system – calcium Phosphate (Promega, Madison, WI #E 12). Also, you can make transfection reagents as follows. Dissolve the mixture listed below in 80 mL of H_2O and adjust the pH exactly to 7.0–7.05 using 5 *M* NaOH or HCl (*see* **Note 2**). Adjust the volume up to 100 mL. Sterilize the 2xHBS solution through 0.22-μm filter. All reagents should be at room temperature prior to use. The following chemicals are used to make 100 mL of 2XHBS solution: 1.6 g NaCl, 0.2 g KCl, 1 g dextrose, 0.027 g HEPES acid, 0.074 g $Na_2HPO_4 \times 2H_2O$.
3. Selection of mammalian cells: Hygromycin B (Sigma, St. Louis, MO #H-3274), Puromycin (Clontech #8052-1, -2).

2.4. Infection of Target Cells

Culture medium for target cells, 0.45 μm Millipore filters, 5- or 10-mL syringes, Polybrene (Hexadimethrine Bromide) Sigma #H9268.

2.5. Regulation of Gene Expression

1. Tetracycline hydrochloride (Sigma #T3383): Dissolve in 1 mg/mL in 70% ethanol, filter sterilize (0.22-Millipore filter), and store at –20°C (stable within 2 mo).

2. Doxycycline hydrochloride (Sigma # D9891): Dissolve in 1 mg/mL in H_2O, filter sterilize (0.22-22 mm filter) and store at 4°C in dark (stable within 1 mo). Small aliquots can be frozen at –20°C.

2.6. Measurement of Infection Using Reporter Gene (LacZ)

1. Reagents for β-gal assay: Staining kit is available from Strategene, La Jolla, CA (#200284). You can also make the solution by using the following chemicals for 5 mL of staining solution: 0.008 g potassium ferricyanide, 0.01 g potassium ferrocyanide, 1.5 mL 2 *M* $MgCl_2$, 75 μL X-gal (40 mg/mL in DMF stock solution).
2. Reagents for fixation of cells: PBS containing 1% glutaraldehide (Sigma #G-5882).

2.7. Cloning and Plasmid Preparation

1. Cloning: All standard reagents for cloning, restriction enzymes, T4 ligase, alkaline phosphatase, Klenow DNA polymerase.
2. *E. coli* competent cells: DH5α (Gibco, Gaithersburg, MD, #18258-012), Epicurian *Coli*^{Æ} SURE^{Æ} competent cells (Stratagene #200152).
3. Plasmid preparation: L-broth, ampicillin, gel extraction kit (Quagen, Chatsworth, CA, #28704), Mini Prep kit (Quagen #27104), Maxi Prep kit (Quagen #12162), CHROMASPIN+TE-1000 Column (Clontech #K1324-1, 2).

3. Methods

3.1. Preparation of Recombinant Retroviral Vector

1. Choose one of the desirable retroviral vectors (*see* **Subheading 2.**). For a large-scale preparation of plasmid, transform a plasmid into a suitable *E. coli* strain (e.g., DH5α or SURE™) in accordance with manuals. Perform a large-scale DNA preparation using Maxi Prep Kit or CHROMASPIN+TE-1000 column. Also, it is possible to isolate and purify DNA by CsCl gradient centrifugation (Sambrook, et al., 1989).
2. Digest the retroviral vector with suitable restriction enzymes and check the digestion of the vector on agarose gel. If a blunt-ligation is required, both 5' and 3' ends of the vector can be filled up by addition of 1 U of the Klenow large fragment of DNA polymerase and dNTPs at 50 μ*M*. Incubate a reaction mixture for 15 min at room temperature and inactivate the enzyme at 75°C for 10 min. To prevent self-ligation of the vector, incubate the digested retroviral vector with the calf intestine (CIP) or the shrimp (SAP) alkaline phosphatase and purify DNA using phenol/chloroform/isoamylalcohol (24/23/1) solution. Isolate the insert from an agarose gel according to manufacturer's specification (Quiagen Gel Extraction kit).
3. Isolate an insert using similar procedures described above. It is important that the 5' and 3' overhang sequences of the insert and the retroviral vector are compatible for ligation. If available restriction sites are not compatible, a blunt ligation can be carried out after filling the overhangs of both insert and vector.

4. Set up the ligation reaction. Calculate the amount of vector and insert needed for ligation reaction in accordance with the size and molar ratio of vector and insert. For example, 1:1 molar ratio (vector/insert) requires 100 ng of 5 kb vector and 20 ng of 1 kb insert.

   For blunt ligation:

   X μL of vector (100 ng)
   Y μL of insert
   1 μL 5 m*M* ATP
   1 μL of 10X buffer
   1 μL 25% PEG
   1 μL T4 ligase
   Z μL H_2O

   10 μL

5. Transform the ligation mixture to a competent *E. coli* strain according to the manufacturer's instruction (*see* **Note 2**). Isolate and purify plasmid DNA from many bacterial colonies using Mini Prep kit and identify the desirable recombinant plasmid by restriction digestion, PCR, and sequencing.
6. Perform a large-scale plasmid preparation of a recombinant retroviral vector. We routinely use the Maxi Prep kit for plasmid preparation and found it satisfactory for transfection.

3.2. Preparation of Packaging Cells

A detailed description of Φ*NX* cells is available from the web site from Dr. Nolan's laboratory (http://www-leland.stanford.edu/group/nolan/NL-Homepage.html). The Φ*NX* cells are maintained in growth media containing DMEM, 10% FBS, 50,000 U/500 mL penicillin, 50 mg/500 mL streptomycin, 1% glutamine and splitted 1:4 or 1:5 every 3–4 d. The Φ*NX* cells are seeded at 3 million cells per 10 cm plate in growth medium, 18–24 h prior to transfection. A high-transfection efficiency is achieved when cells are confluent 40–50% at the time of transfection. Significant reduction in transfection efficiency results when cells are confluent.

3.3. Transfection

1. Change the medium with a solution containing 0.1% of BSA instead of serum 3 h prior to transfection. For the Tet-off system, add tetracycline (1–2 μg/mL) or doxycycline (20 ng/mL–1 μg/mL) at this point (*see* **Notes**).
2. Put 0.5 mL of 2XHBS into one Eppendorf tube.
3. Into another Eppendorf tube, add H_2O + DNA + $CaCl_2$. We suggest using 2 *M* $CaCl_2$ (62.5 μL/mL of reaction mixture). The best results were obtained using 20 μg of DNA/1 mL of the reaction mixture. Dilute DNA with H_2O and add $CaCl_2$ dropwise. The final volume of this mixture should be 0.5 mL.

4. Add the DNA/$CaCl_2$ mixture to 0.5 mL 2XHBS dropwise with vigorous bubbling using an automatic pipetter (keep eject button depressed) for 15–45 s (the length of bubbling time depends on each batch of 2XHBS).
5. Add HBS/DNA solution dropwise onto media, gently and quickly, by spreading across cells in media.
6. Observe the cells under a microscope. Good transfection efficiency is expected when the particle size is small and uniform. Big or aggregated particle usually indicates that the pH is not optimized, resulting in a poor transfection efficiency.
7. Put plates in 37°C incubator and shake plates back and forth to distribute DNA/CaPhosphate particles evenly.
8. After 24 h, change the media to 5 mL fresh DMEM containing 10% FBS (virus is more stable if incubation is carried out at 32°C).

3.4. Infection of Target Cells

1. Split the appropriate cells (which you would like to infect) at 200,000 cells per 60 mm plate in 2 mL of appropriate culture media. For suspension cells, they should be growing in a log phase at the time of infection (for Jurkats, ideal density at time of infection is 5×10^5/mL).
2. Collect supernatant from transfected Φ*NX* cells 24–48 h after transfection. Centrifuge the supernatant at 1000*g* for 5 min to pellet cell debris and filter through 0.45 Millipore filter to remove cells as well. Add Polybrene (4 μg/mL). Virus containing supernatant can be frozen at –80°C for later infection (*see* **Notes 5**).
3. Remove 1 mL of media from each plate with cells you are going to infect.
4. Add 4 μg/mL polybrene (stock solution is 4 mg/mL) to each plate them with gentle shaking.
5. Add 1 mL viral supernatant to cells plated for infection and place them at 37°C with gentle shaking. For suspension cells, pellet 5×10^5 suspension cells and resuspend cell pellets in 1 mL of viral supernatant containing 4 mg/mL polybrene. Spin cells and wash away virus supernatant after incubation for 8–24 h. Suspension cells, especially some B-cells and T cells, are sensitive to polybrene and it may be necessary to titrate polybrene.
6. Remove virus 24 h after infection by feeding cells with fresh medium. For suspension cells, spin down cells and resuspend in 2 mL of fresh media.
7. After 24–48 h of infection cells are ready to assay for biochemical event of interest. The reverse transcription and integration take place within 36 h, depending on cell growth. Expression can be observed at 24 h, usually reaching its maximum at 48 h. For the Tc-regulatable retroviral system, add or remove tetracycline or doxycycline depending on Tet-on or Tet-off viral vector, respectively.

3.5. Determination of the Viral Titer

The viral titer produced by packaging cells is determined as follows.

1. Prepare the target cells by plating one day before infection into a six-well plates, $5 \times 10^4 – 1 \times 10^5$ cells per well)

2. Collect virus-containing medium from the ΦNX cells transfected with a retroviral vector.
3. Centrifuge the supernatant at 1000*g* for 5 min to spindown cell debris and filter through a 0.45-μm filter to remove cells as well. Add polybrene to a final concentration of 4 μg/mL and filter medium through a 0.45-μm filter.
4. Prepare serial dilution of a virus-containing media using a fresh culture medium containing 4 μg/mL of polybrene (six 10-fold serial dilutions are usually prepared).
5. Infect target cells by adding virus-containing medium to the wells.
6. Change the media with virus to a normal one and add appropriate antibiotic (depending on the drug selection marker) 24 h postinfection. The optimal concentration should be determined for each cell line. Puromycin selection takes less than 1 wk at 0.5–1.5 μg/mL while G418 takes 2–3 wk at 0.1–1.0 mg/mL.
7. The titer of virus corresponds to the number of colonies present at the given dilution multiplied by the dilution factor. For example, the presence of 4 colonies in the 10^5 dilution would represent a viral titer of 4×10^5 (*see* **Notes**).

3.6. Measurement of Transduction Using a Reporter Gene (LacZ)

In order to measure an efficacy of retroviral transduction, a retroviral vector containing a reporter gene (alkaline phosphatase, β-galactosidase, luciferase, or green fluorescent protein) should be utilized in parallel experiments. Transfection and infection of retrovirus containing a reporter gene can be carried out together. The following procedure describes the procedure for detection of cells transduced by the retroviral vector containing the β-galactosidase gene.

1. Remove media from adherent cells.
2. Add 2 mL of fixative solution to a 60-mm plate of adherent cells at 4°C for 5 min.
3. Remove medium and wash 3 times with PBS.
4. Transfer 1 mL of prepared staining solution into cells.
5. Optimal staining will occur 12–18 h later.

4. Notes

1. The viral supernatants produced by these methods might contain potentially hazardous recombinant virus. The user of these systems must exercise caution in the production, use, and storage of recombinant retroviral virions, especially those with amphotropic host ranges. This consideration should be applied to all genes expressed as amphotropic and polytropic retroviral vectors. The user is strongly advised NOT to create retroviruses capable of expressing known oncogenes in amphotropic or polytropic host-range viruses. NIH guidelines require that retroviral production and transduction be performed in a Biosafety Level 2 facility.
2. For efficient transfection using calcium phosphate method, pH of 2XHBS solution should be 7.0–7.05. Because pH is so important it would be useful to make and to test 2XHBS with pH 6.95, pH 7.0, and pH 7.05.
3. We found frequent rearrangements (mainly large deletions) in the retroviral vector when the ligation mixture was transformed into DH5α cells. Thus, it is neces-

sary to screen many colonies to find a clone containing a correct insert. Usually, smaller colonies had a higher probability of maintaining a desirable insert. Other *E. coli* strains, which are deficient in recombination, UV repair and SOS repair (SURE cells) may be used in order to stabilize the insert.

4. For the Tet-off system, *Tc* or *Dox* needs to be maintained throughout transfection and infection. The presence of *Tc* does not interfere with transfection or infection. In order to investigate gene regulation by *Tc*, infected cells can be split into two plates and maintained in the absence and presence of *Tc*.
5. Virus containing supernatant can be frozen at –80°C for later infection, although viral titer is decreased to 50% when the virus is frozen and thawed.
6. Retroviral titers vary widely depending upon different retroviral vector and packaging cells. In general, constitutive vectors are likely to yield a high-titer virus (10^6 infectious particles/mL) in comparison to a self-inactivating or a Tc-regulatable vector (10^4 infectious particles/mL). In order to avoid multiple infections, which increase the number of integration events per cell, transduction is usually performed at the multiplicity of infection of 0.1 (for example, 10^4 infectious particles per 10^5 cells). This condition is likely to yield one viral integration per genome.

References

1. Weiss R., Teich N., and Coffin J., (1984 and 1985). RNA tumor viruses. Cold Spring Harbor Laboratory, Cold Spring Harbor, N. Y.
2. Cone, R. D. and Mulligan, R. C. (1984) High-efficiency gene transfer into mammalian cells: generation of helper-free recombinant retrovirus with broad mammalian host range. *Proc. Natl. Acad. Sci. USA* **90,** 8033–8037.
3. Miller. A. D. and Chen. F. (1996) Retrovirus packaging cells based on 10A1 murine leukemia virus for production of vectors that use multiple receptors for cell entry. *J. Virol.* **70,** 5564–5571.
4. Warren, S. P., Nolan, G. P., Scott, M. L., and Baltimore, D. (1993) Production of high-titer helper-free retroviruses by transient transfection. *Proc. Natl. Acad. Sci. USA* **90,** 8392–8396.
5. Kinsella, T. M. and Nolan, G. P. (1996) Episomal vectors rapidly and stably produce high-titer recombinant retrovirus. *Hum. Gene Ther.* **7,** 1405–1413.
6. Yu, S. F., Ruden, T., Kantoff, P. W. Garber, C. Seiberg, M., Ruther, U., et al. (1986) Self-inactivating retroviral vectors designed for transfer of whole genes into mammalian cells. *Proc. Natl. Acad. Sci. USA* **83,** 3194–3198.
7. Deng, H., Lin, Q., and Khavari, P. A. (1997) Sustainable cutaneous gene delivery. *Nature Biotech.* **15,** 1388–1391.
8. Hoeben R. C., Migchielisen, A. A., van der Jagt, R. C., van Ormondt, H., and van der Eb, A. J. (1991) Inactivation of the Molony murine leukemia virus long terminal repeat in murine fibroblast cell lines is associated methylation and dependent on its chromosomal position. *J. Virol.* **65,** 904–912.
9. Yarranton, G. T. (1992) Inducible vectors for expression in mammalian cells. *Curr. Opin. Biotechnol.* **3,** 506–511.

10. Gossen, M. and Bujard, H. (1992) Tight control of gene expression in mammalian cells by tetracycline-responsive promoters. *Proc. Natl. Acad. Sci. USA PNAS* **89,** 5547–5551.
11. Gossen, M., Freundlies, S., Bender, G., Muller G., Hillen, W., and Bujard. H. (1995) Transcriptional activation by tetracyclines in mammalian cell. *Science* **268,** 1766–1769.
12. Hoffman, A., Nolan, G. P., and Blau, H. M. (1996) Rapid retroviral delivery of tetracycline-inducible genes in a single autoregulatory cassette. *Proc. Natl. Acad. Sci. USA* **93,** 5185–5190.
13. Hillen, W. and Berens, C. (1994) Mechanisms underlying expression of Tn10-encoded tetracycline resistance. *Annu. Rev. Microbiol.* 48, 345–369.
14. Sambrook J., Fritsch E. F., and Maniatis T. (1989) *Molecular Cloning*, 2nd ed. Cold Spring Harbor Laboratory, Cold Spring Harbor, NY.

IV

Cell Biology of the Extracellular Matrix

17

Using Self-Assembled Monolayers to Pattern ECM Proteins and Cells on Substrates

Christopher S. Chen, Emanuele Ostuni, George M. Whitesides, and Donald E. Ingber

1. Introduction

We present a method that uses microcontact printing of alkanethiols on gold to generate patterned substrates presenting "islands" of extracellular matrix (ECM) surrounded by nonadhesive regions such that single cells attach and spread only on the adhesive regions. We have used this micropatterning technology to demonstrate that mammalian cells can be switched between growth and apoptosis programs in the presence of saturating concentrations of growth factors by either promoting or preventing cell spreading *(1)*. From the perspective of fundamental cell biology, these results suggested that the local differentials in growth and viability that are critical for the formation of complex tissue patterns may be generated by local changes in cell–ECM interactions. In the context of cell culture technologies, such as bioreactors and cellular engineering applications, the regulation of cell function by cell shape indicates that the adhesive microenvironment around cells can be carefully optimized by patterning a substrate in addition to using soluble factors *(2)*. Micropatterning technology will play a central role both in our understanding how ECM and cell shape regulate cell physiology and in facilitating the development of cellular biosensor and tissue engineering applications *(3–5)*.

Historically, investigations of cellular responses to various adhesive environments were limited by a lack of control over the interfacial properties and the topology of available substrates. It was particularly difficult to generate substrates patterned with adjacent adhesive and nonadhesive regions. In the past decade, the technology to engineer the properties of a surface with molecular-level control and to pattern these substrates with ligands suitable for

From: *Methods in Molecular Biology, vol. 139: Extracellular Matrix Protocols*
Edited by: C. Streuli and M. Grant © Humana Press Inc., Totowa, NJ

biological experiments has advanced rapidly. This progress was obtained as a result of the modification of microfabrication techniques used in the electronics industry in conjunction with polymer science and surface science. This powerful class of "micropatterning" techniques makes it possible to pattern surfaces with defined reactivity and topography with varying degrees of precision, depending on the methods used.

Surfaces with spatially patterned chemical functionalities can be formed using several techniques: vapor deposition, photolithography, and microcontact printing. Vapor deposition of metals through a patterned grid onto polyhydroxyethyl methacrylate (pHEMA) produces a substrate that presents complementary patterns of metal and pHEMA *(6,7)*. Cells attach selectively to the metallic regions because they adhere to the metal (or more properly, to proteins adsorbed on the metals), but not the pHEMA; however, the edge resolution of this method is low (5 μm). Photolithography, which uses ultraviolet light to illuminate photosensitive materials through a patterned mask, can routinely produce patterns of defined chemical properties with resolutions better than 1 μm. To generate surfaces with only selected regions that promote cell attachment, various investigators have lithographically photoablated proteins preadsorbed to a silicon or glass surface *(4)*; uncovered protein-adsorbing regions of a substrate previously coated with photoresist *(8)*; or covalently linked proteins onto photoactivatable chemicals on the surface *(9)*. A major limitation of these approaches is that the "nonadhesive" regions of the pattern are usually surfaces that actually promote the adsorption of protein, and require passivation (blocking of adhesive sites) with a protein such as albumin that prevents the adsorption of ECM proteins and the adhesion of cells. Over a period of days, however, cells are able to migrate onto these regions, probably as a result of degradation of the albumin and deposition of ECM by cells. Several investigators have tried to avoid these problems by using photolithography to pattern monolayers of trichloroalkylsilanes chemisorbed on the surface of SiO_2. Self-assembled monolayers (SAMs) of alkylsiloxanes which present regions of perfluoro- and amino-terminated moieties promote cells to attach in patterns onto the surface; the amino-terminated siloxane promotes the preferential adhesion of cells and the perfluoro-terminated regions resist adhesion without passivation with albumin *(5,10–12)*. Several technical issues remain in using this approach. This type of SAM is not easy to form, and a variety of biologically relevant organic functional groups (e.g., peptides and carbohydrates) are not compatible with the conditions used for its formation, thus limiting the range of surface chemistries available. The mechanism by which cells adhere to SAMs of alkylsiloxanes terminated with amine groups has not been elucidated, although, again, adsorption of proteins from the culture medium onto the charged surface is a plausible first step. Despite these shortcomings, this approach is still a viable one to be used for patterning the adhesion of cells.

Recent advances in the study of SAMs of alkanethiolates on gold surfaces has provided a more versatile approach to the patterning of cells. These SAMs are highly ordered molecular assemblies that chemisorb on the surface of gold with nearly crystalline packing to produce a new interface whose properties are determined solely by those of the head-group of the alkanethiol ***(13)***. This system makes it possible to control the interfacial properties of surfaces exposed to cells with greater molecular-level detail than other methods, and it affords the chance to influence cellular adhesion with greater specificity than with other methods ***(14,15)***. The synthetic procedures used to make alkanethiols are compatible with complex ligands that interact biospecifically on the cell ***(15,16)***. Alkanethiols can be patterned easily on a gold surface using microcontact printing (μCP), a technique in which a flexible polymeric stamp is used to print the alkanethiols in a specified pattern; the size of the stamped regions can be defined arbitrarily with dimensions from 500 nm (or, with greater experimental difficulty from 200 nm) and up ***(17)***. After printing a hydrophobic alkanethiol, the remaining bare surface of the gold is exposed to an alkanethiol that presents tri(ethylene glycol) groups (e.g., $HS(CH_2)_{11}O(CH_2CH_2O)_2CH_2CH_2OH$) that resist the adsorption of proteins. Thus, a pattern of these two SAMs presented on a substrate defines the pattern of ECM that adsorbs from solution onto the substrate ***(3,18)***. The hydrophobic SAMs created on flat gold substrates pattern the otherwise nonspecific adsorption of ECM proteins (fibronectin, fibrinogen, vitronectin, collagen I, and laminin) that promote the adhesion of different cell types (bovine capillary endothelial and rat hepatocytes) to the surfaces, whereas the tri(ethylene glycol) SAMs resist protein adsorption and cell adhesion ***(1,3,18–22)***.

Here, we describe how to use μCP to fabricate substrates that present patterned SAMs with features >500 nm; features as small as 200 nm can be obtained in special cases, but they are not necessary for most conventional biological applications ***(23)***. This technique uses an elastomeric stamp with bas-relief to transfer an alkanethiol to the surface of gold in the same pattern defined by the stamp. The stamps are usually fabricated by pouring a prepolymer of polydimethylsiloxane (PDMS) onto a master relief pattern, which is often formed by photolithographic methods. Because μCP relies on self-assembly of an alkanethiol, it does not require a dust-controlled laboratory environment, and can produce patterned gold substrates at relatively low cost.

2. Materials for Microcontact Printing

2.1. Glass Substrates Coated with Titanium then Gold

Microscope slides (Fisher, Pittsburgh, PA, no. 2) are loaded on a rotating carousel in an electron-beam evaporator (most of these are partially home built). Evaporation is performed at pressures $<1 \times 10^{-6}$ torr. Occasionally, during the evaporation of titanium, the pressure increases above 1×10^{-6} torr, but decreases after allowing the chamber to stabilize for approximately 2 min. Allow the metals to reach evaporation rates of 1 Å/s. Allow 400–500 Å of each metal to evaporate before opening the shutters and exposing the glass slides to 15 Å of titanium (Aldrich, Milwaukee, WI, 99.99+% purity) and 115 Å of gold (Materials Research Corporation, 99.99 + % purity).

2.2. PDMS Stamp with Patterns Molded from a Photolithographic Master

Basic lithographic techniques, concepts, and terminology are described by Madou ***(24)***. Procedures that result in thicker features are available from the manufacturers of other types of photoresists; for the sake of brevity, we do not describe them here. Use test grade N type, 9–13-mils thick silicon wafers (Silicon Sense, Nashua, NH), with <100> orientation, phosphorus dopant, and 1–10 resistivity.

2.2.1. Generating Silicon Master with Desired Pattern Using Photolithography

In a clean room (preferably Class A), clean the wafers by sonicating for 5 min successively in trichloroethylene, acetone, then methanol. Bake at 180°C for 10 min to dry thoroughly. Spin coat (40 s @ 4000 rpm) the wafers with approx 1–2 mL hexamethyldisilazane (Shipley) followed by Shipley 1813 positive photoresist (40 s @ 4000 rpm produces a layer of 1.3 μm); bake the resist at 105°C for 3.5 min. Expose the wafer on a mask aligner (typically a Karl Suss model) through a photomask with features etched in chrome deposited on quartz (Advance Reproductions Corp., North Andover, MA) for 5.5 s at 10 mW/cm^2. Develop the features by immersing in Shipley 351 for 45 s, then rinse with distilled water and dry with a stream of nitrogen. The proper development of the features should be checked under a microscope using a red filter in front of the light source to avoid unwanted exposure of the photoresist. Place the wafers in a desiccator under vacuum for 2 h with a vial (approx 1–2 mL) of (tridecafluoro-1, 1, 2, 2,-tetrahydro-octyl)-1-trichlorosilane (United Chemical Technologies, Bristol, PA).

2.2.2. Molding PDMS Stamp

PDMS (Sylgard 184, Dow Chemical Co.) prepolymer is made by mixing 10 parts of monomer and 1 part of initiator thoroughly in a plastic container and degassing it under vacuum for about 1 h until air bubbles no longer rise to the top. Pour the prepolymer in a Petri dish that contains the patterned silicon wafer, and cure for at least 2 h at 60°C. Peel the PDMS from the wafer and cut the stamps to the desired size with a razor blade.

2.3. Synthesis and Purification of Alkanethiols

The progress of reactions is monitored by thin layer chromatography (TLC) using 0.25-mm silica gel plates (Merck, Mannheim, Germany, or VWR). Column chromatography is performed under nitrogen with silica gel (60–200 mesh, Mallinckrodt). Reactions in nonacqueous solvents are carried out under nitrogen or argon. Organic solvents, unless specified, are HPLC grade (Mallinckrodt, Phillipsburg, NJ). Tetrahydrofuran (THF) that was used as a reaction solvent is distilled freshly on a still that contains benzophenone (1 g/L, Aldrich) and sodium (1 g/L, Aldrich). Dichloromethane used as a reaction solvent is distilled freshly on a still that contains calcium hydride (1 g/L, Aldrich). NMR spectra were collected on samples dissolved in chloroform-*d* (Cambridge Isotope Laboratory). General synthetic procedures are described; the specified quantities of material can be varied by keeping the molar ratios constants to fit the needs of each laboratory. More detailed descriptions of basic organic laboratory techniques are found in Zubrick ***(25)***.

2.3.1. Purification of Hexadecanethiol (HDT; 2 mM in ethanol)

Hexadecanethiol (HDT) (Aldrich) is purified by flash chromatography using hexane as the eluant or by distillation at reduced pressure. The major impurity is a disulfide. The R_f of the product is approx 0.4. The typical ^{1}H NMR spectrum has the following peaks: δ 1.25 (broad singlet, 29 H), 1.6 (quintet, 2 H), 2.5 (quartet, 2 H).

2.3.2. (1-mercaptoundec-11-yl)tri(ethylene glycol) (EGT; 2 mM in ethanol) *(26–28)*

Reaction mixtures are concentrated by rotary evaporation at reduced pressure. The purification of the final product and of the intermediates is carried out using flash column chromatography with silica gel and 98:2 dichloromethane:methanol as the eluant; typical values of R_f are provided.

2.3.2.1. Undec-1-en-11-yl(triethyleneglycol)

Mix 0.34 mL (4.3 mmol) of 50% aqueous sodium hydroxide with 3.2 g (21 mmol) of tri(ethylene glycol) (Aldrich) and stir for 30 min in an oil bath at 100°C. Add 1 g (4.3 mmol) of 11-bromoundec-1-ene (Pfaltz and Bauer) and stir at 100°C for 24 h under argon. After cooling, the reaction mixture is extracted six times with hexane (50–100 mL aliquots), and dried with sodium sulfate (Aldrich). Combine the hexane portions, concentrate them, and purify the resulting yellow oil (R_f = 0.3): a typical yield is approx 70%. ^{1}H NMR (250 MHz) δ 1.2 (broad singlet, 12 H), 1.55 (quintet, J = 7 Hz), 2.0 (quartet, 2 H, J = 7 Hz), 2.7 (broad singlet, 1 H), 3.45 (triplet, 2 H, J = 7 Hz), 3.5–3.8 (multiplet, 12 H), 4.9–5.05 (multiplet, 2 H), 5.75–5.85 (multiplet, 1 H).

2.3.2.2. [1-[(Methylcarbonyl)thio]undec–11yl]tri(ethylene glycol)

Dissolve 0.6 g (2 mmol) of the previous compound in 20 mL of freshly distilled THF; add 10 mg of recrystallized 2, 2'-azobisisobutyronitrile (Aldrich) and 1.4 mL (20 mmol) of thiolacetic acid (Aldrich) and irradiate for 6–8 h with a 450-W medium-pressure mercury lamp (Ace Glass) filtered through Pyrex. Check that the reaction has reached completion before work-up. Take out a 0.1-mL aliquot, reduce under pressure and take an NMR spectrum. The signal from the protons of the alkene group at $\delta = 4.8–6$ ppm should disappear if the reaction has gone to completion. Concentrate the reaction mixture and purify ($R_f = 0.3$): a typical yield is approx 80%. ^{1}H NMR (250 MHz) δ 1.2 (broad singlet, 14 H), 1.6 (multiplet, 4 H), 2.3 (singlet, 3 H), 2.85 (triplet, 2 H, $J = 7$ Hz), 3.45 (triplet, 2 H, $J = 7$ Hz), 3.5–3.75 (multiplet, 12 H).

2.3.2.3. (1-Mercaptoundec–11-yl)tri(ethylene glycol)

Dissolve 0.4 g (1 mmol) of the previous compound in 2 mL of freshly distilled dichloromethane and 8 mL of degassed (argon or nitrogen for 30 min) methanol. Add 0.9 mL (1.2 mmol) of 1.3 *M* sodium methoxide (Aldrich) in degassed methanol. After 45 min, bring the reaction mixture to neutral pH using DL-camphor-10-sulfonic acid (Aldrich), concentrate and purify ($R_f = 0.25$); a typical yield is 50 %. ^{1}H NMR (250 MHz) δ 1.1 (broad singlet, 14 H), 1.2 (triplet, 1 H, $J = 7$ Hz), 1.5 (multiplet, 4H), 2.3 (singlet, 3 H), 2.5 (quartet, 2 H, $J = 7$ Hz), 3.0 (broad singlet, 1 H), 3.4 (triplet, 2 H, $J = 7$ Hz), 3.5–3.75 (multiplet, 12 H).

3. Methods

3.1. Micropattern Stamping Procedure

1. Lay substrate on clean and flat surface, with gold facing upwards. Take care not to scratch the surface with sharp forceps, or by placing substrate upside down. If there is dust visible on the substrate, blow gently with pressurized air or nitrogen.
2. Rinse the PDMS stamp with ethanol and blow off vigorously with a stream of pressurized air or nitrogen for at least 10 s. If any dust remains on the stamp, repeat this procedure.
3. Dip a Q-tip into a 2-m*M* solution of hexadecanethiol in ethanol and gently paint a layer of the solution onto the PDMS stamp. Use a stream of air or nitrogen to gently evaporate the ethanol off the stamp.
4. Gently place the stamp face down onto the gold-coated substrate. Allow the stamp to adhere. This step may require putting gentle pressure on the stamp to press it against the substrate. Let the fully adhered stamp remain on the substrate for at least 10 s.
5. Using forceps, gently peel away the stamp from the substrate, being certain not to smear the stamp against the substrate or to let the stamp re-adhere to the substrate.

6. Return to **step 2** to continue stamping more substrates. After all substrates are stamped, proceed to **step 7**.
7. Using a Pasteur pipet, deliver a solution of EGT dropwise onto each substrate until the liquid covers it entirely. This usually requires approx 0.5–1 mL per square inch of substrate. Incubate with EGT for 30 min.
8. Using forceps cleaned with ethanol and blown dry, grasp the corner of the substrate and rinse with a stream of ethanol on both sides of the pattern for 20 s. Place the substrate on a clean surface, and rinse the forceps with ethanol. Grasp the substrate again in a different location and rinse again with ethanol to wash the area previously masked by the forceps.
9. Blow the ethanol off the substrate with pressurized air or nitrogen.
10. The stamped substrates should be placed into containers taking care not to allow the patterned surface to rub against coarse surfaces. They are stored under nitrogen gas in a cool, dark location. Place the containers in ziplock bags that are filled with nitrogen.

3.2. Coating Stamped Substrates with ECM Proteins and Plating Cells

1. To coat the substrates with ECM, make a solution of the protein (25 μg/mL) in PBS. Typically, 0.25-mL solution per square inch of substrate is sufficient.
2. Place a 0.25-mL drop of ECM solution onto bacteriological Petri dishes or another disposable hydrophobic surface. Float each substrate, with patterned side face down, onto the drops. Let sit 2 h at room temperature.
3. After 2 h, add a large amount (5–15 mL) of 1% bovine serum albumin (BSA, Fraction V, Sigma, St. Louis, MD) dissolved in PBS directly to dish. Remove substrates and place directly into plating medium (remember to flip slide so pattern is facing up again).
4. Plate cells directly on substrates using standard experimental technique.

4. Notes

1. Gold Slide: Choice of glass: We find No. 2 glass cover slips to be less likely to break than No. 1, although at the same time not too heavy to pick up with forceps. We have successfully used standard histology mounting slides as well.
2. Gold Slides: Evaporation: We recommend using an evaporator rather than a sputter coating system to coat the substrates for several reasons. Most sputterers are single source, and are impractical for coating two different metals (Ti and Au) on a substrate. Sputtering also gives less homogeneous films that would require an additional annealing step to correct. And last, sputtering systems generally produce films with higher quantities of metal oxides and other impurities that would interfere with the generatin of the SAM surface.
3. Gold Slides: Storage: Typically, gold-coated substrates become "mottled" after 4–5 wk and are no longer deemed suitable for experiments; streaks with heterogenous transparency develop (they are obvious to the naked eye). This may be

caused by rearrangements in the thickness of the gold layer that are related to impurities present on the glass before evaporation of the gold. Gold substrates that are stamped immediately after evaporation are generally more stable over time (about 3 mo) perhaps because the SAM acts as a resist against impurities ***(29)***.

4. Wafers: Rinse only with water and avoid all contact with organic solvents.
5. PDMS Stamp: Pouring and curing: Often small air bubbles form in the PDMS after it is poured on the master. Cover the dish and gently tap it to allow the bubble to diffuse out of the prepolymer. Typically, stamps are 0.5–1 cm tall.
6. PDMS Stamp: Peeling off of silicon: During curing, a layer of PDMS forms underneath the wafer and holds it to the dish. Invert the dish and gently press on the bottom side of it until the cured PDMS dewets from the surface of the dish. Invert the dish and use a dull edge to trace the contour of the PDMS so as to lift it off the dish. Often the PDMS remains attached to the wafer. Carefully cut the layer of PDMS found under the wafer and gently peel the two surfaces away from each other.
7. Thiol Storage: Typically, alkanethiols that are kept in ethanolic solutions for more than three month become oxidized and form significant amounts of disulfides. Disulfides of EGT are detected by TLC as spots with an R_f of approx 0.15, whereas the thiol has an R_f of 0.25 (using 98:2 CH_2Cl_2:CH_3OH as the eluant). By NMR, disulfides can be distinguished from alkanethiols by the presence of a triplet of peaks (from the methylene group adjacent to the sulfur atom) at approx 2.6 ppm instead of a quartet at 2.5 ppm. Although disulfides are known to form SAMs with interfacial properties similar to those formed with alkanethiols, their assembly is 75 times slower ***(30)***.
8. Thiol Stamping: Over and understamping. Observe the play of light, at an angle, on the micropattern to ensure that the stamp has fully adhered to the substrate. Usually, a pink color will ensure that full adhesion has occurred. Both under and overstamping results in a loss of this interference pattern. Make sure that no patches of nonadhesion remain. Always stamp the hexadecanethiol; stamping the EGT results in less efficient pattern transfer, incomplete formation of SAM, and nonspecific adsorption of proteins in the EGT regions.
9. Protein coating: When adding the BSA solution to bring the substrates out of fibronectin coating, the slides sink onto the dish and adhere to it; since the substrates face the bottom of the dish, the pattern may be damaged. To avoid this, add the BSA solution around the edges of the substrate so that the slides remain afloat.
10. Cell culture: For cell culture, cells should ideally be plated without serum (at least for 1 h), before washing serum media back in. The serum activates certain cell types to attach where they should not. This is cell-type specific, so some exploration may be necessary.
11. Cell culture: The actual surface area that is adhesive on patterned slides is a fraction of that of a regular substrate; therefore, cells should be accordingly seeded at seemingly very low concentrations.

Acknowledgment

The authors would like to thank Drs. Rahul Singhvi, Milan Mrksich, and Laura Dike for their dedication to the development of these techniques. This work was supported by NIH grants CA55833, HL57699, the Defense Advanced Research Projects Agency, and the National Science Foundation.

References

1. Chen, C. S., Mrksich, M., Huang, S., Whitesides, G. M., and Ingber, D. E. (1997) Geometric control of cell life and death. *Science* **276,** 1425–1428.
2. Chen, C. S., Mrksich, M., Huang, S., Whitesides, G. M., and Ingber, D. E. (1998) Micropatterned surfaces for control of cell shape, position, and function. *Biotech. Prog.* **14,** 356–363.
3. Singhvi, R., Kumar, A., Lopez, G., Stephanopoulos, G. N., Wang, D. I. C., Whitesides, G. M., and Ingber, D. E. (1994) Engineering cell shape and function. *Science* **264,** 696–698.
4. Hammarback, J. A., Palm, S. L., Furcht, L. T., and Letourneau, P. C. (1985) Guidance of neurite outgrowth by pathways of substratum-adsorbed laminin. *J. Neurosci. Res.* **13,** 213–220.
5. Healy, K. E., Lom, B., and Hockberger, P. E. (1994) Spatial distribution of mammalian cells dictated by material surface chemistry. *Biotech. Bioeng.* **43,** 792–800.
6. Letourneau, P. C. (1975) Cell-to-substratum adhesion and guidance of axonal elongation. *Devel. Biol.* **44,** 92–101.
7. O'Neill, C., Jordan, P., and Riddle, P. (1990) Evidence for two distinct mechanisms of anchorage stimulation in freshly explanted and 3T3 swiss mouse fibroblasts. *J. Cell Sci.* **95,** 577–586.
8. Bhatia, S. N., Toner, M., Tompkins, R. G., and Yarmush, M. L. (1994) Selective adhesion of hepatocytes on patterned surfaces. *Ann. NY Acad. Sci.* 745 , 187–209.
9. Matsuda, T. and Sugawara, T. (1995) Development of surface photochemical modification method for micropatterning of cultured cells. *J. Biomed. Mat. Res.* **29,** 749–756.
10. Stenger, D. A., Georger, J. H., Dulcey, C. S., Hickman, J. J., Rudolph, A. S., et al. (1992) Coplanar molecular assemblies of amino-and perfluorinated alkylsilanes: characterization and geometric definition of mammalian cell adhesion and growth. *J. Am. Chem. Soc.* **114,** 8435–8442.
11. Britland, S., Clark, P., Connolly, P., and Moores, G. (1992) Micropatterned substratum adhesiveness: a model for morphogenetic cues controlling cell behavior. *Exp. Cell Res.* **198,** 124–129.
12. Kleinfeld, D., Kahler, K. H., and Hockberger, P, E. (1988) Controlled outgrowth of dissociated neurons on patterned substrates. *J. Neurosci.* **8,** 4098–4120.
13. Whitesides, G. M. and Gorman, C. B. (1995) Self-assembled monolayers: Models for organic surface chemistry, in *Handbook of Surface Imaging and Visualization* (Hubbard, A. T., ed.), CRC, Boca Raton, FL, pp. 713–732.

14. Tidwell, C. D., Ertel, S. I., and Ratner, B. D. (1997) Endothelial cell growth and protein adsorption on terminally functionalized, self-assembled monolayers of alkanethiolates on gold. *Langmuir* **13,** 3404–3413.
15. Roberts, C., Chen, C., Mrksich, M., Martichonok, V., Ingber, D., and Whitesides, G. M. (1998) Using mixed self-assembled monolayers presenting RGD and $(EG)_3OH$ groups to characterize long-term attachment of bovine capillary endothelial cells to surfaces. *J. Am. Chem. Soc.* **120,** 6548–6555.
16. Rao, J. and Whitesides, G. M. (1997) Tight binding of a dimeric derivative of vancomycin with dimeric l-Lys-d-Ala-d-Ala. *J. Am. Chem. Soc.* **119,** 10,286–10,290.
17. Wilbur, J. L., Kim, E., Xia, Y., and Whitesides, G. (1995) Lithographic molding: a convenient route to structures with sub-micrometer dimensions. *Adv. Mat.* **7,** 649–652.
18. Mrksich, M., Dike, L. E., Tien, J., Ingber, D. E., and Whitesides, G. M. (1997) Using microcontact printing to pattern the attachment of mammalian cells to self-assembled monolayers of alkanethiolates on transparent films of gold and silver. *Exp. Cell Res.* **235,** 305–313.
19. López, G. P., Biebuyck, H. A., Härter, R., Kumar, A., and Whitesides, G. M. (1993) Fabrication and imaging of two-dimensional patterns of proteins adsorbed on self-assembled monolayers by scanning electron micorscopy. *J. Am. Chem. Soc.* **115,** 10774–10781.
20. López, G. P., Albers, M. W., Schreiber, S. L., Carroll, R. W., Peralta, E., and Whitesides, G. M. (1993) Convenient methods for patterning the adhesion of mammalian cells to surfaces using self-assembled monolayers of alkanethiolates on gold. *J. Am. Chem. Soc.* **115,** 5877–5878.
21. Mrksich, M., Chen, C. S., Xia, Y., Dike, L. E., Ingber, D. E., and Whitesides, G. M. (1996) Controlling cell attachment on contoured surfaces with self-assembled monolayers of alkanethiolates on gold. *Proc. Natl. Acad. USA* **93,** 10,775–10,778.
22. DiMilla, P., Folkers, J. P., Biebuyck, H. A., Harter, R., Lopez, G., and Whitesides, G. M. (1994) Wetting and protein adsorption of self-assembled monolayers of alka nethiolates supported on transparent films of gold. *J. Am. Chem. Soc.* **116,** 2225–2226.
23. Xia, Y., Mrksich, M., Kim E., and Whitesides, G. M. (1995) Microcontact printing of siloxane monolayers on the surface of silicon dioxide, and its appliction in microfabrication. *J. Am. Chem. Soc.* **117,** 9576–9577.
24. Madou, M. (1997) In, *Fundamentals of Microfabrication*, 1st ed. CRC, New York, pp. 1–52.
25. Zubrick, J. W. (1992) *The Organic Chem Lab Survival Manual*, 3rd ed. Wiley, New York.
26. Pale-Grosdemange, C., Simon, E. S., Prime, K. L., and Whitesides, G. M. (1991) Formation of self-assembled monolayers by chemisorption of derivatives of oligo(ethylene glycol) of structure $HS(CH_2)_{11}(OCH_2CH_2)_mOH$ on gold. *J. Am. Chem. Soc.* **113,** 12–20.
27. Prime, K. L. and Whitesides, G. M. (1991) Self-assembled organic monolayers: model systems for studying adsorption of proteins at surfaces. *Science* **252,** 1164–1167.

28. Prime, K. L. and Whitesides, G. M. (1993) Adsorption of proteins onto surfaces containing end-attached oligo(ethylene oxide): a model system using self-assembled monolayers. *J. Am. Chem. Soc.* **115,** 10714–10721.
29. Zhao, X. M., Wilbur, J. L., and Whitesides, G. M. (1996) Using two-stage chemical amplication to determine the density of defects in self-assembled monolayers of alkanethiolates on gold. *Langmuir* **12,** 3257–3264.
30. Bain, C. D., Biebuyck, H. A., and Whitesides, G. M. (1989) Comparison of self-assembled monolayers on gold: coadsorption of thiols and disulfides. *Langmuir* **5,** 723–727.

18

Methods for Preparing Extracellular Matrix and Quantifying Insulin-like Growth Factor-Binding Protein Binding to the ECM

Bo Zheng and David R. Clemmons

1. Introduction

The insulin-like growth factor-binding proteins (IGFBPs) are carrier proteins that are present in all extracellular fluids *(1)*. Similarly, insulin-like growth factor-I (IGF-I) and -II are synthesized by connective tissue cells that are present within the stroma of all tissues, and therefore their presence is ubiquitous in all organs *(2)*. IGF receptors are present in all tissues, which accounts for balanced growth of organs, and deletion of the *IGF-I* gene results in a balanced decrease in organ growth, whereas overexpression of *IGF-I* results in a symmetric increase in growth *(3)*.

The principle role of the IGFBPs is to determine the distribution of IGF-I and -II among cells and tissues and to provide controlled access to receptors *(1)*. Because there is excess binding capacity in all interstitial fluids, all the IGF that circulates is in a bound form. Because protein-bound IGF-I and -II cannot bind to their receptors, mechanisms need to exist to release IGF-I and -II under controlled conditions. Three events have been described that result in controlled release. Because in all cases, the affinity of unmodified forms of IGFBPs in solution is higher than the receptor, these modifications must be ones that result in a reduction in IGFBP affinity in order to result in a more favorable equilibrium. These include proteolysis, binding to cell surfaces or extracellular matrix (ECM), and phosphorylation or dephosphorylation. This chapter concerns association of IGF binding proteins with ECM. ECM binding results in an 8–15-fold reduction in affinity of IGFBP-5 for IGF-I or -II, and therefore it is physiologically significant *(4)*.

From: *Methods in Molecular Biology, vol. 139: Extracellular Matrix Protocols*
Edited by: C. Streuli and M. Grant © Humana Press Inc., Totowa, NJ

When the tissue distribution of IGFs is analyzed by immunohistochemistry, a substantial amount of IGF-I and -II are found bound to ECM. Therefore, it was of interest to determine whether IGFBPs were also present in the ECM. Preparation of fibroblast ECM with immunohistochemical staining showed that, of the three forms of IGFBPs synthesized by human fibroblasts (e.g., IGFBP-3, -4, and -5), only IGFBP-5 was abundant within the ECM *(4)*. IGFBP-3, while being a principle-cell synthetic product (i.e., these cells synthesize approximately 20-fold more IGFBP-3 than IGFBP-5), was barely detectable within the ECM; in contrast, IGFBP-5 was abundant. Other connective tissue cells, such as osteoblasts, chondrocytes, and smooth muscle cells, have also been shown to deposit IGFBP-5 in their ECM. Incubation of IGFBP-1, -2, -4, and -6 with fibroblast ECM showed that they did not adhere to it. The same appears to be true for chondrodyte and osteoblast ECM *(5)*. Under specialized conditions, IGFBP-2 can adhere to fibroblast and/or smooth muscle cell ECM *(6)*.

Because of its abundance by immunohistochemical staining, the affinity of IGFBP-5 for ECM was determined and was calculated to be approx 3×10^{-8} L/M. Because IGFBP-5 was so much more abundant than other IGFBPs within ECM, we determined the physiologic significance of this interaction. ECM was prepared and IGFBP-5 layered onto the ECM. Its affinity for IGF-I was shown to be reduced by approximately eightfold, and enrichment of the ECM with IGFBP-5 resulted in a 2.1-fold potentiation of the cellular growth response to IGF-I *(4)*.

Because of its importance in modulating IGF actions, we wished to determine the structural features of IGFBP-5 that mediated binding to ECM. To do this, ECM was prepared, and synthetic peptides containing regions of IGFBP-5 with large numbers of charged residues were screened for their capacity to compete with the native peptide for binding to ECM. It was shown that a region, termed "peptide A," that contained amino acids 201–218 had the greatest binding activity *(5)*. A second region, between amino acids 131 and 141, was also active in mediating binding to either fibroblast or smooth muscle cell ECM. To further identify the specific amino acids that were important, site-directed mutagenesis was utilized to convert the charged residues within each of these peptides to neutral residues, and then the ability of each of these mutants to bind to ECM was ascertained. This technology was also used to determine constitutive incorporation of IGFBP-5 into matrix during matrix assembly by transfecting cDNAs that encoded ECM binding deficient mutant forms, preparing ECM and quantifying the abundance of IGFBP-5 within that ECM *(6)*. These studies showed that amino acids R207 and R214 were the most important in terms of ECM binding. However, K217 and R218 also mediated binding, as did K202.

In summary, IGFBP-5 association with ECM is an important means of potentiating the cellular response to IGF-I. Sequestering large amounts of IGF-I within the ECM may be an important component of the response to injury when an advanced proliferative activity under IGF-I stimulation is required *(7)*. Likewise, during normal, balanced growth, provision of large quantities of IGFBP-5 in the ECM could provide an important source of growth factor during times of decreased synthesis, when balanced growth is still required. Furthermore, because epithelial cells do not have a large amount of ECM, connective-tissue-cell matrix may provide an important source of IGF for regenerative-epithelial tissue that is rapidly dividing, such as gastrointestinal epithelium. The techniques that have been worked out to isolate ECM *(8)* and then quantify IGF binding to ECM that are detailed below have been instrumental in proving these points.

Under specialized circumstances, IGFBP-2 has also been shown to associate with ECM. Preparation of ECM and layering of IGFBP-2 shows that it is not preferentially bound. However, if high concentrations of IGF-I are added that result in a confirmational change in IGFBP-2, this altered conformation can bind to ECM. This may be important in focally concentrating large amounts of IGF in ECM of tissues that synthesize IGFBP-2. Because IGFBP-2 is an important product of epithelial cells, and because epithelial-cell basement membranes often contain IGFBP-2, this may be a particularly important means for focally concentrating IGFs near epithelial tissue. This has been shown to be very important in the olfactory tissue within the brain, wherein large amounts of IGF-I and IGFBP-2 are synthesized and large amounts can be seen focally concentrated in the ECM of glial cells.

2. Materials

2.1. Preparation of Porcine Smooth Muscle ECM

1. Dulbecco's modified Eagle's medium with high glucose (DMEM-H) (Life Technologies, Grand Island, NY).
2. Fetal bovine serum (FBS) (Life Technologies).
3. Cell scrapers.
4. Laemmli sample buffer without glycerol.

2.2. ^{125}I-IGF BP-5 Binding Assay to pSMC ECM

1. Immobilon membranes (Millipore, Bedford, MA).
2. 30 m*M* sodium phosphate containing 0.2% Triton X-100, pH 7.4.
3. Slot blotter apparatus model 2643 from Bethesda Research Laboratories (Gaithersburg, MD).
4. 0.1 *M* Tris, pH 7.4.
5. Methanol.

6. Tris-buffered saline containing 3% BSA.
7. IGFBP-5.
8. Iodobeads (Pierce, Rockford, IL).
9. 0.25 *M* NA_2PO_4, pH 7.2.
10. Sephadex G-100.
11. Gamma spectrometer.

2.3. Procedures to Verify that the ECM has been Extracted

1. Biomax membranes (Milipore).
2. Glycerol.
3. SDS-PAGE equipment.
4. Vitronectin.
5. Anti-vitronectin antiserum (Sigma).

2.4. Preparation of Fibroblast (GM-10) ECM

1. Human dermal fibroblasts (GM-10) from Coriell Institute (Camden, NJ).
2. Primaria 24-well plates (Falcon Plastics, Becton Dickinson, Rutherford, NJ).
3. Lysis buffer: 0.5% triton X-100 in PBS, 10 m*M* EDTA, pH 7.4.
4. 25 m*M* ammonium acetate, pH 9.0.
5. 1 *M* HEPES, pH 7.3.
6. 30 m*M* sodium phosphate buffer with 1% BSA, pH 7.4.

2.5. Binding of ^{125}I-IGFBP-5 on Unlabeled IGFBP-5 to Fibroblast ECM

1. IGFBP-5.
2. 30 m*M* sodium phosphate buffer, 0.1% BSA, pH 7.4.
3. 0.3 *N* NaOM.
4. Immobilon membranes.

3. Methods

3.1. Preparation of Porcine Smooth Muscle Cell ECM

pSMCs do not synthesize abundant ECM; therefore, the ECM cannot be seen under the microscope. Extracting the ECM is thus a difficult and delicate procedure. Bear in mind that ECM does not remain adherent to culture plates for more than a few weeks, even in optimal storage. For more information about the properties of pSMC ECM, *see* **Note 1**.

1. Porcine aortic smooth muscle cells prepared from 3-wk-old pigs as described by Ross *(9)*.
2. Porcine smooth muscle cells (pSMC) in DMEM-H supplemented with 10% FBS, penicillin at 100 U/mL, and streptomycin at 100 μg/mL (Life Technologies).
3. Subculture the cells in 10-cm plates (Falcon #3001) until confluent.
4. Grow pSMC in 10-cm culture dishes (Falcon 3001) until they are densely confluent. Change the media the day before the ECM is prepared.

5. Wash the plates four times with serum-free DMEM-H. Incubate the plates at room temperature with 2 mL/plate of 2 *M* urea (freshly prepared) for about 15–20 min. Check the plates under the microscope: The cells should be visible against a dark surface. The cells should have a shrunken and collapsed morphology.
6. The media is then aspirated. Use a cell scraper to gently remove the cells from the plate, and pay special attention to the cells around the edges. It is important to remove as many cells as possible, but it also important to be gentle enough to leave the ECM intact.
7. Wash the plate 2–3 times gently with serum-free DMEM-H. Add 0.5 mL 0.5X Laemmli sample buffer without glycerol to each 10-cm plate. Swirl the plate to get complete exposure. Remove the ECM with a cell scraper and transfer it to a 1.7-mL Eppendorf microfuge tube. Wash the plate again with 0.5 mL 0.5X Laemmli buffer to get a second extract, which can be used to test whether or not the ECM has been successfully extracted with urea.
8. Spin the ECM in a microfuge at 14,000*g* for 15 min at –4°C to remove cellular debris. Measure the protein concentrations and store the samples in liquid nitrogen until they are used (preferably less than 3 wk).

3.2. ^{125}I-IGFBP-5 Binding Assay to pSMC ECM

This method is used to determine the binding affinity of ECM proteins for IGFBP-5 and the capacity of various test substances to alter total IGFBP-5 binding to pSMC/ECM (*see* **Note 2**).

1. Dilute the ECM that has been prepared as previously described 1:5 with distilled water.
2. Prepare a piece of Immobilon P membrane, 11 × 3 cm, and cut a piece of 3M Whatman filter paper to the same size.
3. Wet the membrane in methanol and wash in 30 m*M* sodium phosphate containing 0.2% Triton X-100, pH 7.4. Wash the filter paper in the same buffer. Load the slot blot apparatus with the Immobilon P membrane and filter paper.
4. Load 170 µL of diluted ECM into each well. Suction transfer the ECM onto the membrane. Wash each well with 0.1 *M* Tris, pH 7.4 and suction transfer again until the wells are empty.
5. Remove the membrane from the blotting apparatus and allow it to dry on a piece of filter paper. Wet the blots with methanol and wash in TBS for 5 min at room temperature. Block the membranes for 5 h at room temperature in TBS containing 3% BSA. Rinse the membrane in 30 m*M* sodium phosphate with 0.2% Triton X-100 twice for 50 min. Cut the membrane into pieces that each contain four separate ECM extracts.
6. Dilute unlabeled IGFBP-5 with a 30 m*M* sodium phosphate buffer . The recommended concentrations are 0, 50, 100, 500, and 1000 ng/mL. Add to a final volume of 2 mL of 30 mM sodium phosphate buffer. If testing mutants, prepare tubes with identical concentrations. Prepare ^{125}I-labeled IGFBP-5 (specific activity 30 µCi/µg) by mixing 0.5 mCi of NaI, 2.0 µg of IGFBP-5, and one iodobead for 10 min in 0.5 mL of 0.25 *M* Na_2PO_4 (pH 7.2). After 10 min at 22°C,

the bead is removed and the mixture purified by Sephadex G-100 chromatography. Add about 80,000 mL ^{125}I-IGFBP-5 to the solutions containing each concentration of unlabeled IGFBP-5 in a final incubation volume of 2.0 mL.

7. Incubate the membranes with the appropriate treatments overnight on a shaker at 4°C. Wash the membranes three times for 10 min each with 30 m*M* sodium phosphate containing 0.2% Triton X-100, pH 7.4, at 4°C. Allow membranes to dry.
8. Cut the two slot sections from the membranes and place each 2 slot section in 12 × 75 mm polystyrene tube. Count the bound ^{125}I in a gamma spectrometer. Plot the results to determine competition. Nonspecific binding is determined by subtracting the counts per minute bound in the presence of 50 µg/mL unlabeled IGFBP-5. This value should be <15% of the total binding and should be subtracted from each data point.

3.3. Procedures to Verify that the ECM has been Extracted

Save the primary urea extract and secondary Laemmli sample buffer extract and test them in the following manner.

1. Western ligand blotting of IGFBP-5: Concentrate 0.5 mL of sample on a Millipore Ultrafree 0.5-centrifuge filter Biomax 10k NMWL membrane to 40 µL. Add the 40-µL sample to 10 mL of 100% glycerol, and then add to a 12.5% SDS polyacrylamide gel. After the separation, transfer the proteins to an Immobilon membrane. Incubate the filter for 14 h with ^{125}I-IGF-I (specific activity 125 µCi/µg). Add 500,000 cpm to 4 mL of blotting buffer (*see* **ref. *(10)*** for the buffer composition). The filter is washed three times, and the bound ^{125}I-IGF-I visualized by autoradiography. A band with an Mr estimate of 30–32 kDa should be seen. That this is IGFBP-5 can be confirmed by immunoblotting ***(11)***.
2. To test the ECM extract binding to IGFBP-5, using a slot blotter apparatus, concentrate 170 µL of the ECM extract on an Immobilon filter, as described previously. Unlabeled IGFBP-5 should compete with ^{125}I-IGFBP-5 for binding (e.g., >50% reduction in binding) when 500 ng/mL is added. The ^{125}I-IGFBP-5 binding assay protocol is outlined above.
3. Immunoblot the concentrated ECM for vitronectin (*see* **Note 3**). pSMC ECM contains a high concentration of vitronectin. Concentrate 170 µL of ECM extract onto an Immobilon membrane using the slot-blot apparatus. Be sure to run a vitronectin standard (0.2 µg) with this procedure. Incubate the membrane with a 1:1000 dilution of antirabbit and antihuman vitronectin antiserum. Visualize the immune complex as previously described ***(11)***.

3.4. Preparation of Fibroblast (GM-10) ECM

1. Human fetal dermal fibroblasts (GM-10) are maintained in Eagles Minimal Essential Media (EMEM, Life Technologies) supplemented with 10% fetal calf serum (Life Technologies), 110 µg/mL pyruvate, 30 µg/mL asparagine, 21 µg/mL serine, 100 U/mL penicillin, and 100 µ/mL streptomycin.
2. The cells are plated on positively charged 24 Primaria well plates and grown for 7–10 d until confluence.

3. Aspirate the media and rinse the plates twice with PBS.
4. Incubate the cells with 0.3 mL lysis buffer for 2–5 min until the cells lyse and nuclei are no longer apparent by microscopy. Usually this does not take more than 5 min. Aspirate the lysis buffer, gently rinse the cells twice with PBS, and aspirate.
5. Incubate the cells in each well with 0.3 mL of 25 m*M* amonium acetate, pH 9.0, for 3–5 min to remove the remaining nuclei and cytoskeletal elements. Observe the plates microscopically and do not extract for too long a time period.
6. Add 3 drops 1 *M* HEPES, pH 7.3, to neutralize pH and then aspirate. Gently rinse the cells twice with PBS, once with 30 mM sodium phosphate buffer with 1% BSA, pH 7.4.
7. Check the plate under the microscope. Intact ECM should be visible, densely striated with some degree of variation.

To avoid matrix lifting, use cold reagents and keep ECM on ice at all times.

3.5. Binding of ^{125}I-IGFBP-5 or Unlabeled IGFBP-5 to Fibroblast ECM

The purpose of this method is to determine the binding of IGFBP-5 to fibroblast ECM, to quantify the ability of the substances to compete with IGFBP-5 for binding, and to determine the effect of test substances added to cultures to alter the binding of IGFBP-5 to ECM.

1. Prepare ECM according to the previous protocol. Add unlabeled IGFBP-5 peptide in increasing amounts (0.04–2.0 µg/mL in 30 m*M* sodium phosphate buffer, 0.1% BSA, pH 7.4, total volume 500 µL) to duplicate culture wells. Incubate on ice for 30 min to allow IGFBP-5 to bind to ECM. Add 250,000 cpm of IGFBP-5 prepared as described previously to each well. After 14 h at 4°C with gentle agitation, aspirate the media and rinse each well with 30 mM sodium phosphate buffer twice.
2. Extract the ECM off the plate with 0.5 mL of 0.3 *N* NaOH twice and determine the amount of ^{125}I-IGFBP-5 directly by gamma counting. In the experiments measure the capacity of unlabeled IGFBP-5 to bind to the ECM, prepare the fibroblast ECM extracts as described above, add the substances to be tested to duplicate wells containing 0.5 µL of the same buffer described previously. Add 40 ng of unlabeled IGFBP-5 to each well and incubate overnight. Wash the plates as described previously. Extract the ECM with 50 µL of 1X Laemmli sample buffer. Remove insoluble material by centrifuging at 15,000*g* for 10 min. Heat the supernatants to 95°C for 10 min.
3. Electrophorese the supernatants through 12.5% SDS polyacrylamide gel, transfer to Immobilon PDQ (Millipore) membranes, and then analyze the protein extracts by Western ligand blotting, using ^{125}I-IGF-I. Determine band intensity by Phosphor Image analysis (Molecular Dynamics, Sunnyvale, CA). To confirm that the band that binds ^{125}I-IGF-I is IGFBP-5, the extracts can be analyzed by immunoblotting using a 1:1000 dilution of anti-IGFBP-5 antiserum.

4. Notes

1. Experience gained with these technologies has shown that several steps in the methodology require close observation in order to obtain optimal results. In preparing fibroblast ECM, the monolayers need to be observed during the removal of cells and cytoskeletal elements. Failure to observe the cultures microscopically will result in either inadequate removal of cell bodies and nuclei or extensive removal such that most of the ECM is lost. This technique is an approximation at best, and therefore one is always attempting to strike a balance between adequate removal of the non-ECM components and retention of sufficient ECM to complete the studies. The aim is to be able to obtain a preparation of ECM that retains a protein composition that is similar to the native ECM as well as protein folding and conformational states such that binding activities are not completely destroyed. The success to which this can be undertaken is dependent upon the amount of ECM produced per cell. For example, in tissues such as fibroblasts in culture, there is a large amount of ECM produced per cell, so a large amount of matrix can be obtained with successful removal of cell bodies and cytoskeletal elements. However, for smooth muscle cells, there is much less matrix produced per cell; therefore, no method will entirely remove all the cells without removing some of the ECM, and the cultures need to be observed during the extraction procedure. In summary, one must always strike a balance between removal of cellular material and retention of ECM. This can only be done by close, microscopic observation during the extraction procedures.
2. In some cases, such as analysis of binding properties of smooth muscle cell ECM proteins, it may be necessary to perform the binding assays using a matrix other than a plastic culture dish. The reason for this is that all tissue culture plastic is prepared with highly charged polyanionic substances in order to facilitate cell attachment. These charged substances will bind proteins that are added exogenously quite avidly. Therefore, if a large quantity of ECM is present, it will nearly cover and neutralize the polyanionic charges. However, if part of the ECM is removed during cell removal, such as in preparation of smooth muscle cell ECM, then the nonspecific binding to the tissue culture plastic may be so great that one cannot accurately quantify binding to ECM. In order to obviate this problem, we developed a method of filter binding assays. While this method accurately quantifies binding to specific ECM proteins, it does not allow maintenance of these proteins in their natural, three dimensional conformation in which they are maintained in the ECM, since they must be extracted completely and then readhere to the membranes. However, using this technology, one can block nonspecific binding to the filter by pretreating the filters with high concentrations of BSA. Therefore, one can definitively demonstrate valid binding to ECM proteins by showing that only the slot that is blotted on the filter with ECM proteins will bind to a radiolabeled antigen (in our case radioiodinated IGFBP-5). Therefore, this methodology gives an approximation of the amount of ECM protein that is present and the binding capacity of those proteins that can be extracted for a particular radiolabeled substrate. Furthermore, it is quite is easy to correct for

nonspecific binding using this technology, and therefore the residual binding that can be measured is generally a valid assay. Attempts should be made during the binding assay to use appropriate concentrations of divalent cations, salt, and so on, to approximate physiologic conditions that would be present in the normal ECM *in situ*. Again, some assurance that the extracted matrix represents the true matrix as deposited by the cells can be obtained by comparing ratios of different matrix proteins within the extract to determine that they approximate the ratios of those proteins that are present *in situ* as the matrix is synthesized.

3. When conducting binding studies *in situ* using tissue culture plates, the remaining ECM after tissue extraction is often densely adherent to the plate. Therefore, in addition to adding harsh detergents such as SDS, one often needs to scrape the plate with a mechanical scraper in order to remove the remaining ECM proteins. To ensure that removal is complete, a reasonable technique is to re-extract the plate with a high concentration of SDS and then conduct immunoblotting for a common matrix protein such as vitronectin. This will ensure that the method is adequate for complete extraction of the proteins of interest.

Reference

1. Jones, J. I. and Clemmons, D. R. (1995) Insulin like growth factor and their binding proteins: biologic actions. *Endocrine. Rev.* **16,** 3–34.
2. Han, V. K. M., D'Ercole, A. J., and Lund, P. K. (1987) Cellular location of somatomedin (insulin-like growth factor) messenger RNA in the human fetus. *Science* **236,** 193–197.
3. Liu, J. P., Baker, J., Perkins, A. S., Robertson, E. J., and Efstratiadis, A. (1993) Mice carrying small mutations of the genes encoding insulin like growth factor I (IGF-I) and type 1 IGF receptor (IGF/r). *Cell* **75,** 73–82.
4. Jones, J. I., Gockerman, A., Busby, W. H., Camacho-Hubner, C., and Clemmons, D. R. (1993) Extracellular matrix contains insulin-like growth factor binding protein-**5,** Potentiation of the effects of IGF-I. *J. Cell Biol.* **121,** 679–687.
5. Mohan, S., Nakao, Y., Honda, Y., Landale, E. C., Leser, U., Dony, C., Lang, K., and Baylink, D. J. (1995) Studies on the mechanisms by which insulin-like growth factor (IGF) binding protein-4 (IGFBP-4) and IGFBP-5 modulate IGF actions in bone cells. *J. Biol. Chem.* **270,** 20,424–20,431.
6. Arai, A., Busby, W. H., and Clemmons, D. R. (1996) Binding of insulin-like growth factor I or II to IGF binding protein–2 enables it to bind to heparin or extracellular matrix. *Endocrinology* **137,** 4571–4575.
7. Khorsondi, M. J., Fagin, J. A., Ginnella-Neto, Forrester, J. S., and Cercek, B. (1992) Regulation of insulin-like growth factor I and its receptor in rat aorta after balloon degradation: Evidence for local bioactivity. *J. Clin. Invest.* **90,** 1926–1931.
8. Knutson, B. S., Haysel, P. C., and Nachman, R. (1988) L. Plasminogen activator is associated with the extracellular matrix of cultured bovine smooth muscle cells. *J. Clin. Invest.* **80,** 1082–1088.
9. Ross, R. (1971) The smooth muscle cell: Growth of smooth muscle in cultures and formation of elastic fibers. *J. Cell Biol.* **50,** 172–186.

10. Hossenlopp, P., Seurin, D., Segovia-Quinson, B., Hardouin, S., and Binoux, M. (1986) Analysis of serum insulin-like growth factor binding proteins using Western blotting: Use of the method for titration of the binding proteins and competitive binding studies. *Anal. Biochem.* **154,** 138–143.
11. Camacho-Hubner, C., Busby, W. H., McCusker, R. H., Wright, G., and Clemmons, D. R. (1992) Identification of the forms of insulin-like growth factor binding proteins produced by human fibroblasts and the mechanisms that regulate their secretion. *J. Biol. Chem.* **267,** 11,949–11,956.

19

Measuring Interactions Between ECM and TGFβ-Like Proteins

Sarah L. Dallas

1. Introduction

Recent advances in developmental biology have highlighted the importance of the synergistic relationships between structural components of the extracellular matrix (ECM) and growth factors that play fundamental roles in tissue morphogenesis and repair. Increasing evidence suggests that the binding of growth factors to the ECM is a major mechanism for regulation of growth factor activity (for review *see* **ref.** *(1)*. Association of growth factors with the ECM allows storage of large quantities of growth factors in a readily mobilized form, which could allow extracellular signaling to proceed rapidly in the absence of new protein synthesis. This may be particularly important in situations such as tissue repair following injury. Storage of growth factors in the ECM may also facilitate communication between cells that are widely separated in time, by allowing transmission of signals from one cell to a different cell that comes in contact with the same matrix later. Another intriguing possibility is that matrix-bound growth factors could act as a "memory," which provides information about the history of cellular activity of the tissue. Finally, it is possible that matrix-bound growth factors may actually signal differently than their soluble counterparts, for example by escaping internalization after binding to receptors, thus producing an extended signal. Thus matrix-bound growth factors may provide an important new dimension to cell–cell communication by greatly increasing the diversity of action of growth factors.

An example of a matrix protein that is important in growth factor regulation is the latent transforming growth factor beta binding protein-1 (LTBP1). LTBP1 is a member of an emerging superfamily of ECM proteins, which includes LTBPs 1, 2, 3, and 4 *(2–7)*. These proteins have homology to the

From: *Methods in Molecular Biology, vol. 139: Extracellular Matrix Protocols*
Edited by: C. Streuli and M. Grant © Humana Press Inc., Totowa, NJ

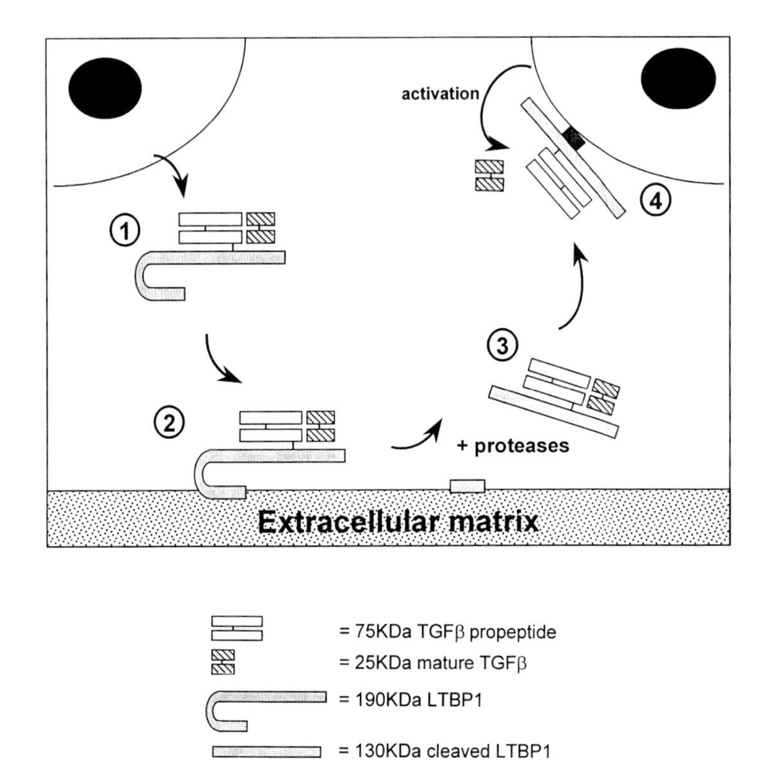

Fig. 1. Proposed model showing the role of LTBP1 in regulating TGFβ activity and availability. (1) Latent TGFβ consists of the mature 25 KDa TGFβ homodimer, which is cleaved from, but remains non-covalently associated with the 75 KDa propeptide. LTBP1 (190 KDa) is complexed with latent TGFβ inside the cell and facilitates its secretion. (2) LTBP1 acts as a carrier to transport latent TGFβ to the matrix for storage. (3) LTBP1 facilitates release of latent TGFβ from matrix by undergoing proteolysis, releasing an approx 130 KDa cleaved fragment of LTBP1 complexed to latent TGFβ. (4) This cleaved fragment of LTBP1, which is protase-resistant, protects the latent complex from activation until it reaches an appropriate target cell. LTBP1 may then play a further role in activation of latent TGFβ.

fibrillins and three of the family members are known to bind transforming growth factor betas (TGFβs) and influence their availability ***(2–4,6)***. LTBP1 is complexed with latent TGFβ prior to secretion and tightly regulates the activity of this growth factor at multiple levels (summarized in **Fig. 1**). LTBP1 facilitates secretion of latent TGFβ from the cell ***(8)***, and then provides a vehicle for storage of the growth factor in the matrix ***(9–11)***. Matrix-bound LTBP1 is highly insoluble because of the fact that it is crosslinked via the action of transglutaminase ***(12)***. However, LTBP1 has a proteinase sensitive "hinge" region, rich in proline and basic amino acids, suggesting that proteolytic cleavage may be the mechanism for release of LTBP1 from matrix. A number of in vitro studies have confirmed that LTBP1 (together with its bound latent TGFβ) can be released from matrix by various proteases ***(9–12)***. However, these studies have focused on examining release by addition of purified exogenous proteases. To date, no technique has been described for measuring release of LTBP1/latent TGFβ by cells during the process of matrix remodeling. Because the release of growth factors from matrix-bound stores may be a key step in regulating their activity, a method for examining this would be of great benefit to researchers in the field. The following account details an immunoprecipitation method for examining the release of matrix-bound LTBP1/latent TGFβ by cells during matrix remodeling. The example described examines the release of LTBP1/latent TGFβ by osteoclasts from matrix laid down by osteoblasts. However, the assay could readily be adapted to examine release of LTBP1/latent TGFβ by any one cell type from the ECM of any other cell type. Furthermore, as long as a good source of antibodies for immunoprecipitation is available, the assay could be modified to examine the cell-mediated release of other matrix-bound growth factors or matrix proteins from ECM.

The principle of the assay is summarized in **Fig. 2**. Primary osteoblast cells are grown in long-term culture to a stage when they are producing an extensive layer of ECM (10–14 d). The cells are then metabolically labeled for 48 h using ^{35}S-cysteine. This labels intracellular and secreted proteins, but also the ECM proteins. The cells are then lysed, leaving the insoluble matrix adhered to the culture plate. This radiolabeled matrix is extensively washed and dried. Avian osteoclast cells are then seeded onto the radiolabeled matrix. Release of LTBP1 and latent TGFβ by the osteoclasts during matrix remodeling is monitored by immunoprecipitating the labeled LTBP1 from the culture media (the LTBP1 antibodies will precipitate both free LTBP1 and LTBP1 complexed to latent TGFβ). Because the osteoclast cells were never exposed to ^{35}S-cysteine, it is assumed that any radiolabeled proteins present in the culture media must have been released from the matrix. Any labeled LTBP1 remaining bound in the matrix (i.e., not released by the cells) can also be measured by performing a plasmin digestion of the matrix to release LTBP1, followed by immunoprecipitation using anti-LTBP1. In this way, it should be possible to correlate release of LTBP1 and latent TGFβ by the osteoclasts into the culture media with a concomitant reduction in LTBP1 and latent TGFβ remaining in the matrix (for example, *see* **Fig. 3**).

2. Materials

Note: All chemicals used should be at least analytical grade. Where a specific manufacturer is used for a particular reagent, it is mentioned in the text.

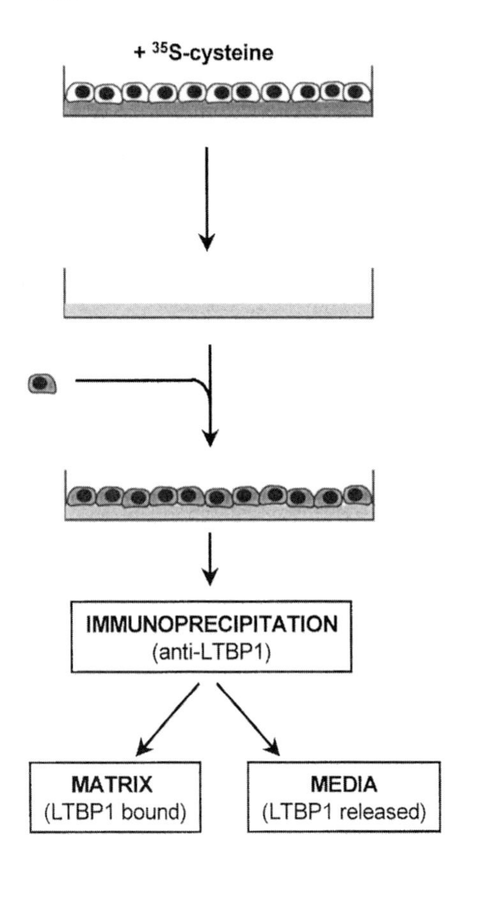

Fig. 2. Immunoprecipitation assay for examining release of matrix-bound LTBP1 and latent TGFβ.

2.1. Preparation of ^{35}S-Labeled ECM from Calvarial Osteoblasts

1. Fetal rat calvarial osteoblast (FRC) culture medium: To 500 mL minimal essential medium, alpha modification (α-MEM) (Gibco Life Technologies, Gaithersburg, MD) add 1.1 g $NaHCO_3$ (if medium is not provided with $NaHCO_3$), 10% heat-inactivated fetal calf serum (FCS) (serum lots are individually checked from several suppliers; heat-inactivate by heating serum to 56°C for 30 min), 2 m*M* L-glutamine (LG) (commercially available as 100X stock; Gibco), and 100 U/mL penicillin/streptomycin (P/S) (commercially available as 100X stock; Gibco). After the cells become confluent, the serum concentration is reduced to 5% and 100 µg/mL ascorbic acid and 5 m*M* β-glycerophosphate are added to the medium (make each as a 100X stock in α-MEM, filter sterilize then make 5-mL aliquots

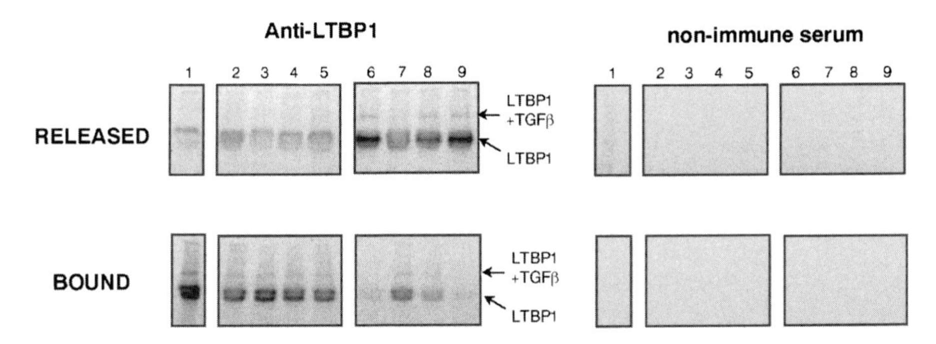

Fig. 3. Immunoprecipitation showing release of LTBP1/latent TGFβ from the ECM of FRC osteoblasts by avian osteoclast cells. ^{35}S-labeled ECM was prepared from fetal rat calvarial osteoblasts grown in six-well culture plates. Avian osteoclast precursor cells were then seeded onto the matrix at 3×10^6 cells per well. The upper panels show the labeled LTBP1/latent TGFβ immunoprecipitated from the culture medium of avian osteoclasts during days 3–4 of culture (released). The lower panels show the labeled LTBP1/latent TGFβ immunoprecipitated from the matrix following plasmin digestion at the end of the experiment on day 4 (bound). Nonimmune serum controls on the right indicate negligible precipitation of non-specific bands. Lane 1 shows a control in which no avian osteoclasts were seeded onto the matrix. A small amount of LTBP1 is released passively from the matrix, however, the majority remains bound. Lanes 2–5 show matrix seeded with avian osteoclasts that were not stimulated with 1,25-dihydroxyvitamin D_3. A small amount of LTBP1 was released into the medium by these cells (lane 2). This release was inhibited by aprotinin (20 μg/mL, lane 3), but not by leupeptin (2 μ*M*, lane 4) or pepstatin (1 μ*M*, lane 5). Lanes 6–9 show matrix seeded with avian osteoclasts that were cultured in the presence of 10^{-8} *M* 1,25 dihydroxyvitamin D3. These vitamin D-stimulated osteoclasts released essentially all the LTBP1/latent TGFβ from the matrix (lane 6). Release of LTBP1/TGFβ was partially inhibited by aprotinin (20 μg/mL, lane 7) and leupeptin (2 μ*M*, lane 8), but not by pepstatin (1 μ*M*, lane 9). Note that in each case, release of LTBP1/latent TGFβ into the culture medium is associated with a concomitant reduction in the amount bound to the matrix.

and store at –20°C, protected from light). These should be added fresh to the media on the day of use.

2. Phosphate-buffered saline (PBS): dissolve 8 g NaCl, 2 g KCl, 1.44 g $Na_2HPO_4 \cdot 2\ H_2O$, 2 g KH_2PO_4 in 800 mL distilled water. Make volume up to 1 L. The pH should be 7.4 without further adjustment. Filter sterilize and store at 4°C.
3. Radiolabeling media: to 80 mL cysteine and methionine-free DMEM (Gibco) add 10 mL α-MEM (Gibco), 5 mL dialyzed FCS (dialyze 50 mL heat-inactivated FCS 4 times against 2 L PBS using a 12–14 kDa molecular weight cut off membrane, then filter sterilize and store at –20°C), 2 m*M* LG (use 100X stock as above), 100 U/mL P/S (use 100X stock as above), 1 mL L-methionine (commercially available as 100X stock, Gibco), 100 μg/mL ascorbic acid (use 100X stock

as above), 3 m*M* β-glycerophosphate (use 100X stock as above). Add 100 μCi ^{35}S-cysteine per mL labeling media immediately prior to use (NEN Life Science Products, Boston, MA).

4. Lysis buffer: to 500 mL PBS, pH 7.4 add 5 mL Triton-X-100. Filter sterilize and store at 4°C.
5. 50 m*M* ammonium acetate pH 9.0: dissolve 1.93 g ammonium acetate in 500 mL distilled water. Adjust pH to 9.0 using glacial acetic acid. Filter sterilize and store at 4°C.

2.2. Analysis of Release of LTBP1 and TGFβ from ^{35}S-Labeled ECM by Avian Osteoclast Cells

1. Avian osteoclast culture medium: To 500 mL α-MEM (Gibco) add 1.1 g $NaHCO_3$ (if medium is not provided with $NaHCO_3$), 5% heat-inactivated FCS, 5% heat-inactivated chicken serum (CS) (serum lots are individually checked from several suppliers; heat-inactivate by heating serum to 56°C for 30 min), 2 m*M* LG (use 100X stock as above) and 100 U/mL P/S (use 100X stock as above). Add 6 μg/mL cytosine-β-D-arabino-furanoside fresh to the medium on the day of use (make a 1 mg/mL stock in PBS, filter sterilize and store for up to 1 wk protected from light at 4°C). For reduced serum avian osteoclast culture media, reduce serum to 2.5% FCS and 2.5% CS.
2. Vitamin D_3 solution: 1,25-dihydroxyvitamin D_3 (BioMol Research Labs, Plymouth Meeting, PA) is diluted to give a 10^{-3} *M* stock in 95% ethanol. This stock is extremely light sensitive and should be stored at –20°C under argon in a light-protected vial. The 1,25-dihydroxyvitamin D_3 is added to avian osteoclast culture media to give 10^{-7} to 10^{-8} *M*. The stock solution should be diluted immediately before use and this should be done under reduced lighting conditions. This reagent is not very stable in aqueous solution and any unused diluted solutions should be discarded.
3. Protease inhibitors: (Boehringer-Mannheim Corp, Indianapolis, IN) (a) aprotinin- make a 10-mg/mL stock in distilled water (stable for 6 mo stored at –20°C in aliquots), add to samples to give a final concentration of 10 μg/mL, (b) leupeptin- make a 2-m*M* stock in distilled water (0.95 mg/mL) (stable for 1 mo stored at –20°C in aliquots) add to samples to give a final concentration of 2 μ*M*, (c) pepstatin- make a 1 m*M* stock in methanol (0.686 mg/mL) (stable for 1 mo stored at –20°C in aliquots) add to samples to give a final concentration of 1 μ*M*, (d) prefabloc (AEBSF)- make a 100-m*M* stock in distilled water (23.95 mg/mL) (stable for 1 mo stored at –20°C in aliquots) add to samples to give a final concentration of 2 m*M*.

2.3. Immunoprecipitation

1. 1 *M* Tris-HCl, pH 8.0: dissolve 60.55 g Tris in 400 mL distilled water. Adjust pH to 8.0 using concentrated HCl. Bring volume to 500 mL.
2. Plasmin digestion buffer: (0.1% 1-o-*n*-octyl-β-glucopyranoside, 3 m*M* $MgCl_2$, 3 m*M* $CaCl_2$, 150 m*M* NaCl, 10 m*M* Tris-HCl, pH 8.0). For preparation of 500 mL; add

0.5 mL 1-o-n-octyl-β-glucopyranoside (Boehringer-Mannheim) to 400 mL distilled water and mix well until dissolved. Add 0.305 g $MgCl_2 \cdot 6H_2O$, 0.166 g $CaCl_2$, 4.38 g NaCl, 5 mL 1 *M* Tris-HCl, pH 8.0. Make volume up to 500 mL. Filter sterilize and store at 4°C.

3. Plasmin solution: dissolve plasmin (Boehringer-Mannheim) in plasmin digestion buffer to make a 1 U/mL stock (stable for at least 1 mo in aliquots at –70°C). Dilute to 0.2 U/mL in plasmin digestion buffer immediately prior to use.
4. 5% BSA solution: dissolve 2.5 g bovine serum albumin (BSA) in 50 mL PBS. Check that pH is neutral, as some BSA preparations are acidic. Filter sterilize and store in aliquots at –20°C (stable for several mo).
5. Antibodies: anti-LTBP1 antibody is a rabbit polyclonal antiserum against purified human platelet LTBP1 (Ab39). Characterization and specificity of this antibody is described in ***(2)***.
6. 0.5 *M* EDTA: dissolve 84.06 g Na_2-EDTA in 400 mL distilled water. Adjust pH to 8.0 with 1 *M* sodium hydroxide (NaOH) and adjust final volume to 500 mL with distilled water.
7. Radio immunoprecipitation assay (RIPA) buffer: 50 m*M* Tris-HCl, 150 m*M* NaCl, 0.5% sodium deoxycholate (DOC), 1% Nonidet P40. For preparation of 500 mL; dissolve 3.03 g Tris, 4.4 g NaCl in 450 mL dH_2O, bring to pH 7.3 with HCl. Add 5 mL nonidet P40, 12.5 mL 20% DOC, slowly bring pH to 7.2 with HCl, bring volume to 500 mL, filter sterilize and store at 4°C (stable for several months).
8. Immunoprecipitation wash buffer 1 (low salt): 1% Triton X-100, 1% DOC, 0.1% SDS, 50 m*M* Tris-HCl, 150 m*M* NaCl, 10 m*M* EDTA, pH 7.5. For preparation of 1 L; add 10 mL Triton X-100 to 800 mL dH_2O, mix well until dissolved. Add 6.06 g Tris, 8.8 g NaCl and when dissolved bring to pH 7.5 with HCl. Add 10 mL 10% SDS (this is commercially available as a 10% stock. Otherwise it can be prepared dissolving 10 g SDS (Bio-Rad Laboratories, Hercules, CA) in distilled water to a final volume of 100 mL. Note: care should be taken when weighing out this reagent, gloves and a face mask should be worn to avoid contact with the skin and respiratory tract), add 20 mL 0.5 *M* EDTA. Bring volume to 1 L, filter through 0.45 μ*M* filter and store at 4°C (stable for several months).
9. Immunoprecipitation wash buffer 2 (high salt): 1% Triton X-100, 0.5 *M* NaCl, 20 m*M* Tris-HCl (pH 7.5). For preparation of 1 L; add 10 mL Triton X-100 to 800 mL dH_2O, mix well until dissolved. Add 2.42 g Tris, 29.3 g NaCl and when dissolved bring to pH 7.5 with HCl. Bring up volume to 1 L. Filter through 0.45 μ*M* filter and store at 4°C (stable for several months).
10. 3X SDS-PAGE sample buffer (nonreducing): (Electrophoresis quality reagents are purchased from Bio-Rad Laboratories). Dissolve 2.28 g Tris in 50 mL distilled water, add 3 g SDS, 1.5 mg bromophenol blue. Adjust pH to 6.8 with HCl. Make volume up to 70 mL, then add 30 mL glycerol. For 3X reducing sample buffer add 3 mL 2-β-mercaptoethanol. 3X sample buffer should be filtered through a 0.45-μm filter and then stored in aliquots at –20°C (stable for several mo). To make 1X sample buffer, dilute 3X sample buffer 1:3 with distilled water (note: SDS will precipitate at cool temperatures, therefore make sure that SDS is thoroughly dissolved when thawing out frozen aliquots).

11. Gel fixing solution 1 (20% TCA, 50% isopropanol): to 30 mL distilled water, add 20 mL 100% trichloroacetic acid (TCA) solution (100 g TCA made up to 100 mL with distilled water, stable for several mo), add 50 mL isopropanol. This solution should be made fresh on the day of use.
12. Gel fixing solution 2 (10% TCA, 25% isopropanol): make up gel fixing solution 1 and dilute 1:1 with distilled water. This solution should be made fresh on the day of use.

3. Methods

Note: the procedures for isolation of FRC osteoblasts ***(11,14–15)***, and avian osteoclast precursor cells ***(16)*** are well-established techniques and will not be described in this account. The reader is referred to the relevant publications to obtain details of these procedures. The procedures described below are modified from protocols originally described in ***(9–13)***. The reader is referred to these articles for further information and variations on these techniques.

3.1. Preparation of ^{35}S-Labeled ECM

1. Plate FRC osteoblasts in a six-well culture plate at 8×10^4 cells/well in 4 mL α-MEM supplemented with 10% FCS, 100 U/mL P/S, 2 m*M* LG. Culture at 37°C in a 5% CO_2 incubator.
2. When the cells reach confluence (approximately 3 d later) change the media to 4 mL α-MEM supplemented with 5% FCS, 100 U/mL P/S, 2 m*M* LG, 100 μg/mL ascorbic acid, 3 m*M* β-glycerophosphate.
3. Continue culturing the cells, changing the media every 3 d. Between days 6–12 a thick layer of ECM is laid down and multilayered "nodules" of cells should be observed. These then become mineralized, forming bone-like structures.
4. On day 12 of culture, aspirate the media, wash the cells twice in PBS and then radiolabel the cells for 48 h using 2 mL radiolabeling media containing 100 μCi ^{35}S-cysteine per mL (*see* **Note 1**).
5. After labeling, aspirate the media and wash the cells twice in ice-cold PBS. Add 2 mL ice-cold lysis buffer and incubate for 15 min at 4°C. Once the cell membranes are disrupted, this should leave a thick, membrane-like layer of ECM adherent to the culture plate. Extreme care should be taken with subsequent washes to avoid disturbing the adherent layer of matrix. Gently aspirate the lysis buffer and wash the insoluble matrix layer 5 times with 4 mL ice-cold PBS (*see* **Note 2**).
6. Aspirate the final PBS wash and air-dry the plates in the sterile laminar flow hood (*see* **Note 3**).

3.2. Analysis of Release of LTBP1 and TGFβ from ^{35}S-Labeled ECM by Avian Osteoclast Cells

1. Bring ^{35}S-labeled matrix to room temperature (if stored frozen) and then rehydrate for 5 min with 4 mL PBS per well. Discard PBS, wash once more with

4 mL PBS, and then seed avian osteoclast precursor cells onto matrix at 3×10^6 cells/well in 4 mL avian osteoclast culture medium (*see* **Notes 4** and **5**). It is important to include a control well cultured with media alone (i.e., without cells) to determine the background level of release from the matrix and also to indicate the amount of LTBP1 in the matrix that is available for release by the avian osteoclasts.

2. Allow cells to adhere for 2 h at 37°C in a 5% CO_2 incubator then change media to 2 mL reduced serum avian osteoclast culture media. Culture for 48 h. For avian osteoclasts, experiments are usually performed in the presence and absence of 1,25-dihydroxyvitamin D_3, (10^{-7} to 10^{-8} *M*) which stimulates the formation of mature osteoclasts and stimulates their bone resorptive activity. At this stage any test treatments, such as protease inhibitors, antibodies, growth factors, and so on, can be added to the media to determine their effects upon release of LTBP1 from the matrix by the cells.
3. Harvest the 0–48 h culture media and transfer to 1.5-mL vials. Add protease inhibitors (aprotinin 10 µg/mL, leupeptin 2 µ*M*, pepstatin 1 µ*M* and prefabloc 2 m*M*) to the media samples. Microfuge at 16,000*g* for 10 min at 4°C and transfer supernatant to a clean tube. Store at –70°C until ready for immunoprecipitation. Add 2 mL fresh media (with or without test factors) to the cells and culture for a further 48 h. This long culture period is used because optimal release of LTBP1 and TGFβ by avian osteoclast occurs on days 3–4 (*see* **Note 4**).
4. Harvest the 48–96 h culture media as described above.
5. Prepare the ECM by washing the cultures twice with ice-cold PBS then adding 2 mL ice-cold RIPA buffer and incubating at 4°C for 15 min (protease inhibitors should not be included at this stage because a plasmin digestion will subsequently be performed to release any LTBP1 that is still bound to the matrix). Carefully aspirate the RIPA buffer and wash a further 4 times in ice-cold PBS. Aspirate the final PBS, and wash and air-dry the plates (*see* **Note 6**). Plates can be stored at –70°C for 1–2 d if required, prior to plasmin digestion and immunoprecipitation.

3.3. Immunoprecipitation of LTBP1 and TGFβ

In this procedure, labeled LTBP1 is immunoprecipitated from the media samples to indicate how much was released into the culture medium by the avian osteoclasts. To measure the LTBP1 which is still bound to the matrix, a plasmin digestion is required since LTBP1 is crosslinked to the matrix and requires proteolytic cleavage for release. Note: unless stated otherwise, samples should be kept on ice for all steps described in the following procedure. A refrigerated microfuge should be used for all centrifugation steps.

1. Digest matrix preparations with 0.5 mL plasmin solution (0.2 U/mL) to release LTBP1 and TGFβ that is still bound to the ECM. Incubate for 1 h at 37°C with gentle shaking.
2. Transfer plasmin digest to a 1.5 mL vial and then add protease inhibitors (aprotinin 10 µg/mL, prefabloc 2 m*M*). Microfuge at 16,000*g* for 10 min and transfer the supernatant to a clean tube.

3. Preclear media samples and plasmin digests with nonimmune rabbit serum (This step reduces background nonspecific bands, which may bind to the protein G agarose or to other serum proteins). Do this by adding 5 µL nonimmune rabbit serum to each sample, then incubating for 30 min at room temperature on an end over rotator.
4. Add 30 µL protein G agarose (Boehringer Mannheim) and incubate for 2 h at 4°C on rotator.
5. Microfuge at 14,000 rpm for 5 min then transfer the supernatant to a clean 1.5-mL tube and discard the pellet. Add 5% BSA solution to the plasmin digests to give a final concentration of 0.2% (This provides excess protein which may help reduce nonspecific binding of radiolabeled proteins to the protein G agarose).
6. Divide each sample and use one half for immunoprecipitation with the specific antiserum (anti-LTBP1) and the other half for a nonimmune rabbit serum control (this control is important as it indicates any bands that are nonspecific). Do this by adding 5 µL of the appropriate serum to each tube. Incubate 30 min at room temperature on a rotator.
7. Add 30 µL protein G agarose and incubate for 2 h at 4°C on a rotator (*see* **Note 7**).
8. Microfuge at 16,000*g* for 5 min, discard the supernatant and wash the pellet as follows; (a) 3 times in 1 mL wash buffer 1 (low salt), (b) 3 times in 1 mL wash buffer 2 (high salt), and (c) one time in 1 mL PBS.
9. Aspirate PBS from the pellet, then centrifuge again and aspirate more if possible to leave the pellet as dry as possible.
10. Add 30 µL 1X nonreducing sample buffer to pellet and store samples at –20°C until ready to run SDS-PAGE.
11. Boil samples for 10 min, then separate proteins on a 4–20% gradient SDS-PAGE gel (load 25 µL sample per lane). SDS-PAGE analysis is a standard procedure performed in most laboratories and details of this technique will not be described here. For a standard protocol and discussion of this technique please refer to ***(17)***.
12. After performing SDS-PAGE, place the gel in a tray and fix with gel fixing solution 1 for 30 min at room temperature with gentle shaking.
13. Change the solution to gel fixing solution 2 and incubate for a further 30 min at room temperature with gentle shaking (*see* **Note 8**).
14. Pour off this solution and add sufficient En^{3}Hance solution (Du Pont, NEN) to cover the gel. Incubate for 1 h at room temperature with gentle shaking. This step as well as the subsequent washes should be performed in a fume hood.
15. Pour off En^{3}hance solution and wash the gel three times in distilled water for 20 min at room temperature.
16. Dry the gel on a gel drier and then expose dried gel using Kodak X-OMAT AR5 film.

Figure 3 shows an example of how this method was used to examine release of LTBP1 and latent TGFβ from bone matrix by avian osteoclasts.

4. Notes

1. The radiolabeling media contains one-tenth the cysteine content of normal culture media. This is necessary because of the long labeling period used. If a shorter

labeling period of only a few hours were found to be sufficient with a particular cell type, it may be preferable to use media that is totally free of cysteine for labeling.

2. An incubation with 50 m*M* ammonium acetate, pH 9.0 can be used following the lysis step to remove nuclear matrix proteins (as recommended in the procedure described by Jones et al. ***(18)***. However, owing to the alkaline pH of this buffer, it is not recommended for analysis of TGFβ, because TGFβ can be activated by pH extremes.
3. ^{35}S-labeled ECM can be used directly without drying, however, it is our experience that if the matrix is dried first, there are less problems with loss of adherence later on when the cells are seeded onto the matrix.
4. Any cell type could be used to examine its ability to release LTBP1/latent TGFβ from the labeled ECM at this step. The seeding densities and culture periods should be optimized for each cell line to determine the conditions for maximal release of LTBP1/latent TGFβ from the labeled ECM. We have found the optimum seeding density for avian osteoclast precursors to be 3×10^6 cells per well. This is because in order to form mature osteoclasts, these precursor cells must first fuse in culture. A higher density of cells allows more efficient fusion. In avian osteoclast cultures, maximal release of LTBP1/latent TGFβ is seen on days 3–4. This corresponds with the appearance of mature multinucleated (fused) osteoclasts. We have found that a good seeding density for MDA-MB-231 human breast cancer cells is 5×10^5 cells/well and that maximal release of LTBP1/latent TGFβ from FRC matrix by these cells occurs by 24 h.
5. It is extremely important when preparing cells for these experiments to ensure that any trypsin or other enzymes that have been used to prepare the cell suspension be removed prior to seeding the cells onto the matrix. It is recommended that cell preparations which have been trypsinized be treated with one-tenth volume of serum to inactivate the enzymes. Additionally the cell suspension should be washed two times in PBS before plating them onto the labeled matrix.
6. The ECM produced by long-term FRC cultures is usually present as a thick membrane-like layer which is still adherent to the culture plate. If ECMs from other cell types are used and are found not to adhere very well to the culture plate, the entire lysate, including insoluble (matrix) material can be transferred to a 1.5 mL tube. This can then be microfuged and the DOC insoluble pellet washed and used for plasmin digestion.
7. If a high level of background bands is obtained, even in the nonimmune control samples, this background can be reduced by using preadsorbed protein G agarose at this step. This is prepared by incubating the protein G agarose in unlabeled cell lysate from the same cells as those used in the immunoprecipitation (this blocks nonspecific binding sites on the protein G agarose with unlabeled protein and thereby reduces background caused by nonspecific adsorption of irrelevant proteins). For 1 mL agarose beads (bed volume) incubate with 4 mL unlabeled cell lysate for 1 h at room temperature on a rotator. Centrifuge the protein G agarose, aspirate the lysate, then wash agarose four times in PBS. Resuspended agarose in PBS to give a 50% slurry.

8. At this stage, if more convenient, fixation in gel fixing solution 1 can be omitted and the gel can be placed straight into gel fixing solution 2 for overnight fixation.

Acknowledgments

The author would like to thank Dr. Lynda F. Bonewald for critical review of this manuscript and Frances J. Ramirez for expert secretarial help. Financial support from the National Institutes for Health and the National Osteoporosis Foundation is also gratefully acknowledged.

References

1. Taipale, J. and Keski-Oja, J. (1997) Growth factors in the extracellular matrix. *FASEB J.* **11,** 51–59.
2. Kanzaki, T., Olofsson, A., Morén, A., Wernstedt, C., Hellman, U., Miyazono, K., et al. (1990) TGFβ1 binding protein: A component of the large latent complex of TGFβ1, with multiple repeat sequences. *Cell* **61,** 1051–1061.
3. Tsuji, T., Okada, F., Yamaguchi, K., and Nakamura, T. (1990) Molecular cloning of the large subunit of transforming growth factor type β masking protein and expression of the mRNA in various rat tissues. *Proc. Natl. Acad. Sci. USA* **87,** 8835–8839.
4. Morén, A., Olofsson, A., Stenman, G., Sahlin, P., Kanzaki, T., Claesson-Welsh, L., et al. (1994) Identification and characterization of LTBP-2, a novel latent transforming growth factor β binding protein. *J. Biol. Chem.* **269,** 32,469–32,478.
5. Bashir, M. M., Han, M., Abrams, W. R., Tucker, T., Ma, R., Gibson, M., Ritty T., et al. (1996) Analysis of the human gene encoding latent transforming growth factor-β-binding protein-2. *Int. J. Biochem. Cell* **28,** 531–542.
6. Yin, W., Smiley, E., Germiller, J., Mecham, R. P., Florer, J. B., Wenstrup, R. J., and Bonadio, J. (1995) Isolation of a novel latent transfomring growth factor-β binding protein gene (LTBP–3) *J. Biol. Chem.* **270,** 10,417–10,160.
7. Giltay, R., Kostka, G., and Timpl, R. (1997) Sequence and expression of a novel member (LTBP–4) of the family of latent transforming growth factor-β binding proteins. *FEBS Lett.* **411,** 164–168.
8. Miyazono, K., Olofsson, A., Colosetti, P., and Heldin, C. H. (1991) A role of the latent TGFβ1-binding protein in the assembly and secretion of TGFβ1. *EMBO J.* **10,** 1091–1101.
9. Taipale, J., Miyazono, K., Heldin, C. H., and Keski-Oja, J. (1994) Latent transforming growth factor-β1 associates to fibroblast extracellular matrix via latent TGF-β binding protein. *J. Cell Biol.* **124,** 171–181.
10. Taipale, J., Lohi, J., Saharinen, J., Kovanen, P. T., and Keski-Oja, J. (1995) Human mast cell chymase and leukocyte elastase release transforming growth factor-β1 from the extracellular matrix of cultured human epithelial and endothelial cells. *J. Biol. Chem.* **270,** 4689–4696.
11. Dallas, S. L., Miyazono, K., Skerry, T. M., Mundy, G. R., and Bonewald, L. F. (1995) Dual role for the latent TGFβ binding protein (LTBP) in storage of latent

TGFβ in the extracellular matrix and as a structural matrix protein. *J. Cell Biol.* **131,** 539–549.

12. Nunes, I., Gleizes, P., Metz, C. N., and Rifkin, D. B. (1997) Latent transforming growth factor-β-binding protein domains involved in activation and transglutaminase-dependent cross-linking of latent transforming growth factor-β. *J. Cell Biol.* **136,** 1151–1163.
13. Saharinen, J., Taipale, J., and Keski-Oja, J. (1996) Association of the small latent transforming growth factor-β with an eight cysteine repeat of its binding protein LTBP1. *EMBO J.* **15,** 245–253.
14. Bellows, C. G., Aubin, J. E., Heersche, J. N. M., and Antosz, M. E. (1986) Mineralized bone nodules formed in vivo from enzymatically released rat calvaria cell population. *Calcif. Tissue Int.* **38,** 143–154.
15. Harris, S. E., Bonewald, L. F., Harris, M. A., Sabatini, M., Dallas, S. L., Feng, J., et al. (1994) Effects of TGFβ on bone nodule formation and expression of bone morphogenetic protein-2, osteocalcin, osteopontin, alkaline phosphatase and type I collagen mRNA in prolonged cultures of fetal rat calvarial osteoblasts. *J. Bone Miner. Res.* **9,** 855–863.
16. Alvarez, J. I., Teitelbaum, S. L., Blair, H. C., Greenfield, E. M., Athanasou, N. A., and Ross, F. P. (1991) Generation of avian cells resembling osteoclasts from mononuclear phagocytes. *Endocrinology* **128,** 2324–2335.
17. Scopes, R. K. and Smith, J. A. (1998) Electrophoretic separation of proteins, in *Current Protocols in Molecular Biology* (Ausubel, F. M., Brent, R., Kingston, R. E., Moore, D. D., Seidman, J. G., Smith, J. J., and Struhl, K., eds.), Wiley, pp. 10.0.1–10.2.30.
18. Jones, J. I., Gockerman, A., Busby, W. H., Camacho-Hubner, C., and Clemmons, D. R. (1993) Extracellular matrix contains insulin-like growth factor binding protein 5: potentiation of the effects of IGF-I. *J. Cell Biol.* **121,** 679–687.

20

Using Organotypic Tissue Slices as Substrata for the Culture of Dissociated Cells

Daniel E. Emerling and Arthur D. Lander

1. Introduction

Dissociated cell culture has proved extremely valuable for studying the functional relationship between specific cell surface and extracellular matrix (ECM) molecules and a variety of cellular behaviors such as proliferation, differentiation, migration, and neuronal process outgrowth. When cells are cultured on substrata made from purified cell surface or ECM components, the direct actions of these molecules on cells can be examined easily. However, for some in vitro studies, it is desirable to examine the functions of such molecules in an environment that more closely approximates that which exists in vivo. Such studies might include, for example, investigations in which the endogenous distribution and heterogeneity of such molecules is important, screens for uncharacterized factors that effect a particular cell behavior, or studies examining the combinatorial effects of the many cell surface and ECM components that occur in intact tissue.

We devised an assay in which dissociated cells are plated onto living slices of tissue, much as they can be plated onto tissue culture plastic. This procedure allowed us to examine cell behaviors within the organotypic environment of a tissue substratum. Primary neurons were plated onto slices of living mouse forebrain, cocultured at 37°C, fixed, and then examined. The degree of attachment of the cells and the orientation of process outgrowth from them was found to be highly dependent upon the anatomical region of the forebrain slice onto which the cells settled when plated (**Fig. 1** and **ref.** ***1***). When the same cells were plated onto cryostat sections of the same tissue, neither the degree of overall attachment and outgrowth nor the region-dependent effects were as dramatic *(1)*.

From: *Methods in Molecular Biology, vol. 139: Extracellular Matrix Protocols*
Edited by: C. Streuli and M. Grant © Humana Press Inc., Totowa, NJ

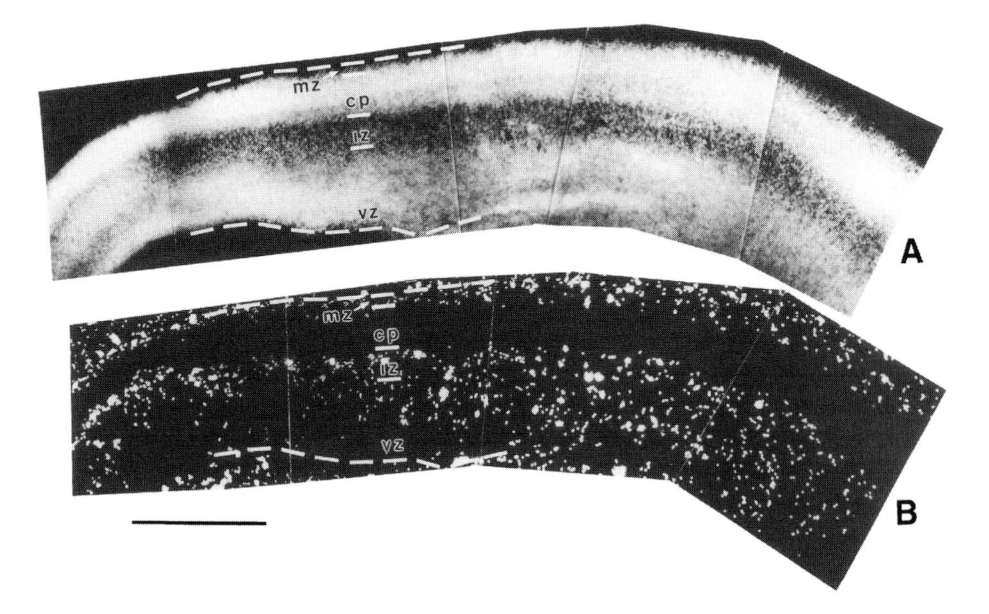

Fig. 1. Laminar specific attachment of thalamic neurons to a living slice of embryonic mouse cerebral neocortex. Dissociated and fluorescently labeled embryonic thalamic neurons were plated onto sagittal slices of embryonic day 15 cortex and were cocultured for 3 h at 37°C. Cocultures were then rinsed to remove nonattached thalamic cells, fixed, counterstained, and visualized. **(A)** Cortical laminae of a bisbenzamide-stained slice can be seen under UV fluorescence as differences in nuclear density. Rostral is to the right, and caudal is to the left. **(B)** The same slice viewed under rhodamine optics shows the distribution of attached thalamic cells. The pial and ventricular edges of the cortex are demarcated by dotted white lines at the top and bottom of each figure. Lines are positioned in the two photos to reference the same points in the two views of the slice. Very few thalamic cells attach to the cortical plate (CP), whereas more attach to the intermediate zone (IZ), marginal zone (MZ), and ventricular zone (VZ) as well as to the culture support off the slice (seen at the edges of the photos). Density of attached cells is greatest on the intermediate zone just subjacent to the cortical plate (i.e., the subplate). Scale bar is 500 μm.

The living-slice method of coculture is amenable to the same experimental manipulations that are performed on typical dissociated cell cultures (e.g., pharmacological treatment, enzymatic digestions, antibody blockade, etc.). For example, we used this approach to investigate the functions of chondroitin sulfate in neuronal attachment and axon pathfinding. We treated slices with chondroitinase ABC to enzymatically remove endogenous chondroitin sulfate, and then compared the behaviors of primary neurons plated on treated and control slices (**Fig. 2** and **ref.** ***2***).

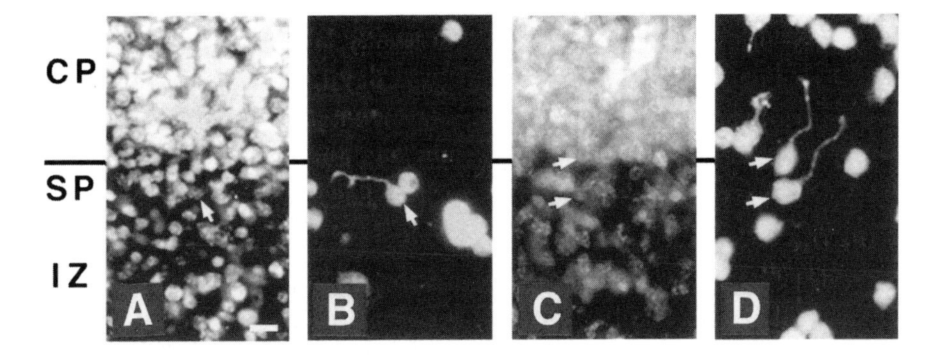

Fig. 2. Cell attachment and neurite extension on slices of embryonic mouse cerebral cortex treated with chondroitinase ABC. Cocultures of cortical slices and thalamic neurons were prepared as in **Fig. 1**, except that slice cultures were treated with carrier (complete medium), or 1 U/mL chondroitinase, for 3 h at 37°C, prior to the addition of dissociated and labeled thalamic neurons. High-magnification, fluorescence micrographs show the bisbenzamide-stained cortical plate (CP), subplate (SP), and intermediate zone (IZ) of control **(A)** and treated **(C)** tissue slices. **(B,D)** The same fields viewed under rhodamine optics reveal the attached, thalamic cells and their neurites. Arrows indicate neurite-bearing thalamic cells and their corresponding nuclei. In control cocultures **(A,B)**, cells attach well to the subplate, but poorly to the cortical plate. Neurites that extend on the subplate and intermediate zone tend to orient parallel with the cortical layers, and neurites that cross from the subplate onto the cortical plate are extremely rare. In cocultures treated with chondroitinase (C, D), cells attach well to the cortical plate, neurite outgrowth on the cortical plate is enhanced, and processes that originate on the subplate often cross onto the cortical plate. Scale bar is 10 μm.

This chapter will focus on how to cut tissue slices for use as culture substrata and how to make and fluorescently label a dissociated cell suspension from tissue. It is assumed that the investigator already has procedures to obtain the tissue(s) of interest. The first four parts of the protocol involve preparing the slices and dissociated cells for coculture and include: 1. embedding the tissue in agarose and slicing it on a vibratome, 2. mounting the tissue slices onto supports and transferring them to culture, 3. dissociating tissue into a cell suspension, and 4. labeling dissociated cells with a vital dye (**Fig. 3**). The fifth section deals with culturing, fixing, and counterstaining the cocultures, as well as how to mount them onto microscope slides for visualization.

The choice of technique for visualizing the cells on the slices (e.g., immunofluorescence, *in situ* hybridization, transgenic marker, etc.) will, of course, depend upon the experimental needs of the investigator. The method described below involves labeling the cells in suspension, before plating, with a fluorescent, fixable, cytoplasmic dye that makes it easy to discern plated cells from

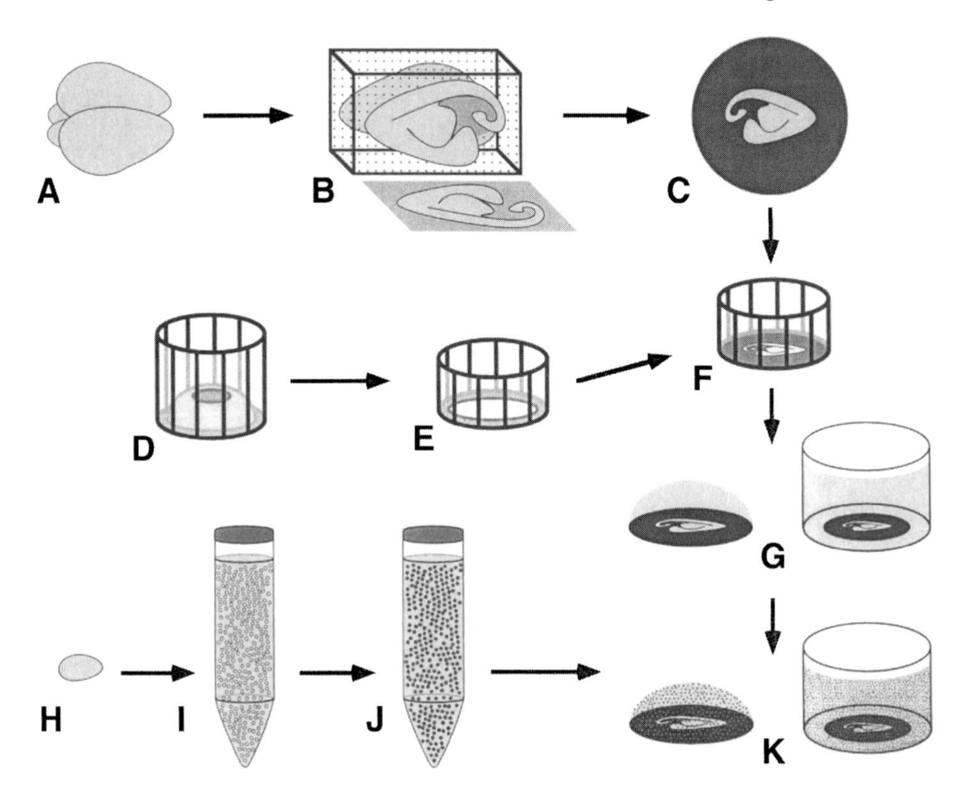

Fig. 3. Steps to making cocultures of dissociated cells plated onto living tissue slices. The tissue to be sliced is dissected **(A)**, embedded in agarose and sliced on a vibratome **(B)**, and then placed onto nitrocellulose disks for support **(C)**. Disks with mounted slices are transferred from the vibratome using a holder made from the cap of a round-bottom tube (**D**, *see* **Subheading 2.2.3.**). A hole is punched in the top of the cap, leaving a rim, and the cap is shortened **(E)** so that the disk can be placed in the bottom of the cap, forming a well **(F)**. After transfer from the vibratome, disks are removed from the holder (*see* **Subheadings 3.2.4.** and **3.2.5.** for details), and placed either on parafilm under a bubble of medium or in tissue culture wells **(G)**. To obtain a cell suspension, the tissue is dissected **(H)**, dissociated **(I)**, and labeled with a fluorescent vital dye **(J)**. Cells are then allowed to settle onto the slice cultures **(K)**.

those in the slice. Cells can be similarly labeled with membrane-intercalating vital dyes such as DiI *(3)*. Bisbenzamide, a blue-fluorescent, nuclear stain, is useful for making visible the overall anatomy of cultured slices. Protocols for immunostaining such slices can be found elsewhere *(1,2)*.

We optimized this method for making living slices of perinatal mouse forebrain to use in acute (<24 h) cocultures with dissociated primary neurons. The sectioning protocol should be applicable for use with most tissues, but may require some modifications depending upon the properties of the particular tissue used. Similarly, the dissociation protocol was optimized for neural tissue and may require modification for other tissues or for particular experimental requirements.

We found that, after 1 d in culture, the tissue slices may begin to flatten. This process alters anatomical structure and borders. The particular questions and needs of the investigator will determine whether longer culture periods are needed and if such in vitro alterations could affect the interpretation of data. Regardless of the culture period, investigators should take care to ensure that particular results are not because of artifacts such as folds, tears, or other incongruities in the tissue slice. A wise approach is to examine all fixed, mounted slices for such defects prior to examining the behaviors of the labeled dissociated cells plated on them.

2. Materials

All solutions should be sterile (except paraformaldehyde). Unless noted, all chemicals are from Sigma, St. Louis, MO. All media and balanced salt solutions are from Gibco, Gaithersburg, MD. Procedures should be performed aseptically and, when possible, in a laminar flow hood.

2.1. Embedding and Slicing Tissue

1. Phosphate Buffered Saline (PBS): 137 m*M* NaCl, 2.68 m*M* KCl, 7.83 m*M* Na_2HPO_4, 1.47 m*M* KH_2PO_4.
2. PBS/glucose: PBS supplemented with 0.45% glucose and 1 m*M* sodium pyruvate (from 100X stock, Gibco). Filter sterilize.
3. Low melting point (LMP) agarose solution: 2% LMP agarose (Gibco-BRL) in PBS. Autoclave to dissolve agarose and to sterilize. When cool, but still liquid (<50°C), supplement to 0.3 % final glucose with sterile 30% glucose stock. Stable at 4°C in sealed bottle at least 6 mo.
4. Vibratome tray (Technical Products International, Inc., St. Louis, MO) and teflon coated vibratome blades (Ted Pella).
5. Tissue adhesive, such as Locktite™ (Ted Pella) or any comparable glue may work.

2.2. Mounting Slices onto Supports for Culture

We chose to mount tissue slices onto black nitrocellulose disks because this material provided the rigidity we needed for our experimental manipulations. Black was chosen to minimizes reflection and scattering of light during examination by fluorescence microscopy. To ensure that slices attach well to the support, the disks are pretreated with the lectin concanavalin A, however, other proteins (e.g., laminin, fibronectin, etc.) may work as well or better for a particular tissue. For this treatment and for coculturing, disks can be placed into 12- or 24-well flat bottom tissue culture plates. Alternatively, disks can be placed onto Parafilm™ (*see* **Subheading 2.2.1.**). Holders (*see* **Subheading 2.2.3.**) are made to transfer disks with mounted tissue slices out of the vibratome tray without damaging the slices.

The preparation and use of nitrocellulose disks as supports is somewhat laborious (*see* below and **Subheading 3.2.**), and they are not suitable for transmitted light microscopy. As an alternative, slices can be mounted onto commercial culture well inserts. Several suppliers (Millipore, Bedford, MA, Costar, Cambridge, MA, and Nunc, Wilsbaden-Biebrich, Germany) manufacture these inserts with growth surfaces made from a variety of translucent or transparent polymer materials (e.g., polycarbonate, cellulose, etc.) and treated with a variety of ECM molecules (e.g., collagen, laminin, etc.). Characteristics such as autofluorescence, transparency, rigidity, and porosity vary greatly among the different membranes and, based on the particular experimental needs, should be taken into account when making a choice of insert to use.

1. Parafilm™ dishes: When pipeted onto nitrocellulose disks that are on Parafilm™ (American National Can), small volumes (≈100 µL) of solutions remain in a bubble over the disk because of the hydrophobic properties of the Parafilm™. Cut a circle of Parafilm™ (≈10–12 cm diameter), remove backing, and place down in a 150-mm plastic Petri dish. Using both hands, hold the sheet flat and, with the back edge of a forceps or other blunt metal object, scratch several short lines, perpendicular to the edges of the Parafilm™, around its perimeter. This will score the Parafilm™ into the plastic of the dish. Sterilize under ultraviolet (UV) light (>1 h).
2. Nitrocellulose disks: Sterilize the cutting edge of a cork borer (12 mm diameter) with flame and/or ethanol. Place a sheet of sterile, black nitrocellulose (0.45 µm; Sartorius, Bohemia, NY) into a plastic Petri dish and press the borer into it to cut out disks. Transfer the cut disks with a sterile forceps to another Petri dish for storage.
3. Holders for nitrocellulose disks: A holder is made from the cap of a 6-mL round bottom tube (Falcon, #2063). The cap is inverted and used as a well, on the bottom of which the tissue-bearing nitrocellulose disks will be placed. Take a plastic cap and place it onto the cutting end of an 11-mm cork borer, as if to cap the end of the cork borer as a tube. The indent at the top of the cap should just fit into the hole of the borer. Using a large rubber stopper as a backing, punch a hole out of the cap. This should leave the top of the cap with just a rim. Discard the rest of the cap top that was cut out. Using a scalpel or razor blade, cut the length of the cap in half, discarding the half that contains indents on the inner side of the cap and saving the half that has the rim (**Fig. 3E**). Sterilize by storing in ethanol.
4. Concanavalin A: 1.3 mg/mL in DMEM (or other medium or balanced salt solution). Filter sterilize and store aliquots at –20°C.
5. Complete medium: DMEM (glutamine free, 4.5 g/L glucose) supplemented with 10 µg/mL transferrin, 5 mg/mL crystalline grade bovine serum albumin (BSA;

ICN Biochemicals), 20 n*M* progesterone, 30 nM sodium selenite, 100 μ*M* putrescine, 10 μg/mL bovine insulin, 1 m*M* sodium pyruvate, 50 I.U./mL penicillin, 50 μg/mL streptomycin, and 25 m*M* HEPES (pH 7.2). Filter sterilize. Store at 4°C. Use preferably within 24 h. Other basal medium and supplements may be more appropriate for the particular tissue and dissociated cells used.
6. Sterile filter paper or gauze. Bleached (white), filter paper from a sealed pack should be suitable.

2.3. Dissociating Tissue into a Cell Suspension

1. Balanced Salt Solution (BSS): calcium–magnesium-free, Hank's Balanced Salt Solution. This can be supplemented with 10 m*M* HEPES, pH 7.2, if good pH control is needed (BSS will become somewhat basic in room air).
2. Trypsin stock: 5 mg/mL of trypsin from bovine pancreas in BSS. Sterile filter. Store as aliquots at –20°C.
3. DNase stock solution: 1 mg/mL DNase I (grade II, Boehringer Mannheim, Germany) in DMEM or other basal medium. Sterile filter. Store as aliquots at –20°C.
4. Trypsin inhibitor: 5 mg/mL trypsin inhibitor (Type I-S) from soybean in DMEM or other basal medium. Sterile filter. Store as aliquots at –20°C.
5. 9" Pasteur pipets with cotton plug. Autoclave to sterilize.
6. 4% BSA: 4% bovine serum albumin (crystalline grade; ICN Biochemicals) in BSS; bring to pH ≈ 7.3 with 1 M NaOH. Sterile filter. Store as 2.0–2.5 mL aliquots at –20°C.

2.4. Labeling Dissociated Cells

Dye stock solution: Use Cell Tracker™ (Molecular Probes, Eugene, OR) dyes, such as CMTMR (rhodamine derivative) or CMFDA (fluorescein derivative). Make 10 m*M* stock solution in DMSO and store as aliquots at –20°C.

2.5. Coculture, Fixation, Counterstaining, and Mounting

1. Paraformaldehyde: 4% paraformaldehyde (J. T. Baker, Phillipsburg, NJ) in 0.1 *M* sodium phosphate buffer (pH 7.5) or PBS. Paraformaldehyde is toxic, so perform this procedure in a fume hood. Add paraformaldehyde to water (one half of the final solution volume), stirring at 65°C. Add 5 *M* NaOH dropwise until paraformaldehyde goes into solution. Bring to the final volume with a 2X stock of buffer (0.2 M sodium phosphate or 2X PBS). Filter to remove particulate. For paraformaldehyde/sucrose, make fix with 0.12 *M* final sucrose. Add sucrose with the 2X buffer. Good for 2 wk at 4°C.
2. Bisbenzamide (Hoechst 33258): 1 mg/mL stock solution (100X) in water, then filter and store at 4°C. For use, make fresh by diluting in PBS to 10 μg/mL.
3. Mounting solution: Saturated sucrose solution containing 0.1% sodium azide (Fluka AG). Store at room temperature. Other aqueous mounting media may work as well.
4. Glass microscope slides, 22 × 22 mm coverslips, super glue, and clear nail polish.

3. Methods

3.1. Embedding and Slicing Tissue

1. Autoclave or ethanol flame the vibratome tray to sterilize it and then insert it in the vibratome.
2. Melt the LMP agarose solution, then cool to and keep at 37°C in a water bath.
3. Dissect tissue for slicing and store on ice in PBS/glucose.
4. Fill a 35-mm plastic Petri dish or similar container with the agarose and place the dissected tissue in the dish (*see* **Note 1**), being careful not to transfer too much liquid with it (*see* **Note 2**). Place the dish on ice to speed the hardening of the agarose.
5. Cut a block (sides ≈1.5 cm in length) containing the tissue out of the agarose in the dish using a sterile scalpel or razor blade. In another dish, you can further cut the block so that the tissue is oriented in the desired plane of sectioning and is at the correct height for the blade when the block is placed in the vibratome tray (*see* **Note 3**).
6. Place a drop of adhesive on the dry tray surface and immediately place the agarose block on the glue in the proper orientation. Once the glue has set (*see* **Note 4**), fill the tray with ice-cold PBS/glucose so that the block is submerged and surround the tray with an ice bath.
7. Insert a sterile vibratome blade in the holder on the vibratome (*see* **Note 5**).
8. Section the tissue into slices 150–200μm thick (*see* **Note 6**). The optimum vibratome settings for speed and vibration of the blade will depend upon the size and consistency of the particular tissue and should be empirically determined (*see* **Note 7**). Once cut, allow the slice to settle to the bottom of the tray and separate it from any agarose (*see* **Note 8**).

3.2. Mounting Slices onto Supports for Culture

1. Treat the nitrocellulose disks with the concanavalin A solution for 4–18 h at 4°C (*see* **Note 9**) in a culture well plate or on Parafilm™ in a humidified atmosphere. Rinse well in PBS before use.
2. Place a disk in the vibratome tray and position it near the slice on the bottom of the tray. Carefully manipulate the tissue slice onto the disk (*see* **Note 10**). Multiple slices can be placed on the disk, if they can fit.
3. Take a sterile disk holder (*see* **Subheading 2.2.3.**) and, with forceps, hold it on the tray bottom next to the disk (*see* **Note 11**). With another forceps in the other hand, grab the edge of the disk with mounted slice(s) and place it into the holder so that the disk forms the bottom of a well (**Fig. 3F** and *see* **Note 12**).
4. Lift the holder and disk out of the tray so that the well they form remains filled with the PBS/glucose (*see* **Note 13**). Place the holder onto the sterile filter paper or gauze so that the buffer is drawn down, through the nitrocellulose, by capillary action, and rapidly lift the holder off the paper just before the slice gets dry (*see* **Note 14**).
5. Using forceps, remove the disk from the holder and transfer it to parafilm or a culture well plate. Add complete medium before the tissue dries out (≈100 μL for

disks on parafilm). Slices can be stored for short periods (≈1.5 h) in medium on ice until all slices collected. Culture in a humidified atmosphere at 37°C (*see* **Note 15**).

3.3. Dissociating Tissue into a Cell Suspension

1. Dissect tissue for dissociation and place in 1 mL BSS in a 15-mL polypropylene, conical tube (*see* Note 16).
2. Add 45 µL of the trypsin stock and 100 µL DNase (final concentrations of 0.2 mg/mL and 0.09 mg/mL, respectively). Incubate for ≈10 min at 37°C with occasional, gentle mixing (*see* **Note 17**).
3. Add another 250 µL DNase, 300 µL of complete medium, and 100 µL of the trypsin inhibitor. These additions raise the final DNase concentration to 0.2 mg/mL and the inhibitor to 0.3 mg/mL. Incubate for another 2–5 min at 37°C.
4. Make a trituration pipet by flame-polishing the tip of a Pasteur pipet so that it tapers to a hole ≈0.2–0.7 mm in diameter (*see* **Note 18**). To prevent loss of cells on the glass, coat the inside of the pipet with 4% BSA or serum just before use.
5. Gently triturate the tissue by drawing and ejecting the entire tissue suspension in and out of the pipet (*see* **Note 19**). To avoid introducing bubbles into the suspension, eject it from the pipet along the wall of the tube at a height above the final liquid level. Repeat until the most of the tissue has dissociated (about 5X, *see* **Note 20**).
6. Add ice cold BSS to bring the final volume to 10–12 mL, mix, and allow the undissociated tissue to settle by gravity, while keeping the tube on ice (5–10 min). Transfer the cell suspension to a new tube and discard the settled chunks.
7. If cells are to be labeled, go to **Subheading 3.4.1.** If cells are to be plated directly, go to **Subheading 3.4.4.**

3.4. Labeling Dissociated Cells

1. Make fresh labeling medium by diluting the dye stock in warm complete medium to a final concentration of 1–25 µM (*see* **Note 21**). Centrifuge the labeling medium to remove undissolved dye and filter sterilize before use.
2. Centrifuge the cell suspension (100–300*g*, 10 min, 4°C) and gently resuspend in 0.5–2 mL labeling medium. Incubate at 37°C for 30–50 min with occasional mixing to resuspend settled cells (*see* **Note 22**).
3. Add cold BSS to bring final volume to 10–12 mL and mix.
4. Below the cell suspension, underlay a 2–2.5 mL cushion of cold, 4% BSA. Place the tip of a Pasteur pipet, filled with BSA solution, at the bottom of the tube and slowly eject the solution so as not to disturb the cushion as it forms (*see* **Note 23**). Centrifuge (150–300*g*, 10 min, 4°C).
5. Resuspend the pellet in complete medium and dilute the cell suspension to the appropriate density for plating (*see* **Note 24**).

3.5. Coculture, Fixation, Counterstaining, and Mounting

1. Carefully aspirate or pipet the medium off the slices and add the cell suspension. Alternatively, you can add the cells in a small volume directly to the medium

already on the slices or after removing a portion of it (*see* **Note 25**). Coculture in a humidified atmosphere at 37°C (*see* **Note 15**).

2. To fix cocultures, slowly pipet warm paraformaldehyde/sucrose under the medium of the cocultures. Alternatively, you can transfer disks directly to warm paraformaldehyde (*see* **Note 26**). Fix at room temperature for 15–30 min. Stop the fix by replacing it with PBS (*see* **Note 27**). Fixed cocultures can be stored at 4°C in PBS with 0.02% sodium azide.
3. To fluorescently counterstain the nuclei of cells in the slice, add bisbenzamide (10 μg/mL final in PBS) to live or fixed cocultures at room temperature and incubate for at least 10 min (*see* **Note 28**).
4. To mount cultures on microscope slides, a "stage" is built up on the slide to hold the coverslip over the coculture without crushing it. Use superglue to affix two 22 × 22 mm coverslips on a slide at least 15 mm apart but so that a third coverslip can bridge across them when placed over the gap. Place a disk on the slide in the space between the two stages and fill the area with mounting medium (*see* **Note 29**). Overlay the third coverslip and aspirate off any excess solution. Seal the edges with nail polish and let dry. Fixed and mounted cocultures can be stored at 4°C.

4. Notes

1. The tissue may sink to the bottom of the dish. To prevent this, pour the agarose in two layers. Pour a bottom layer, and when it begins gelling, but is not yet solidified, pour a top layer and add the tissue. It should sink to the top of the solidifying bottom layer.
2. Any space created by liquid trapped between the tissue surface and the hardened agarose can cause uneven slicing. To prevent this, thoroughly coat the tissue by moving it around in molten agarose before it hardens. Avoid making bubbles.
3. To minimize the deflection of the block each time the blade enters it, make cuts to taper the side of the block that the blade will enter. When placed in the tray, the taper of the block should point toward the blade like an arrow.
4. If the glue does not hold well, there may have been too much liquid on the agarose. Aspirate excess liquid before placing the block on the glue. Dry glue on the tray can be removed with a razor blade.
5. Oil placed on the blades by the manufacturer can be removed with ethanol. If flame-sterilizing the blades, minimize exposure to heat since this can dull and distort the blade edge.
6. At this thickness, slices of embryonic and postnatal mouse forebrain were sturdy enough to be manipulated and not tear, but were thin enough so that flattening and distortions of the slice surface were minimal during culture. Some tissues may allow thinner, or require thicker, sectioning.
7. For embryonic and postnatal mouse forebrain, a relatively slow speed and the maximum vibration provided the best settings to begin cutting the tissue. After most of the slice had been cut, however, the vibration intensity was lowered to minimize damage to the delicate, sectioned tissue. In general, the softer the tissue, the slower the speed and the greater the vibration needed.

8. In most cases, the agarose should easily separate from the slice. If the agarose is still attached, separate it from the slice using a pairs of forceps in each hand. Cutting the agarose with a scalpel blade may help. Although the slice can be mounted onto its support with the agarose attached, be aware that the agarose may detach after time in the culture.
9. Culture inserts can also be coated.
10. This may be done with a forceps or by lifting the slice with a flat spatula. To ensure attachment, use a forceps to lightly press the slice onto the disk in an unimportant region of the tissue. Slices can be similarly placed into inserts held under the fluid level.
11. Holders float in water. The rim of the holder should be on the bottom.
12. The slice should always remain under the surface of the liquid. Use a forceps to lightly press the disk down so that it is flush with the rim of the holder.
13. Surface tension at the air/liquid interface can damage the tissue or remove it from the disk if the disk is simply pulled from the PBS/glucose after mounting the slice.
14. This step removes most of the PBS. It also further pushes the slice onto the disk by the force of the capillary action. Alternatively, the holder and disk can be placed on parafilm and the PBS removed by careful pipeting. Inserts can be treated similarly.
15. Complete medium with DMEM is buffered for an 8% CO_2 atmosphere, but other medium may be different.
16. Large tissue chunks should be cut into smaller pieces. Volumes and additions can be scaled up for large amounts of tissue.
17. The optimal protease concentration and length of incubation may vary for different tissue types.
18. Rotate the pipet within the flame to get an even taper. The optimal bore size will depend upon the tissue type. Too small of a bore will kill cells, whereas a hole too large will not effectively dissociate the tissue. We routinely used bore sizes of 0.2–0.4 mm for perinatal mouse nervous tissue. Instead of flame-polished glass, a disposable pipet tip (e.g., 1 mL blue tips) may work as well.
19. If the entire suspension does not fit into the pipet, divide the suspension among more than one tube into volumes that fit the trituration pipet , and triturate each individually.
20. Repeatedly triturating cells that are already dissociated will lower cell viability. If further dissociation is necessary, allow the undissociated tissue chunks to settle, transfer the dissociated cell suspension to another tube, add more medium or BSS to the chunks, and continue to triturate them. Pool all fractions when done. A longer protease digestion (**Subheading 3.3.2.**) may promote easier dissociation during the trituration step, but can also lower cell viability.
21. The optimum concentration for CellTracker™ dyes will depend upon the cell type and length of culture and should be empirically determined. The optimal labeling concentration should not be exceeded because of the toxic effects of DMSO and high dye concentrations. We used 14 µm CMTMR for labeling primary embryonic neurons.

22. If stringy precipitate is seen floating, it may be DNA released from dead cells. Remove it by adding DNase (*see* **Subheading 2.3.3.** for stock solution) to 0.1–0.2 mg/mL final after tripling the volume with complete medium to dilute the dye. Continue the incubation for 5–15 min, until precipitate dissolves.
23. 4% BSA is more dense than BSS. The cushion efficiently separates the cells from dye, enzymes, and debris left from the labeling and dissociation.
24. We routinely plated $1–2 \times 10^6$ cells/disk in 90 μL on parafilm. This gives 1000–2000 cells/mm^2 in the middle portion of the disk. At the edges, densities are lower due to the shape of the medium bubble on the disk.
25. This step can be done immediately after the slices are cut or after a period of slice preculture and/or treatment. We routinely plated dissociated cells 4–5 h after slicing.
26. Transfer of cocultures through an air/liquid interface will remove cells that are not well attached to the slice.
27. After fixation, the cocultures can better withstand manipulation through an air/liquid interface.
28. If immunostaining is performed, bisbenzamide can be added after this procedure, before mounting the slides. Rinsing with PBS after the bisbenzamide may slightly lower the background fluorescence, however, the stain is effective when slices are mounted directly from the dye solution.
29. Cocultures grown in well inserts can also be mounted this way by cutting the membrane out of the insert with a scalpel blade and mounting the membrane like a disk.

Acknowledgments

This work was supported by NIH grant NS26862.

References

1. Emerling, D. E. and Lander, A. D. (1994) Laminar specific attachment and neurite outgrowth of thalamic neurons on cultured slices of developing cerebral neocortex. *Development* **120,** 2811–2822.
2. Emerling, D. E. and Lander, A. D. (1996) Inhibitors and promoters of thalamic neuron adhesion and outgrowth in embryonic neocortex: functional association with chondroitin sulfate. *Neuron* **17,** 1089–1100.
3. Honig, M. G. and Hume, R. I. (1986) Fluorescent carbocyanine dyes allow living neurons of identified origin to be studied in long-term cultures. *J. Cell Biol.* **103,** 171–187.

21

Neuroepithelial Differentiation Induced by ECM Molecules

José María Frade and Alfredo Rodriguez Tébar

1. Introduction

Many important processes in the development of the vertebrate central nervous system (CNS) rely on signals induced by extracellular matrix (ECM) components *(1,2)*. These molecules are known to contain multiple functional domains that deliver distinct signals to the cells with which they interact. We will focus here on a well-known ECM glycoprotein, laminin-1, which plays a variety of roles in the development of the CNS *(2)*.

Laminins are a family of ECM glycoproteins, one of which, laminin-1, was first isolated from the murine Engelbreth-Holm-Swarm sarcoma *(3)*. Laminin-1 has been shown to exist as a heterotrimeric cross-shaped molecule containing a large $\alpha1$ chain and two smaller $\beta1$ and $\gamma1$ chains. Further studies showed that other isoforms of the three chains exist, giving rise to various classes of laminins *(4)* including merosin *(5)* and Schwannoma laminin *(6)*.

One of the biological activities of laminin-1 during the development of the CNS is the induction of neuroblast differentiation (pipetthe exit from the cell cycle and the acquisition of neuronal markers). This activity can be tested in vitro, such an assay being a useful tool to dissect out the inductive forces provided by any extracellular factor. In our case, a bioassay analyzing neurogenesis induced by ECM molecules should mimic the in vivo conditions in which neurons are generated. Therefore, such a bioassay for neuron differentiation would require i) a source of CNS proliferating precursor cells; ii) the presence of soluble factors necessary for neuronal differentiation, and iii) an appropriate ECM substrate which will be the object of this communication.

The chick retina is a region of the CNS that has been extensively studied and possesses multiple advantages as a model system. Its morphology is quite

From: *Methods in Molecular Biology, vol. 139: Extracellular Matrix Protocols*
Edited by: C. Streuli and M. Grant © Humana Press Inc., Totowa, NJ

simple compared with other regions within the CNS, it is completely isolated from other CNS structures, a feature that simplifies its dissection, and it is completely avascular, therefore, there is no contamination from nonnervous tissues during the process of dissection. The embryonic day 5 (E5) chick retina consists of a neuroepithelium mainly composed of precursor cells, which are undergoing active neurogenesis ***(7)***. This structure has been the object of active study and conditions that permit neurogenesis in cultures of dissociated neuroepithelial cells are well known ***(8,9)***. Among these conditions, the presence of at least the long arm of laminin-1 is absolutely required ***(8)***.

Here we will describe a set of protocols that can be used for the analysis of neuronal differentiation induced by ECM molecules using dissociated E5 chick retina neuroepithelial cells in culture.

2. Materials

2.1. Coating with ECM Substrata

1. 10 mm round, coverslips (Menzel-Gläser, Germany).
2. 65% nitric acid.
3. Sterile distilled water.
4. Absolute ethanol.
5. 35/10-mm four-well Petri dishes (Greiner, Fickenhausen, Germany).
6. Poly (D-L) ornithine (Cat. no. P8638) (Sigma, St. Louis, MO) (0.5 mg/mL in 150 m*M* Na borate, pH 8.4). Aliquots may be stored at –20°C for more than 1 yr.
7. Ca^{2+}-, Mg^{2+}-free PBS: 136.89 m*M* NaCl (8 g/L), 2.68 m*M* KCl (0.2 g/L), 6.46 m*M* $Na_2HPO_4 \cdot 2H_2O$ (1.15 g/L), 1.47 m*M* KH_2PO_4 (0.2 g/L), pH 7.3 (autoclaved). All reagents from Merck (Darmstadt, RFA).
8. Laminin-1 (10 µg/mL, final concentration, in Ca^{2+}, Mg^{2+}–free PBS) (Gibco-BRL, Gaithersburg, MD, or Collaborative Biomedical Products, New Bedford, MA).
9. Merosin (10 µg/mL; final concentration) (Gibco-BRL).
10. Schwannoma laminin (10 µg/mL; final concentration) ***(6)***.
11. Fibronectin (10 µg/mL; final concentration) (Gibco-BRL).
12. Laminin-1 proteolytic fragments P1 and E8 ***(10)***. These fragments are used in soluble form, they are added in equimolar amount as that of laminin-1 to cells plated onto fibronectin (P1: 3.22 µg/mL; E8: 2.66 µg/mL; final concentration in Ca^{2+}, Mg^{2+}–free PBS).

2.2. The Differentiation Bioassay

1. Chick embryos staged as described elsewhere ***(11)***.
2. Material for dissection (dissecting scissors, Dumont #5 tweezers, and curved iris forceps).
3. Bovine serum albumin (BSA) (Sigma) (30 mg/mL in sterile Ca^{2+}-, Mg^{2+}-free PBS). Aliquots may be stored at –20°C for more than 1 year.
4. Trypsin (Worthington). The powder should be aliquoted and stored at 4°C under dry conditions for less than 1 year (storage conditions are very important for reproducibility)

5. Soybean trypsin inhibitor (Sigma) (10 mg/mL in sterile Ca^{2+}-, Mg^{2+}-free PBS). Aliquots may be stored at –20°C.
6. DMEM/F12 Ham (1:1) medium (Sigma) with N2 supplements *(see below)*. 100 mL medium should contain 10 mg transferrin (Sigma), 1 mL insulin solution (Sigma) (0.5 mg/mL in sterile Ca^{2+}-, Mg^{2+}-free PBS containing 4 μL/mL concentrated HCl, this solution may be stored at 4°C), 10 μL putrescin solution (Sigma) (160 mg/mL in sterile Ca^{2+}-, Mg^{2+}-free PBS), 10 μL progesterone solution (Sigma) (60 μg/mL in 70%), 10 μL sodium selenite solution (Sigma) (52 μg/mL in sterile Ca^{2+}-, Mg^{2+}-free), 100 μL gentamicin solution (Sigma) (50 μg/mL; this solution is supplied by the manufacturer ready to use and may be stored at 4°C; all other solutions, unless otherwise indicated can be stored at –20°C for more than 1 yr). Ensure that the pH of the medium is in the range of 7.15–7.25.
7. [^{3}H]methyl thymidine (25 Ci/mmol) (Amersham, Arlington Heights, IL).
8. G4 monoclonal antibody ***(12)***, ascitic fluid.
9. Krebs-Ringer-HEPES (KRH): 125 m*M* NaCl (7.30 g/L), 4.8 m*M* KCl (0.36 g/L), 1.3 m*M* $CaCl_2 \cdot 2H_2O$ (0.19 g/L), 25 m*M* free-acid HEPES (5.96 g/L), 1.2 m*M* $MgSO_4 \cdot 7H_2O$ (0.30 g/L), 1.2 m*M* KH_2PO_4 (0.16 g/L), 5.6 m*M* Glucose (1.11 g/L); pH 7.3. All reagents were from Merck, except HEPES, that came from Sigma.
10. Biotinylated sheep antimouse IgG species-specific whole antibody (Amersham).
11. Streptavidin-texas red (Amersham).
12. 4% paraformaldehyde solution in Ca^{2+}-, Mg^{2+}-free PBS (heat 100 mL of this buffer containing 4 g paraformaldehyde and 130 μL 1 *M* NaOH until the solution becomes transparent, this solution should be freshly made).
13. DePeX (Serva, Heidelberg, Germany).
14. NTB-2 photographic emulsion (Kodak)/2% glycerol in water (1:2). In the darkness, add one volume of solid NTB-2 emulsion to two volumes of 2% glycerol prewarmed to 42°C. Incubate at 42°C for 15 min and mix the solution avoiding the production of air bubbles.
15. D19 developer (Kodak) diluted 1:1 in water.
16. 25% sodium thiosulfate ($Na_2S_2O_3$) (Sigma) in water.
17. Ca^{2+}-, Mg^{2+}-free PBS/Glycerol(1:1).

3. Methods

E5 chick retina consists mostly of neuroepithelial cells in active proliferation. These cells can be easily dissociated to obtain single cells *(see below)* and thus the influence of particular ECM molecules in neuron differentiation can be analyzed in vitro. After 20 h in culture, cells labeled with [^{3}H]methyl thymidine and a neuronal marker represent neuroepithelial cells that were still in the process of cell division at the time of plating and that have acquired neuronal properties during the time in culture (pipet in vitro differentiated neurons). Here we will describe how to quantify neuronal differentiation using as neuronal marker the glycoprotein G4. This molecule is a member of the immunoglobulin superfamily of cell adhesion molecules that is expressed mostly by retinal ganglion cells but not neuroepithelial cells ***(12)***.

In principle there are two approaches to this kind of in vitro assay, which will be described below. The main application would be simply a comparison of the response of dissociated retinal cells to different substrata, when dissociated cells are plated directly onto different ECM proteins. This has the disadvantage of lacking a real negative control. This problem can be circumvented by the second approach, analyzing the effect of a particular ECM molecule or ECM proteolytic fragment containing domains of interest compared with a neutral substratum. In this case, the substrata that are the object of study are added in a soluble form to the cells cultured onto fibronectin. These two approaches will be further detailed below.

3.1. Coating with ECM Substrata

1. 10-mm round glass coverslips are pretreated with 65% HNO_3 for 1 h and are rinsed several times with distilled water until reaching a neutral pH. Finally, they are rinsed twice in ethanol and left to dry in sterile conditions (under a hood) at room temperature.
2. Coverslips are then transferred to four-well Petri dishes using sterile forceps and coated with polyornithine solution for at least 2 h at 37°C.
3. Remove the polyornithine solution by aspiration and rinse the coverslips three times with distilled water. Coat with 10 μg/mL (diluted in sterile Ca^{2+}-, Mg^{2+}-free PBS) laminin-1, merosin, or fibronectin and incubate at 37°C for 2 h at least.

3.2. Differentiative Bioassay Using ECM Molecules as a Substrate

1. Two neuroretinas from an E5 chick embryo are used as the source of biological material. The procedure of dissection is as follows (*see* **Fig. 1**). First, remove the eyes using curved iris forceps and peel off the surrounding mesenchyme. Then, transfer the eyes to a Petri dish covered with a thin layer of Ca^{2+}-, Mg^{2+}-free PBS leaving the optic nerve exit facing up. Pick up the pigmented epithelium using Dumont #5 tweezers and peel it off carefully. The dark color of this tissue facilitates its recognition. When the equator is reached, put the eye upside down and remove the rest of the pigment epithelium by pulling up while the rest of the eye is held with one of the tweezers. The neuroretina now can be isolated easily by removing the lens and vitreous body (the transparent, gelatinous mass inside the eye).
2. Mix 800 μL Ca^{2+}-, Mg^{2+}-free PBS with 100 μL 30 mg/mL BSA and 50 μL freshly prepared trypsin solution (10 mg/mL in Ca^{2+}-, Mg^{2+}-free PBS) and incubate the two retinas in this buffer for 12 min at 37°C with gentle shakings every 3 min. After this time the retinas should look soft (*see* **Note 1**)
3. Immediately stop the enzymatic reaction by adding 100 μL soybean trypsin inhibitor (10 mg/mL).
4. Cells are then dissociated by passing the suspension gently through a fire-polished, enlarged bore of a Pasteur pipet five times (*see* **Note 2**). The cell density is then estimated using a hemocytometer.

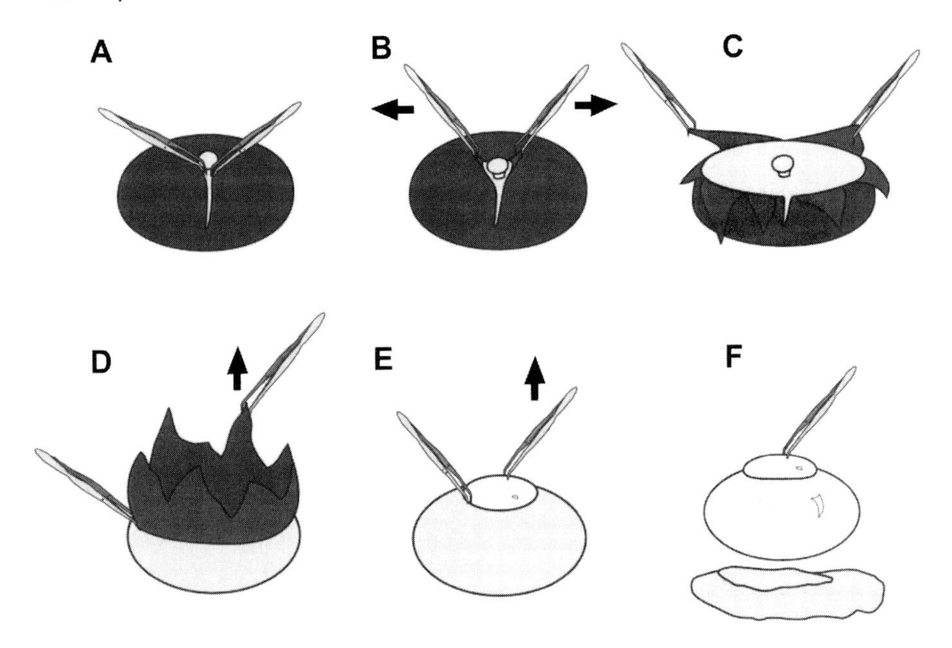

Fig. 1. Dissection of the chick neuroretina (tissue depicted in grey) from the pigmented epithelium (black) and lens/vitreous body (white). *See text* for details.

5. A volume of the cell suspension is diluted in DMEM/F12 Ham (Sigma) medium containing N2 supplement to reach 275,000 cells/mL (after a complete cellular dissociation, around 100 µL of the suspension should be added to 900 µL of medium; never dilute the medium below 50%).
6. Remove the substratum solution from the four-well Petri dishes described in the last section and wash three times with Ca^{2+}-, Mg^{2+}-free PBS. Then, remove this buffer, transfer 55 µL of the cell suspension and incubate for 1 h at 38°C in a water-saturated atmosphere containing 5% CO2. Most cells will attach to either substratum within 20 min of plating. Finally, add 2 mL of previously prewarmed DMEM/F12 Ham medium containing N2 supplement and 0.5 µCi/mL [^{3}H]methyl thymidine and incubate the same conditions for 20 h.
7. After 20 h in vitro, cells are immunolabeled with a neuronal marker (G4). First, the coverslips are picked up with fine forceps and washed three times with KRH. Then they are put in new four-well Petri dishes and incubated for 20 min in 1:100 dilution of G4 mAb ascitic fluid in KRH at room temperature. After this incubation time, coverslips are rinsed again three times with KRH and incubated with biotinylated goat antimouse antibodies diluted 1:100 in KRH for 20 min at room temperature. Finally, the coverslips are rinsed again three times with KRH and incubated in Streptavidin-Texas Red (1:100) in KRH for 20 min at room temperature. After this incubation, the coverslips are rinsed in KRH three times and then the cells are fixed by incubation in 4% paraformaldehyde for 30 min.

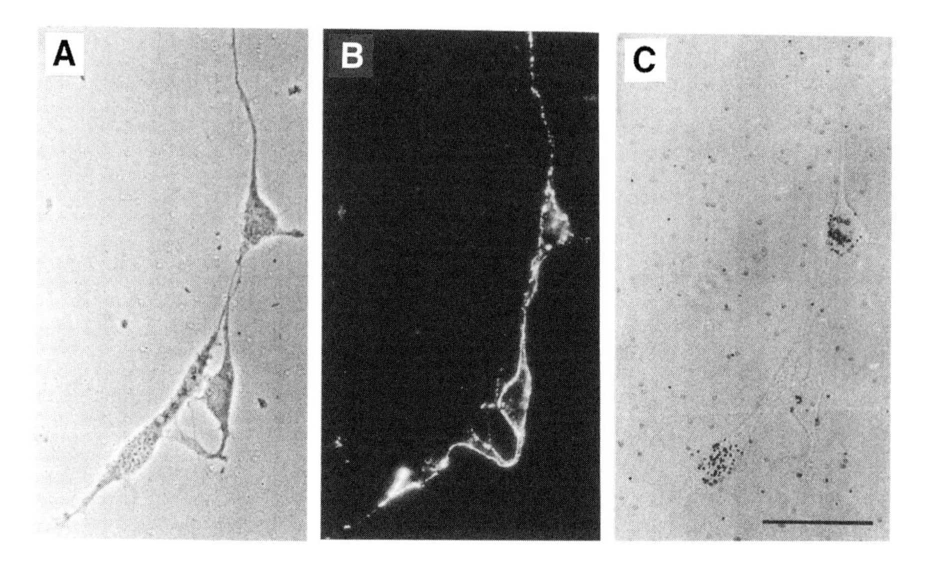

Fig. 2. Effects of laminin-1 on cultured E5 retinal cells. Dissociated cells were cultured on polyornithine/laminin-1 in the presence of 0.5 μCi/**mL** [^{3}H]thymidine. After 20 h in culture, cells were immunolabeled with G4 mAb (specific for neurons) and subjected to autoradiography. Cultured cells were finally observed under phase-contrast (**A**), fluorescence (**B**), and bright-field (**C**) microscopy. Observe the presence of G4-positive, [^{3}H]thymidine-positive neurons. These cells were still in process of cell division at the moment of plating and have acquired neuronal properties during the time in culture (i.e., in vitro differentiated neurons). From *(8)* (with permission).

8. Cultures are then dehydrated by passing the coverslips through 50%-, 70%-, 96%- and absolute ethanol. The coverslips are stuck to a glass slide with DePex and then covered with NTB-2 photographic emulsion and exposed for 2–3 d in the darkness. 100 μL photographic emulsion should be laid onto the coverslip surface containing the cells and quickly removed to leave only a thin film covering the cells.
9. Develop the photographic reaction for 2 min using D19 developer diluted 1:1 in water. Wash once in water, fix in 25% sodium thiosulfate for 5 min and wash extensively with water. Finally, mount coverslips with Ca^{2+}-, Mg^{2+}-free PBS/Glycerol and observe under a fluorescence microscope. Cells double labeled for G4 and [^{3}H]methyl thymidine are considered to be in vitro differentiated neurons (*see* **Fig. 2**).

3.3. Differentiative Bioassay Using Soluble ECM Molecules

Unlike the protocol described previously, which lacks appropriate controls, the protocol described here uses fibronectin as a control substratum, the ECM molecule to be tested is added to the medium subsequently. This protocol is based on the property of fibronectin as a permissive substrate for cell proliferation, but not for neuron differentiation in E5 chick neuroretina cell cultures ***(8,9)***. This protocol is essentially identical to that described in the previous section except that cells are always plated on fibronectin and, after 1 h, medium containing [^{3}H]methyl thymidine and different ECM molecules (10 µg/mL as final concentration) or laminin-1 proteolytic fragments (in equimolar amount compared to the native molecule; P1: 3.22 µg/mL; E8: 2.66 µg/mL) is added.

4. Notes

1. The time used for the incubation of E5 retinas in the presence of trypsin is crucial and should be adjusted whenever a new batch of the enzyme is used.
2. The diameter of the pore of the pipet used for cellular dissociation is crucial for the quality of the cells and should be empirically tested. In our hands a diameter of 1–1.5 mm yields the best results.
3. A control of cell quality should be included. This control is based in the proliferative capability of the dissociated cells cultured on laminin-1 for 20 h. An increase of 2.2–2.6-fold in the number of cells should be obtained under these conditions.

Acknowledgments

We thank C. Bauereiss for help with the preparation of **Fig. 1**.

References

1. Adams, J. C. and Watt, F. M. (1993) Regulation of development and differentiation by the extracellular matrix. *Development* **117,** 1183–1198.
2. Reichardt, L. F. and Tomaselli, K. J. (1991) Extracellular matrix molecules and their receptors: functions in neural development. *Annu. Rev. Neurosci.* **14,** 531–570.
3. Timpl, R., Rohde, H., Robey, P. G., Rennard, S. I., Foidart, J. M., and Martin, G. R. (1979) Laminin - a glycoprotein from basement membranes. *J. Biol. Chem.* **254,** 9933–9937.
4. Tryggvason, K. (1993) The laminin family. *Curr. Op. Cell Biol.* **5,** 877–882.
5. Ehrig, K., Leivo, I., Argraves, W. S., Ruoslahti, E., and Engvall, E. (1990) Merosin, a tissue specific basement membrane protein, is a laminin-like protein. *Proc. Natl. Acad. Sci. USA* **87,** 3264–3268.
6. Davis, G. E., Manthorpe, M., Engvall, E., and Varon, S. (1985) Isolation and characterization of rat schwannoma neurite-promoting factor: evidence that the factor contains laminin. *J. Neurosci.* **5,** 2662–2671.
7. Prada, C., Puga, J., Pérez-Méndez, L., López, R., and Ramírez, G. (1991) Spatial and temporal patterns of neurogenesis in the chick retina. *Eur. J. Neurosci.* **3,** 559–569.
8. Frade, J. M., Martínez-Morales, J. R., and Rodríguez-Tébar, A. (1996a) Laminin-1 selectively stimulates neuron generation from cultured retinal neuroepithelial cells. *Exp. Cell Res.* **222,** 140–149.

9. Frade, J. M., Martí, E., Bovolenta, P., Rodríguez-Peña, M. A., Pérez-García, D., Rohrer, H., et al. (1996b) Insulin-like growth factor-I stimulates neurogenesis in chick retina by regulating expression of the a6 integrin subunit. *Development* **122,** 2497–2506.
10. Beck, K., Hunter, I., and Engel, J. (1990) Structure and function of laminin: anatomy of a multidomain glycoprotein. *FASEB J.* **4,** 148–160.
11. Hamburger, V., and Hamilton, H. (1951) A series of normal stages in the development of chick embryo. *J. Morphol.* **88,** 49–92.
12. Rathjen, F. G., Wolf, J. M., Franck, R., Bonhoeffer, F., and Rutishauser, U. (1987) Membrane glycoproteins involved in neurite fasciculation. *J. Cell Biol.* **104,** 343–353.

22

Migration Assays for Oligodendrocyte Precursor Cells

Emma E. Frost, Richard Milner, and Charles ffrench-Constant

1. Introduction

In the developing central nervous system (CNS), oligodendrocyte precursor cells (OPCs) originate in discreet regions, which may be at a significant distance from their final position. These cells must, therefore, undergo extensive migration in order to reach this final destination *(1)*. The regulation of OPC migration remains to be fully elucidated, with the role of external factors such as extracellular matrix proteins, and growth factors, as well as cell–cell interactions with other neural cells currently being investigated. In order to assess the possible regulation of OPC migration, we have refined two separate assays that present the OPCs with putative regulatory molecules in different forms.

First, a modified Boyden chamber, used to analyze directed migration of cells, and the role of haptotactic, as well as chemotactic stimuli, and their potential synergy, in regulating OPC migration (*see* **refs.** ***2*** and ***3***). In this assay, cells migrate through pores in a filter, from an upper chamber containing a high concentration of cells, to a lower chamber containing a soluble chemoattractant. In addition the filter can be coated on the lower surface and in the pores to form a haptotactic gradient. Second, an agarose drop assay (originally described by Varani et al. 1978 *[4]*) in which cells migrate away from a small drop containing a high concentration of cells, over an appropriate substrate. In this assay cell migration is driven by the innate motility of the cells and the high concentration of cells in the drop, and allows the analysis of chemokinetic or random migration, in the presence or absence of haptotactic stimuli or growth factors, (*see* **refs.** ***5*** and ***6***).

From: *Methods in Molecular Biology, vol. 139: Extracellular Matrix Protocols*
Edited by: C. Streuli and M. Grant © Humana Press Inc., Totowa, NJ

1.1. Advantages and Disadvantages of the Migration Assays

1.1.1. Chemotaxis Chamber

The advantages of the microchemotaxis assay are that it allows for the simultaneous assay of 48 wells per chamber, allowing either several different chemotactic agent concentrations to be assayed in a single experiment, or chemotactic and haptotactic stimuli to be assayed together. It is a quick assay, with significant migration often being seen within 8 h (or less) allowing large numbers of experiments to be conducted in a relatively short period of time. An added bonus is the small volume required for each well, reducing the cost of experiments involving blocking antibodies and growth factors. Although counting the migrated cells is labor-intensive, it is a precise and objective process with low-error rates of less than 10% between operators. The major disadvantage of this assay is that it requires large numbers of cells to be prepared ($1.5–2.0 \times 10^6$ cells per chamber), which may present problems when primary cells are being studied.

1.1.2. Agarose Drop Assay

The advantages of the agarose drop assay are that it allows migratory behavior of the same starting population of cells to be examined under many different conditions, e.g., with different extracellular matrix (ECM) substrates and inhibitors. It also enables the extent of cell migration to be examined at more time-points, e.g., a single experiment with removal of the plate from the incubator and cell migration measured from each drop before the plate is returned to incubator. This assay permits interventions to be made with the introduction and subsequent removal of inhibitors or blocking antibodies after the start of the assay, thus examining the reversible nature of blocking agents during the course of one experiment. Information about the morphology of migrating cells on different substrates and under different conditions, can also be determined using this assay. The disadvantages of this assay are that it takes several days, during which time cells may synthesize and secrete their own ECM in addition to the ECM molecule under investigation. Promigratory agents, which are also mitogenic such as growth factors, will result in an increase in cell numbers, which may be misinterpreted as cells actively migrating out of the drop. Finally, the assay is conducted in larger volumes, i.e., a 24-well plate, and therefore also requires larger volumes of growth factors and or blocking antibodies.

2. Materials

2.1. The Chemotaxis Chamber

1. Nucleo Pore™ 48-well chemotaxis chambers (Corning Costar, Cambridge, MA, Cat. No. AP48) (*see* **Note 1**).

2. Nucleo Pore™ 8-µm pore polypropylene filter beds—Corning Costar (Cat. No. 155846) (*see* **Note 2**).
3. Cells.
4. Growth medium for your cells. For these studies, Sato's modification DMEM is used.*
5. Bacterial grade plastic Petri dishes (i.e., not tissue culture treated).
6. Sterile 10 mL pipets.
7. P1000, P200, P20 variable pipets.
8 Sterile pipet tips.
9. Blunt ended forceps, two pairs.
10. 100% methanol.
11. Eosin, 1% aqueous solution (Merck, Darmstadt, Germany, Cat. No. 35084 4K).
12. Haemotoxylin, Ehrlich's original formulation (Merck Cat. No. 35017 5T).
13. 3 Beakers or staining jars.
14. DePex (Merck Cat. No. 36125 2B).
15. Humidified chamber (e.g., plastic box with wet tissues in the bottom).
16. Phosphate buffered saline (PBS), sterile, tissue culture grade.
17. Protein of interest, e.g., extracellular matrix protein.
18. Chemotactic agent of interest, e.g., growth factor such as platelet-derived growth factor (PDGF).
19. Poly-D-lysine (PDL) (Sigma P-7405).
20. 0.5 *M* acetic acid.
21. Sterile distilled water.
22. Terg-A-Zyme (Alconox, Inc., New York) or other nonaggressive enzyme cleaner. Not Decon.

2.2. Agarose Drop Migration Assay

1. Low melting point agarose (Sigma A-9045).
2. Phosphate buffered saline (PBS) sterile, tissue culture grade.
3. Microwave oven.
4. Water bath set at 37°C.
5. Cells.

*Sato's modification of DMEM for oligodendrocyte precursor migration assay. There are several versions of Sato's modification of DMEM. The following recipe has been used in this lab for 5 yr without problems:

Dulbecco's Modified Eagle's Medium, high glucose with pyruvate (DMEM) (Sigma D-6546); Penicillin/Streptomycin solution f.c. 100 U/mL (Life Technologies - Cat. No. 15140 114); L-Glutamine (Sigma G-6392) f.c. 4 m*M*; Putrescine (Sigma P-5780) f.c. 16 µg/mL; Thyroxine (Sigma (T4) T-1775) f.c. 400 µg/mL; Tri-Iodothyroxine (Sigma (T3) T-6397) f.c. 400 µg/mL; Progesterone (Sigma P-8783) f.c. 66 ng/mL; Sodium Selenite (Sigma S-5261) f.c. 5 ng/mL; Bovine Serum Albumin (BSA) Fraction V (Sigma A-4919) f.c. 100 µg/mL; Insulin (Sigma I-1882) f.c. 5 µg/mL; holo-Transferrin (human) (Sigma T-0665) f.c. 50 µg/mL. T3 and T4 do not dissolve in water/DMEM, they need to be solubilized in 0.1 *M* NaOH, and 0.2 *M* NaOH, or 4.0 *M* NH_4OH in 75% MetOH, respectively.

6. Growth medium for your cells. (*See* **Subheading 2.1.4.**)
7. Variable pipets, P20, P100, and P1000.
8. Sterile pipet tips.
9. 24 Multiwell tissue culture treated plates (e.g., Nunc, Wiesbaden-biebrich, Germany, Cat. No. 143982).
10. Fetal bovine serum (FBS) for rat derived oligodendrocytes, or Horse Serum (HS) for mouse derived cells.
11. Sterile microfuge tubes (1.5 mL volume).
12. Plastic beaker.
13. Protein of interest, e.g., extracellular matrix protein or growth factor.

3. Methods

3.1. The Chemotaxis Chamber

For the purposes of this description, the cell type in use is the oligodendrocyte precursor cell, for which the growth medium would be Sato's modification of DMEM *(7)*.

1. Soak the filter in 0.5 *M* acetic acid for 40 min at 60°C, or overnight at room temperature
2. Carefully, wash the filter twice in sterile distilled water (*see* **Notes 2** and **3**).
3. Precoat with 5 µg/mL PDL for a minimum of 1 h (*see* **Notes 4**).
4. Wash the filter carefully, twice, in sterile distilled water.
5. Dry the filter (*see* **Note 5**).
6. If coating the filter with ECM protein, place shiny side down on a drop of ECM protein in PBS, and on a clean microscope slide in a humidified chamber, at 37°C; for a minimum of 1 h, preferably overnight. Carefully, rinse the filter in PBS before placing onto chamber (*see* **Note 6**).
7. Subtract microglia from shaken cells, by differential adhesion to nontissue culture treated plastic, for 20–25 min (*see* **Note 7**).
8. During which time, prepare the solutions for the chamber, e.g., PDGF dose response curve ± other factors.
9. Pipet-test solutions into the bottom plate of the chemotaxis chamber (*see* **Notes 8** and **9**).
10. Snip a corner off the filter to allow for orientation after staining. Once the filter is stained, it is virtually impossible to reorientate it correctly without snipping off a corner. Orientation of the filter is required to not only identify which side to wipe the cells off, but also to identify which wells are which (*see* **Note 10**).
11. Carefully place the filter shiny side down on the bottom plate making sure to not move the filter once it is placed, otherwise, you might cross-contaminate the test solutions (*see* **Note 11**).
12. Carefully place the silicon gasket on top of the filter (*see* **Note 12**).
13. Place the top plate on the chamber, press down with one hand, and maintain even pressure to avoid bubbles that can form in the lower plate. Then, screw down the thumb nuts, making sure that all are as finger tight as possible (*see* **Note 13**).

14. Place the chamber in a humidified chamber at 37°C until you are ready to put the cells in (*see* **Note 14**).
15. Remove the cell suspension from the Petri dishes and centrifuge at 240*g* for 5 min (*see* also **Note 7**).
16. Remove supernatant and resuspend the cells in growth medium.
17. Triturate the cells for 30 s, rest for 30 s, and retriturate, repeating this three times (*see* **Note 15**).
18. Take a small aliquot and count the cells.
19. Dilute the cells to achieve 35–40,000 cells per 50 μL, and add any test solution to the cells.
20. Pipet 50 μL cells into each of the wells in the upper plate, taking care to avoid any bubbles forming on the top of the filter (*see* **Note 16**).
21. Replace in the humidified chamber, at 37°C, for the requisite period of time, 8 or 16 h (*see* **Note 17**).
22. At the completion of the experiment, carefully remove the thumb nuts, lift the top plate off the chamber, and using forceps place the filter in 100% methanol for 2 min to fix the cells (*see* **Notes 18–20**).
23. Stain with eosin for 2 min.
24. Stain with haemotoxylin for 1 min.
25. Rinse in distilled water, then differentiate the stain in tap water for 10–20 s, then rinse in distilled water.
26. Place the filter shiny side down on a clean microscope slide taking care to avoid the filter folding during handling (*see* **Note 21**).
27. Wipe the unmigrated cells off the top of the filter using a wet cotton bud (*see* **Note 22**).
28. Dry the filter using a dry cotton bud or a tissue.
29. Mount on DePex.
30. Count the cells using a light microscope (*see* **Notes 23** and **24**).

3.2. Agarose Drop Migration Assay

For the purposes of this description the cell type in use is the oligodendrocyte precursor cell, for which the growth medium would be Sato's modification of DMEM *(7)*.

1. Make a solution of 1% agarose, for example, 50 mg of agarose in 5 mL of PBS, and dissolve in a microwave oven. When dissolving agarose, you need to microwave it for a few short periods of time, making sure that you do not overheat the mixture, and mix between each period of heating. Once the agarose is fully dissolved, the tube can be placed in a 37°C water bath to prevent premature gelling of the agarose (*see* **Note 25**).
2. Make up the working solutions for resuspending the cells, e.g., Sato's medium plus 10% FBS, for rat oligodendrocyte precursor cells or horse serum (HS), for mouse oligodendrocyte precursor cells, and warm to 37°C in the water bath prior to the experiment.

3. Obtain the cells for investigation, by overnight mechanical shaking of mixed glial cultures for oligodendrocyte precursors, centrifuge to a pellet, at 240*g* for 5 min.
4. Remove the supernatant and resuspend the pellet in 40 µL of Sato's medium containing FCS (or HS), triturate up and down vigorously several times before transferring exactly 40 µL of the cell suspension to a clean microfuge tube (*see* **Note 26**). The cell density used varied between 20 and 100 × 10^6 cells/mL (*see* **Note 27**).
5. Transfer the tube with the cell suspension in to the 37°C water bath to equilibrate the temperature. After 2 min or so, transfer the tube to a plastic beaker containing 37°C water from the bath, to maintain the temperature during subsequent handling in the TC hood. Add 20 µL of the 1% agarose to the cells and mix thoroughly.
6. Immediately, plate 1.5 µL drops of the cell suspension at the centre of wells in a 24-well tissue culture plate. Place the plate in a refrigerator at 4°C for 15 min to allow the agarose drop to set (*see* **Notes 28** and **29**).
7. As the drops are setting, prepare the solutions to be added around the drop. In the case of migration on PDL, the solution to be added would be Sato's medium alone. However, to examine cell migration on different ECM molecules, stock solutions of ECM should be diluted into Sato's medium to give a final concentration of 10 µg/mL.
8. After the drop has been at 4°C for 15 min, carefully add 50 µL Sato's medium media (+/– ECM) around the drop (*see* **Note 30**). When ECM solution is being used, the dish should then be placed into the incubator at 37°C for 2 h in order for the ECM solution to coat the plastic surface.
9. After 2 h, a further 450 µL of Sato's medium can be carefully added (down the side of the well) before returning the plates to the incubator (*see* **Note 31**).
10. The extent of cell migration can then be measured at daily intervals for 1–5 d using a phase contrast microscope with a calibrated graticule in the eyepiece, in which the width of one grid square represents 100 µm actual distance at a magnification of 10×. Cells migrate out to form a uniform corona around the drop (*see* **Fig. 1**). At any one time point, the distance between the edge of the drop and the leading edge of migrating cells within the corona can be recorded on four sides of the drop. We routinely count at 0, 90, 180, and 270° angles around the drop (*see* **Note 32**). A distribution curve can be established by measuring the number of cells at various distances from the edge of the agarose drop.
11. The cells should be allowed to migrate out of the drop for 24 h before any function blocking reagents are introduced. This allows direct analysis of the inhibition of cell migration per se, as opposed to the reagent interfering with the ability of the cells to exit from the agarose drop. The inhibitors are added carefully, down the side of the wells, in volumes of 5–20 µL.
12. In order to discount cell proliferation as a factor in this assay, aphidicolin at 20 µL/mL can be included in the surrounding media. Aphidicolin inhibits DNA replication, and thus the contribution of cell proliferation to the expanding corona of cells around the drop can be discounted ***(11)***.

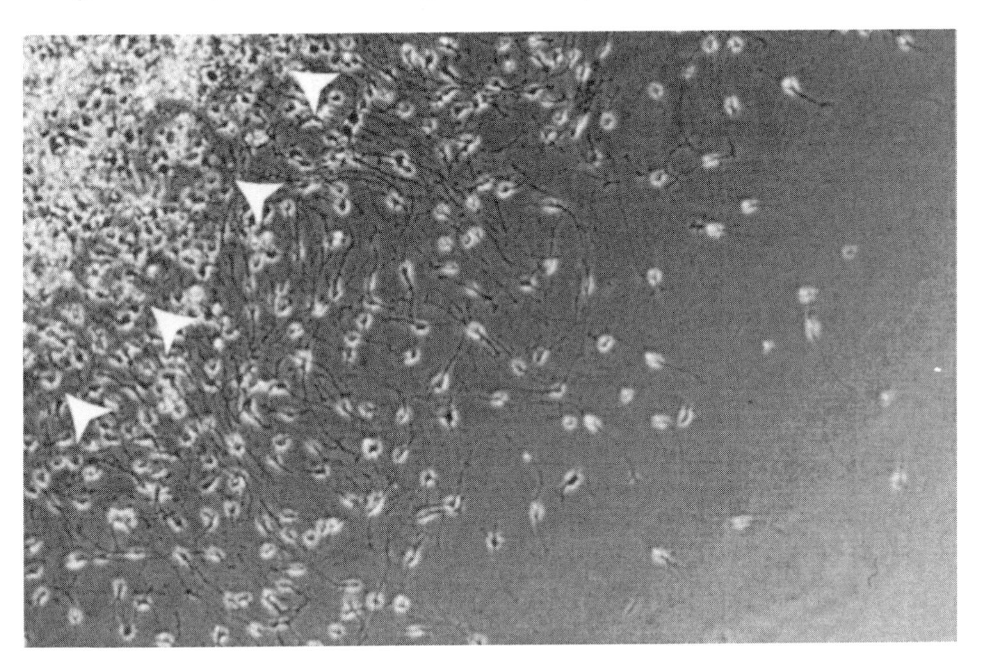

Fig. 1. Migration away from an agarose drop. Arrow heads indicate the edge of the drop.

4. Notes

1. The chemotaxis chambers are made of acrylic, where they scratch, pit, and warp relatively easily. Do not over tighten the thumb nuts. Exercise caution when handling, and never use solvents to clean them. Always take great care not to scratch the surface, which might compromise the water tight seal required for successful assembly of the chamber, keep it clean, and do not allow solutions to dry in the wells. If you need to clean the wells, use cotton buds soaked in a nonaggressive soap solution. Clean the chamber between each experiment, using Terg-A-Zyme, or other proprietary, nonaggressive, enzymatic cleaning solution. Replace the gaskets periodically, as the holes for the retaining pins begin to enlarge with use, and small scratches appear regardless of how carefully you treat them.
2. When handling the filters, do not touch them with bare hands to avoid greasy marks. The filters are very fragile, and tear very easily, they also puncture, stretch, and crease easily. Great care is required during handling. Flat-tipped forceps are the best handling tool, one for each end of the filter, to avoid it folding during placement on the chamber or transfer from solution to solution.
3. When washing and precoating the filters, it is easiest to immerse them in the solution.
4. Prepare the filters for each experiment as you need them. Although you can store PDL-coated filters for several weeks, the quality of the coating deteriorates with time and is noticeable when staining.

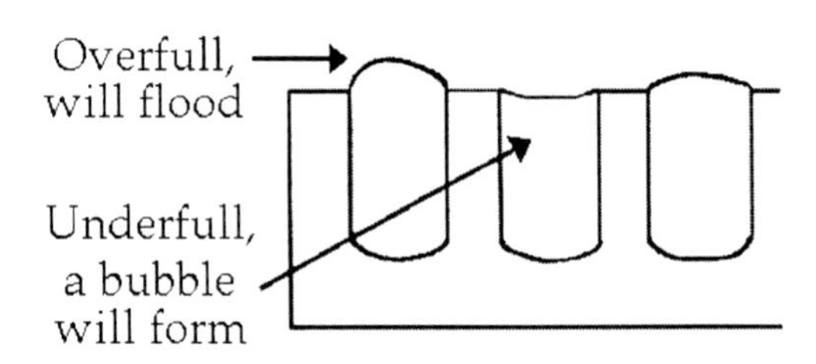

Fig. 2. Schematic representation of filled wells, indicating the level required to avoid flooding or bubbles. This experiment has been set up to assay the effect of blocking antibody on PDGF-stimulated migration. It allows six replicates of each PDGF concentration in the presence or absence of antibody.

5. When drying the filter, suspend it from the roof of a horizontal laminar flow hood by a bulldog clip stuck to the hood with sticky tape. This way the filter is held secure, and dries quickly without touching anything else. When removing from rinse before drying, do not blot the filter to speed up the drying, as this may affect cell migration, either by blocking the pores, or by affecting the homogeneity of the adhesive coating.
6. When coating the filter with ECM proteins, place the filter onto a 100–200 μL drop of protein on a clean microscope slide in the humidified chamber.
7. The microchemotaxis chamber can be used for several different cell types *(8)*, including neutrophils *(9)* and leukocytes *(10)* as well as oligodendrocyte precursor cells *(2)*. OPCs are usually obtained from mixed glial cultures, and contaminating microglial cells need to be subtracted from the cell population prior to the assay. Preparation of each cell type to be used will obviously differ.
8. If the chamber is not quite dry before you use it, aspirate the wells using a yellow Gilson tip, do not use glass which might scratch the chamber.
9. When placing your test solution into the wells of the lower plate, it is important to form a positive meniscus on the top of the well (*see* **Fig. 2**). A negative meniscus will result in a bubble forming on the underside of the filter. However, it is important to not have too large a meniscus, otherwise you may get flooding and cross-contamination from well to well. You will find that each chamber has a different lower well volume, and this can vary by several microliters. It helps to have already ascertained the volume for each chamber before you start your first experiment. The volumes should not change with use of the chambers, however, your pipets might change, and recalibration periodically might also cause slight changes. So check carefully each time, to ensure that you are not under- or overloading the wells.
10. When designing your experiment, it is important to remember to have control and test solutions on each chamber (*see* **Fig. 3**). Although differences between chambers should not occur, it is possible. In order to be able to confidently compare test-against control, it is safest to assume there might be.
11. Place the filter shiny side down onto the chamber. This is according to the manufacturer's instruction, and does not appear to make a difference to migra-

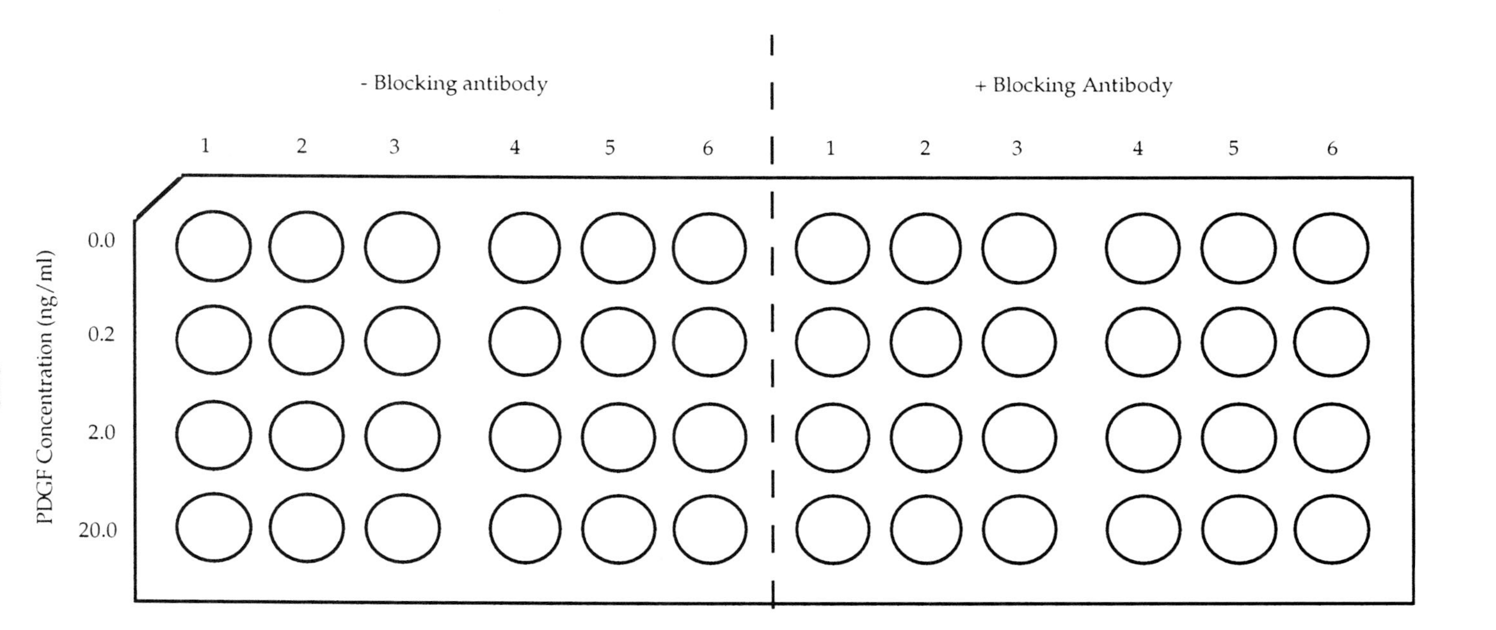

Fig. 3. Example of an experimental design. This experiment has been set up to assay the effect of blocking antibody on PDGF-stimulated migration. It allows six replicates of each PDGF concentration in the presence or absence of antibody.

tion if you get it wrong, but it is as well to be consistent. When you place the filter on the lower plate, first, place the middle of the filter down and allow the ends to "drop" into place. It important to not move the filter after the first placing, to avoid cross-contamination from well to well, or bubble formation on the underside of the filter.

12. The silicon gasket is present to allow formation of a water tight seal between the upper and lower plates of the chamber. It is important to check the gasket for tears, dimples, or other damage which might jeopardize formation of a complete seal, including the little bit of filter you have just cut off the corner.
13. When assembling the chamber, place the silicon gasket onto the pins, without pushing it onto the filter. At this point, if the filter lifts off the lower plate, bubbles might form on its underside. The pins will hold the gasket above the filter until you push the top plate down onto the bottom plate. At this point, you need to keep constant, even pressure on the top plate so that the filter does not lift slightly.
14. Once the chamber is assembled correctly, it is important to place the chamber in an appropriate CO_2 atmosphere, as alkaline conditions will inhibit cell migration, therefore, you need to try to equilibrate the Sato's pH as much as possible. It is also helpful to humidify the chamber, which makes adding the cells easier (*see* **Note 16**).
15. Oligodendrocytes are inclined to clump together, and a single-cell suspension is required for migration. Therefore, it is necessary to disaggregate the cells by trituration prior to plating into the chamber. If you are using a different cell type, this step may be omitted.
16. When adding the cells to the upper well, it is vital that there are no bubbles in the well, which will prevent the cells reaching the filter. There are several ways to do this, but if the chamber has been humidified first it is easier. If the wells are already wet, there is less likelihood of bubble formation when adding the cells, and the small amount of moisture is not sufficient to affect the osmolality of the medium. It helps considerably if the medium in use has been warmed to 37°C and swirled to remove bubbles from the tube. This will allow dissolved gases to come out of the solution before you put it in the well. One way to add the cells is to place the pipet tip on the surface of the filter, then inject the cells in a quick, smooth action. However, this method runs a high risk of puncturing or stretching the filter surface. A better way is to place the pipet tip against the side of the well, angled (*see* **Fig. 4**), injecting the cells in a quick smooth action. If a bubble does form, which you will know because the cell suspension will not fit into the well, the size of the bubble will determine how much above the top of the well the meniscus forms. Simply remove the cell suspension and reinject it, or remove the bubble with the pipet tip, reinjecting any cells that you remove in this process. Sometimes very small bubbles form, which barely affect the meniscus. To check for such bubbles, look directly into the well from above, bubbles can also be seen though the sides of the chamber. It is important to remove any bubbles before the experiment starts. It is also important to remember that if you remove the cell suspension from the well, do not place it back into the stock of cells, because proteins from the filter or lower well may have contaminated the solution, and you would thus cross-contaminate all the remaining cells.

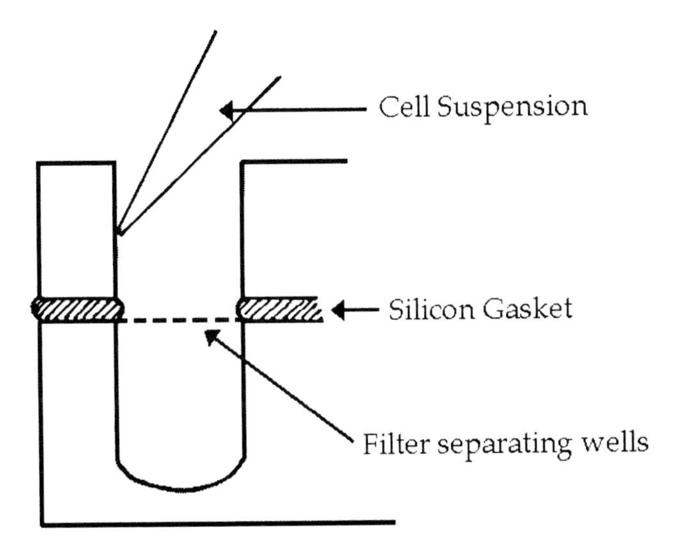

Fig. 4. Schematic diagram indicating the approximate position and angle of the pipet tip when injecting cells into wells. Note that the gasket bulges out, this may cause a bubble to form at the edge of the well.

17. Cells will require different time periods to achieve adequate migration, e.g., microglia will migrate in 4 h *(3)*, as will neutrophils *(9)*, however, 8 h are required for oligodendrocyte migration *(2)*. Therefore, you need to refer back to the literature for your cell type or run preliminary experiments to ascertain the time required by your cells.
18. When disassembling the chamber prior to staining the filters, take care not to touch the underside of the filter where your cells are located.
19. Once fixed in methanol, the filter becomes even more brittle, and has a tendency to fold or roll up, it is important to avoid the filter surfaces touching each other, which runs the risk of cells being transferred from place to place, or being scraped off altogether. This is why two pairs of forceps are useful. The fixing time and eosin-staining time are not critical, but do not over stain in haemotoxylin, otherwise, it becomes difficult to distinguish between cytoplasm and nucleus.
20. Immediately after use, place the chamber in water to avoid the protein solutions drying out on the filter or gasket.
21. This is when snipping the corner off of the filter becomes invaluable as a method of aligning the filters.
22. Take great care when cleaning the unmigrated cells off the upper surface of the filter, as it may tear, fold, or crease at this time. The filters are inclined to lift off the slide and fold over so that you accidentally wipe migrated cells off. It may help to gently wipe the upper surface of the filter with a wet cotton bud to remove most of the cells, then to blot the filter carefully, with a paper towel. This helps the filter stick to the slide better, reducing the risk of it lifting and folding over.

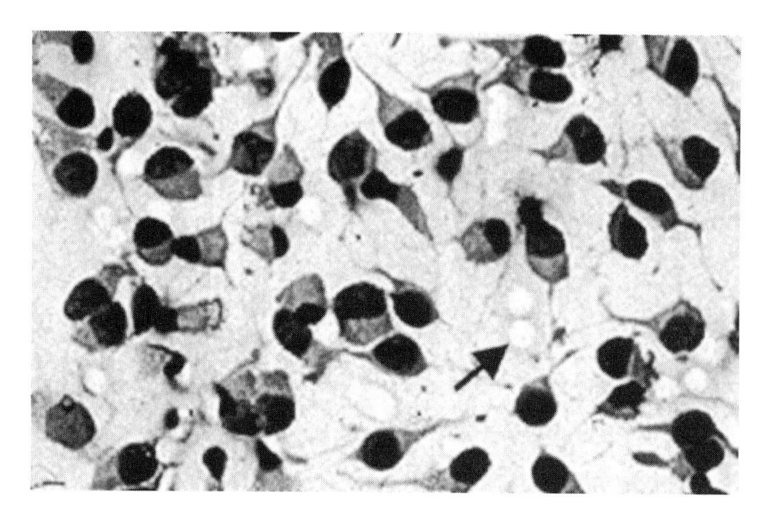

Fig. 5. Photomicrograph of migrated oligodendrocyte precursors on the underside of a filter. A pore in the filter is indicated by the arrow.

23. When counting the cells, count the nuclei only, as some cytoplasm might appear through the pores without the cells migrating through (*see* **Fig. 5**). One method of counting is to use a square graticule, count three or five fields of view per well, depending on magnification (e.g., three fields is adequate at 10×, but five fields are required for 20 or 40×). Your magnification factor will depend on how many cells have migrated, as you will find it impossible to count the cells accurately if you are looking at 600–800 per field of view at 10×. Whereas if you have only 50 cells migrating through your filter, you can count at a lower magnification. As long as you count at the same magnification across each filter, and as long as each filter contains a control against which you compare the results, you can normalize the data and compare across filters/experiments (*see* also **Note 10**).
24. Migration can also be quantified using a spectrophotometric method, based on the extraction of stain from the nuclei of migrated cells (for further details, *see* **ref. *11***).
25. The agarose solution will solidify if taken out of the water bath for any prolonged period of time (>1 min depending on the volume), but can be reliquefied on subsequent days for further experiments by reheating in the microwave oven, or at 65°C in a water bath for 30 min. One preparation can last approximately 1 mo. There is no need to filter either the PBS or the agarose solution because the heating process sterilizes the solution before every experiment.
26. Transfer the cells to a clean tube to obtain a precise volume and avoid drops of solution on the side of the tube running down and adding to the 40 μL volume, thus further diluting the agarose drop, which may prevent it from gelling properly.
27. It is important to optimize the cell density for a particular population of cells; if the density is too low, cells will not survive, and if it's too high, the agarose drop

may disintegrate and fragment. Two ways to do this would be to reduce the absolute volume of cell suspension and agarose, at the same time maintaining the ratio, or by changing the ratio of cells suspension to agarose. Trial and error is required to do this successfully.

28. In order to produce a nice round drop, the pipet tip can be lowered to make contact with the plastic at a perpendicular angle and then the 1.5 µL volume released. The substrate for plating the drops should always be dry; if not the 1.5 µL tends to spread further and unevenly, and does not produce an even drop. The number of drops per dish can be limited, e.g., to 12, to prevent excessive desiccation of the drops while plating.
29. It is very easy to dry out the drop excessively during this procedure. This will manifest an excessive amount of cell death, particularly at the periphery of the drop. This can be reduced by either cutting down on the time the drops are setting in the refrigerator (e.g., from 15 min to 7–10 min) or by plating a smaller number of drops on each plate (instead of 12 per plate, try only four or six).
30. Add the medium to the gelled drop by taking up 50 µL of Sato's medium into the pipet tip and then drawing a circle around the agarose drop by slowly releasing the media. Then, adding the remainder of the 50 µL into the center of the circle, cover the agarose drop. This should be done very carefully in order to avoid excessive mechanical forces detaching the agarose drop from the plastic dish. It is also very easy to detach the drop when adding the solution around the drop. This can be limited by adding the surrounding 50 µL of media in a more gentle fashion, or alternatively, by placing the drop straight onto a PDL substrate (*see* **Note 31**). In addition, the plates should be handled with care and not moved abruptly throughout the experiment.
31. Because the drops may adhere in the early stages, it is not uncommon for them to detach during the course of the experiment. This can be avoided by placing the drops into wells that have been precoated with PDL all over. However, when cell migration on ECM substrates is to be examined it is advisable not to precoat the entire surface with PDL because this tends to mask the difference in the effects of the different ECM molecules. In this case, one approach is to mark the underside of the well with a fine black point and then add 1 µL PDL onto the plastic. This then evaporates, but acts as additional anchorage for the agarose drop to adhere to (the black point allows precise positioning of the agarose drop after evaporation of the PDL).
32. It is essential that the counting procedure remains constant throughout, and between experiments, you need to ensure that the plates are always aligned in a uniform manner on the microscope prior to counting the cells.

Acknowledgments

The authors would like to thank Marion Perryman for her technical advice on the chemotaxis chamber, and Chandike Mallawaarachchi for his helpful comments on the manuscript. This work has been funded by the Multiple Sclerosis Society and the Wellcome Trust.

References

1. Kiernan, B. W. and ffrench-Constant, C. (1993) Oligodendrocyte precursors (O-2A progenitor cells) migration; a model system for the study of cell migration in the developing central nervous system. *Development* **(Suppl.)**, 219–225.
2. Frost, E., Kiernan, B., Faissner, A., and ffrench-Constant, C. (1996) Regulation of oligodendrocyte migration by extracellular matrix: evidence for substrate specific inhibition of migration by tenascin-C. *Dev. Neurosci.* **18**, 266–273.
3. Armstrong, R. C., Harvath, L., and Dubois-Dalq, M. E. (1990) Type-1 astrocytes and oligodendrocyte-Type 2 astrocyte glial progenitors migrate toward distinct molecules. *J. Neurosci. Res.* **27**, 400–407.
4. Varani, J., Orr, W., and Ward, P. A. (1978) A comparison of the migration patterns of normal and malignant cells in two assay systems. *Amer. J. Pathol.* **90**, 159–172.
5. Milner, R., Edwards, G., Streuli, C., and ffrench-Constant, C. (1997) A role in migration for the $\alpha v\beta 1$ integrin expressed in oligodendrocyte precursors. *J. Neurosci.* **16**, 7240–7252.
6. Milner, R., Anderson, H. J., Rippon, R. F., McKay, J. S., Franklin, R. J. M., Marchionni, M. A., Reynolds, R., and ffrench-Constant, C. (1997) Contrasting effects of mitogenic growth factors on oligodendrocyte precursor migration. *Glia* **19**, 85–90.
7. Bottenstein, J. E. and Sato, G. H. (1979) Growth of a rat neuroblastoma cell line in serum-free supplemented medium. *Proc. Natl. Acad. Sci. USA* **76**, 514–517.
8. Grotendorst, G. R. (1987) Spectrophotometric assay for the quantification of cell migration in the Boyden Chamber chemotaxis assay. *Methods Enzymol.* **147**, 144–152.
9. Harvath, L., Falk, W., and Leonard, E. J. (1980) Rapid quantification of neutrophil chemotaxis: use of a polyvinylpyrrolidone free polycarbonate membrane in a multiwell assay. *J. Immunol. Methods* **37**, 39–45.
10. Falk, W., Goodwin, R. H., and Leonard, E. J. (1980) A 48-well microchemotaxis chamber assembly for rapid and accurate measurement of leukocyte migration. *J. Immunol. Methods* **33**, 239–247.
11. Ikegami, S., Taguchi, T., Ohashi, M. Oguro, M., Nagamo, H. and Mano, Y. (1978) Aphidicolin prevents mitotic cell division by interfering with the activity of DNA polymerase-α. *Nature* **274**, 458–460.

23

Cell Adhesion Assays

Martin J. Humphries

1. Introduction

This chapter will outline in detail the two standard assays used in the author's laboratory for quantitating the adhesion of cells to an immobilized substrate. The attachment assay, which employs a colorimetric detection of bound cells, is based on Kueng et al. *(1)*, and the spreading assay, which employs phase contrast microscopy to measure the flattening of adherent cells, is based on the method of Yamada and Kennedy *(2)*.

It is important to realize that cell adhesion is a complex process that involves many different molecular interactions, including receptor-ligand binding, changes in the fluxes through intracellular signaling pathways, and modulation of cytoskeletal assembly. Consequently, adhesion assays not only measure the contacts between a cell and extracellular adhesion proteins, but also provide information about other cellular events. For this reason, care needs to be taken before choosing to perform adhesion assays. The most common uses of adhesion assays are a) to test the ability of a specific type of cell or cell line to adhere to a specific adhesive substrate, and b) to test the sensitivity of a specific cell-substrate interaction to inhibitors, but it is also apparent that adhesion assays can be used to probe the contribution of other cellular processes.

A number of factors will affect the decision whether to use a cell-spreading assay or a cell-attachment assay. Spreading assays take longer to perform, but are less prone to nonspecificity. For example, many molecules can mediate attachment of cells in a nonphysiological manner, but very few of these molecules are able to mediate spreading. In addition, information can be gained about the ways in which the cells respond to the substrate by observing cells in a spreading assay. For example, the morphology of cells can differ on different substrates even if the level of spreading is the same, and now that our under-

From: *Methods in Molecular Biology, vol. 139: Extracellular Matrix Protocols*
Edited by: C. Streuli and M. Grant © Humana Press Inc., Totowa, NJ

standing of the signaling mechanisms that control cell morphology has improved, spreading assays can give indirect indications of the intracellular events that are triggered by certain substrates. A further advantage of spreading assays is that they are more sensitive when used to measure inhibitory activity because the read-out from the assay is more reliant than attachment assays on multiple adhesive interactions and partial disruption by an inhibitor is sufficient to see blockade. Finally, because spreading assays do not need replicate wells, they are more economical in their use of substrates. As described below, both types of adhesion assay are relatively quick to carry out. The actual assays can be performed in half a day, however the quantitation of spreading assays can take a similar length of time.

Sometimes there is no alternative to an attachment assay, because some cells are unable to spread at all, whereas other cells can only spread on specific substrates. Although attachment assays still require multiple cell-substrate contacts to allow the cell to withstand the washing steps in an attachment assay, fewer contacts are needed than for a cell to spread.

Both the spreading and attachment assays are expressed as percent adhesion and it can be expected that the level of adhesion obtained will vary depending upon the cell type and adhesive substrate under study. In spreading assays, a level of 80% is common, and often higher levels can be obtained. Most importantly, the background level of spreading on BSA-coated plastic should be as low as possible. Frequently, this is actually zero, but certainly this should not rise above a few percent. The level of attachment observed by the dye-staining method is usually not as high as for spreading, but 60–70% should be attainable. A low figure for the BSA background is also important for attachment assays, but generally it is difficult to reduce this level below 5% without adversely affecting the experimental signal (*see* **Note 1**).

2. Materials

2.1. Common to Both Assays

1. 10 mg/mL heat-denatured BSA in divalent cation-free Dulbecco's PBS. This is a better blocking agent than native BSA solutions. After dissolving the BSA, filter through a 0.22-µm filter to remove undissolved protein, and incubate in a water bath at 85°C for 10–12 min. The solution should be slightly hazy; it should not be clear, as this will contain insufficiently aggregated BSA, nor should it be white, because here the aggregates will be too large. After cooling, it is ready for use. Occasionally, some cell types find heat-denatured BSA toxic and therefore care should be taken to wash the wells of the microtiter plate after blocking.
2. 96-well tissue-culture-treated microtiter plate (Costar, Cambridge, MA) (*see* **Note 2**).
3. Dulbecco's PBS (Gibco-BRL, Gaithersburg, MD).

4. Dulbecco's MEM with 25 m*M* HEPES (Gibco-BRL).
5. 5% (w/v) glutaraldehyde. This solution is hazardous and skin contact should be avoided.

2.2. For Attachment Assay Only

1. 0.1% (w/v) crystal violet, 200 m*M* MES, pH 6.0. It is important to filter this solution, as its intense blue color can make it difficult to determine whether it has dissolved properly. If the solution is not filtered, specks of solid crystal violet can also be added to the experimental wells, resulting in high absorbance readings. This solution is hazardous and skin contact should be avoided.
2. 10% (v/v) acetic acid.

2.3. For Spreading Assay Only

Dulbecco's divalent cation-free PBS (Gibco-BRL), 0.05% (w/v) sodium azide. This solution is hazardous.

3. Methods

3.1. Attachment Assay

1. Dilute adhesion molecule with Dulbecco's PBS.
2. Add the diluted adhesion molecule to the wells of the microtiter plate (100 µL/well). Leave a blank well or wells for measuring background spreading on blocked plastic (*see* **Note 2**).
3. Incubate at room temperature for 60 min or at 4°C overnight (*see* **Note 4**).
4. Aspirate, add 200 µL 10 mg/mL heat-denatured BSA in divalent cation-free PBS and incubate at room temperature for 30 min.
5. While the blocking is underway, prepare a suspension of the cells to be examined (*see* **Note 5**).
6. Count the cell density on a hemocytometer, resuspend the cells to a concentration of 5×10^5/mL for fibroblasts and similarly sized cells and 10^7/mL for lymphoid cells in warm DMEM/HEPES (gassed with 5% [v/v] CO_2), and incubate at 37°C in a 15-mL polypropylene tube for 10 min. For experiments examining the effects of inhibitory agents, aspirate the BSA, add 50 µL of 2X agent diluted into PBS followed by 50 µL cells. Add 50 µL PBS followed by 50 µL cells for control wells (*see* **Note 6**).
7. The incubation time chosen for attachment assays depends on the cell type, as some cells adhere more quickly than others, but 15–20 min is usually adequate.
8. Fix cells in the wells to be used for determining 100% attachment value by addition of 100 µL 5% (w/v) glutaraldehyde. Assays are usually performed in triplicate or quadruplicate.
9. Remove nonadherent and loosely attached cells by either tapping the plate or gently washing the wells with one or more 100-µL aliquots of PBS (*see* **Note 7–9**).
10. Aspirate the final wash and fix attached cells by addition of 100 µL 5% (w/v) glutaraldehyde for 20 min at room temperature (or at 4°C overnight if necessary).

11. Wash wells with 3 × 100 µL water.
12. Stain with 100 µL 0.1% (w/v) crystal violet, 200 m*M* MES, pH 6.0 for 60 min at room temperature (*see* **Note 10**).
13. Wash wells with 3 × 400 µL water.
14. Solubilize dye in 100 µL 10% (v/v) acetic acid and incubate on orbital shaker at 150 rpm for 5 min at room temperature.
15. Measure absorbance at 570 nm using a plate reader. As described below in **Subheading 4.**, use the data from 20%, 50%, and 100% inocula to determine the value for 100% attachment, and then express experimental data as a percentage.

3.2. Spreading Assay

1. The coating and blocking of microtiter plate wells is exactly as described for the cell-attachment assay.
2. Preparation of cells is also the same, with the exception that the working cell density should be 1 × 10^5/mL (*see* **Note 11**). Again, for experiments examining the effects of inhibitory agents, add 50 µL of 2X agent diluted into PBS followed by 50 µL cells. Add 50 µL PBS followed by 50 µL cells for control wells (*see* **Note 12**).
3. Incubate at 37°C in a CO_2 incubator for 60–90min.
4. Fix cells by direct addition of 10 µL 50% (w/v) glutaraldehyde and leave at room temperature for 30 min.
5. Aspirate fixative and store cells in PBS (without divalent cations), 0.05% NaN_3.
6. Determine the percentage of cells that adopt a spread morphology using an inverted phase contrast microscope (*see* **Note 13**).

4. Notes

1. The major problem likely to be encountered in both assays is that cells do not adhere. Many reasons could explain this, including coating plates with insufficient amounts of adhesive substrate, bad batches of adhesive substrate, use of poor protein-binding microtiter plates or badly constructed plates with uneven wells, squirting liquids too vigorously onto the bottom of the wells, mycoplasma infection, or a poorly growing cell culture, and (for attachment assays) too much washing.
2. Most tissue-culture treated microtiter plates are adequate for adhesion experiments, although we find Costar to be excellent. Immulon 4 plates have higher protein binding capacity and are particularly good for assays involving small proteins. Usually there is no need to carry out spreading assays with replicate wells, because quantitation is performed by counting multiple fields from within the same well.
3. The concentration of adhesion molecule required for coating will depend on a number of factors, including the efficiency with which it coats plastic, the size of the molecule, and the apparent affinity with which it is bound by cellular receptors. In most cases, adhesion assays are used to measure the adhesion of cells to extracellular matrices or purified extracellular matrix molecules. The key com-

ponents of such matrices are usually large macromolecules that coat plastic relatively well; they are also bound with at least moderate affinity by cells. For these reasons, a concentration range between 1–20 µg/mL is usually adequate, although it is advisable to carry out a range-finding dose-response experiment before focusing on a narrow range. If a nonmatrix molecule or a complex mixture is to be tested, a higher concentration should be used. The handling of adhesion molecules prior to dilution will vary; some molecules, such as fibronectin, are best thawed quickly at 37°C, whereas others, such as laminin, are best thawed slowly on ice.

4. Time-course studies have shown substantial coating of proteins onto plastic within an hour at room temperature and this allows the assay to be performed quickly. However, if the adhesion molecule binds weakly to plastic, or if it is more convenient to carry out the experiment the next day, wells can be coated overnight usually without detrimental effects.
5. Trypsin, EDTA, or trypsin/EDTA solutions are commonly used to detach adherent cells. The action of these reagents must be terminated prior to using the cells in spreading assays (e.g., by resuspending the cells in DMEM with 10% [v/v] fetal calf serum), however, the use of these agents usually has no deleterious effect on adhesive activity provided the cells are not overtrypsinized. It is important to guard against clumping or aggregation of cells, therefore, gentle conditions should be used when centrifuging and resuspending cells. All solutions used during the preparation of cell suspensions should be warmed to 37°C.
6. Cells are left upright in a polypropylene tube in a humidified 5% humidified CO_2 atmosphere with the lid off to allow them to recover from the process of detachment. Alternatively, the tube can be capped and left on its side in an incubator to stop the cells from settling and aggregating into a large clump at the bottom of the tube (they should not be left too long or there may be some nonspecific adhesion to the sides of the tube). The cells should be pipeted gently prior to use to ensure dispersion. Finally, gassing of the cell suspension with CO_2 can sometimes give enhanced spreading, although this is not always needed. It is important that the DMEM/PBS mixture has the opportunity to equilibrate as rapidly as possible with gaseous CO_2 in order to reestablish the buffer. This process can be aided by leaving the lid off the microtiter plate in the incubator. It is also our experience that the adhesion of some cell types is improved by raising the concentration of gaseous CO_2 from the usual 5% (v/v) to 7% (v/v). This can be particularly effective at increasing binding to poorly adhesive substrates. It may also be advisable to use a particular incubator and/or time of day when the door to the incubator will not be opened, as this helps prevent alkalinization of the medium. Ensure that the shelves holding the microtiter plates are horizontal, as uneven shelves lead to uneven settling of cells.
7. For attachment assays, the key parameter is the washing protocol as this is the major determinant of the signal:noise ratio. This is the most critical stage in an attachment assay, and needs to be optimized for each cell type used. Different cells respond differently to tapping and washing, and we recommend varying the

number of tapping or washing cycles to obtain the best signal:noise ratio (i.e., attachment to an adhesive substrate compared to attachment to BSA-blocked plastic). Sometimes this can even be judged by eye in a pilot experiment.

8. More specific problems include cell death in the assay, which could be caused by exposure of sensitive cells to heat-denatured BSA, and clumping of cells either in the centre of a well, or around the perimeter, which is caused by swirling of the plate. In addition, large errors in attachment assays can result from a) inaccurate pipeting, which can come from use of multichannel pipetors, or b) suboptimal washing of wells (note that the volume of BSA blocking solution is higher than the volume of adhesive substrate, and that the wash step after crystal violet staining is larger still).
9. It is advisable to use pipet tips that have their ends cut off for attachment assays. This is to prevent coated proteins and/or cells being washed off directly by a fine stream of liquid. Wells should be included to estimate a value for 100% attachment: here, cells are added directly to uncoated plastic or to polylysine-coated plastic and fixed without washing. The most accurate way of determining this value is to add cell aliquots corresponding to 20%, 50%, and 100% of the experimental inoculum and then extrapolate the resulting graph. It is also possible that the absorbance value for 100% may be off the linear range of the plate reader. Attachment assays rely much more on the accuracy of pipeting than spreading assays, and therefore it is advisable to use a P200 pipetor rather than a multichannel pipetor wherever possible.
10. Staining can also be performed overnight without detriment to the final results. Blank wells should be included to subtract the background binding of crystal violet to plastic. Avoid getting crystal violet solution on the rims of the wells, as this dries during incubation and can be difficult to remove by washing.
11. For cell-spreading assays, an important parameter is the health of the cells. Cultures should be actively growing, but should have been passaged more than 24 h previously. We have observed relatively poor spreading responses in cells that were passaged the day before a spreading assay.
12. Spreading can sometimes be increased by incubating the plate containing the initial 50-µL aliquots at 37°C for several min to allow them to warm up prior to addition of cells. To ensure good spacing of cells, guard against swirling, tapping, or shaking the wells once cells have been added. In our experience, a single pipeting of cells down the side of the well into the PBS solution produces good dispersal.
13. Understandably, the optical quality of the plastic that is used to make microtiter plates is not ideal for phase contrast microscopy. However, the observation of adherent cells can be greatly improved by adding sufficient PBS/azide to form an inverted meniscus at the top of the well and then carefully placing a glass cover slip over the plate. Quantitation of percent spreading is usually carried out by counting three separate lots of 100 cells selected from random areas of the well. Both the selection of cells and minimization of double counting are aided by the use of an eyepiece graticule. Different methods can be used to determine whether

a cell is spread or not. Perhaps the most quantitatively accurate way is to use image analysis software to measure average cell area, however, this tends to produce small differences that may be hard to interpret without the application of other criteria relating to cell shape. Instead, we prefer to assign specific criteria to a definition of spreading and apply these to each cell. The usual criteria are that the cell body should be phase-dark and that cytoplasm should be visible around the entire circumference of the nucleus. Different cells adopt different morphologies and therefore these criteria might need to be slightly modified on a case by case basis.

References

1. Kueng, W., Silber, E., and Eppenberger, U. (1989) Quantification of cells cultured on 96-well plates. *Anal. Biochem.* **182,** 16–19.
2. Yamada, K. M. and Kennedy, D. W. (1984) Dualistic nature of adhesive protein function: fibronectin and its biologically active peptide fragments can autoinhibit fibronectin function. *J. Cell Biol.* **99,** 29–36.

24

Tissue Engineering and Cell-Populated Collagen Matrices

Paul D. Kemp

1. Introduction

Tissue engineering seeks to produce living, three-dimensional cellular constructs that can be used as clinical replacements of damaged tissues and organs as well as research tools to study cell and matrix interactions that occur in higher-order systems. To organize the cells into a three-dimensional structure in vitro, a provisional extracellular matrix support is required. The two main methods to achieve this are a) to culture the stromal cells on a three-dimensional synthetic meshwork, or else b) embed the cells within a three-dimensional type I collagen lattice. The contracted collagen lattice can be used for a variety of practical applications including the support of epithelial growth and differentiation in order to produce a skin replacement ***(1–5)***. However, this model system can also be exploited for experiments to study cell–matrix interactions such as the influence of tension on cell phenotype ***(6)***.

2. Materials

2.1. Production of the Collagen Solution

1. PBS:Water: Mix one part phosphate-buffered saline with 2 parts of purified water. Filter sterilize the solution through an appropriate 0.22-μm filter into a sterile, screwcap storage vessel(s). Store the solution at 2–8°C before use (stable for 1 yr).
2. 0.5 *M* Acetic Acid: Carefully make up 286 mL of glacial acetic acid to 10 L with purified water. Filter sterilize the solution through an appropriate 0.22 m filter into a sterile, screwcap storage vessel(s). Store the solution at 2–8°C before use (stable for 1 yr).

From: *Methods in Molecular Biology, vol. 139: Extracellular Matrix Protocols*
Edited by: C. Streuli and M. Grant © Humana Press Inc., Totowa, NJ

3. 8.75 m*M* Acetic Acid: Add 12.5 mL of glacial acetic acid to 25 L of purified water. Mix well and filter the solution through a 0.22-µm filter into sterile, screwcap storage vessel(s). Store the solution at 2–8°C before Use (it is not recommended to store this solution. Large volumes are required and maintenance of sterility can be an issue).

2.2. Production of Fibroblasts

All solutions must be sterilized by passing through an appropriate sterile, 0.22-µm filter before use.

1. Antimicrobial solution: PBS containing 100 µg/mL Gentamicin sulphate; 250 µg/mL amphotericin B (store at –20°C, stable for 1 yr).
2. Enzyme solution: PBS containing: 2 mg/mL trypsin, 5 mg/mL collagenase, 50 µg/mL gentamicin sulphate, and 1.25 meq/mL of amphotericin B (store at –20°C, stable for 1 yr).
3. Trypsin/versene: 500 µg/mL trypsin; 200 µg/mL versene (store at –20°C, stable for 1 yr).
4. Sterile filtered aqueous solution of 95% (v/v) ethanol.
5. Dulbecco's modified Eagle's medium (DMEM) containing 10% newborn calf serum (NBCS). Store at 2–8°C (stable for 1 mo).

2.3. Production of Cell-Populated Collagen Matrix

All solutions must be sterilized by passing through an appropriate sterile, 0.22-µm filter before use.

1. 71.2 mg/mL sodium bicarbonate. Once filter sterilized, the solution can be stored in 50-mL aliquots at –20°C before use.
2. 7.25 m*M* L-glutamine. Once filter sterilized, the solution can be stored in 15-mL aliquots at –20°C before use.
3. 50 µg/mL gentamicin sulphate. Once filter sterilized, the solution can be stored in 15-mL aliquots at –20°C before use.
4. Premix: Add the following in sequence, to a 15-mL conical tube. 2.2 mL of 10X MEM, 0.2 mL of L-glutamine, 0.025 mL of gentamicin sulphate, 0.7 mL of sodium bicarbonate, 2.5 mL of newborn calf serum. The premix must be stored on ice and used on the day of preparation.

3. Methods

3.1. Production of Collagen Solution

The mechanical properties of cell-contracted collagen gels is markedly superior if the type I collagen used contains intact telopeptides. Atelopeptide collagen in which the telopeptides have been enzymatically digested contracts more rapidly and extensively and the resulting lattice has inferior tensile strength *(7)* (*see* **Note 1**). Bovine digital extensor tendons provide a suitable yield of high-purity acid-extracted collagen *(7)*. This solution is stable for a year if stored in a sterile condition at 2–8°C. In order to minimize the bioburden, sterile technique should be used throughout the process and all open transfers of the collagen solution should take place in a tissue-culture hood.

1. Dissect out the digital extensor tendon from calf hooves (*see* **Note 2**) by making two lateral incisions and peeling back the dorsal flap of skin to expose the common digital extensor tendon.
2. Cut the tendon from the surrounding tissue and sheath. Place the tendons in ice water until all are collected. The tendons can be stored at –70°C until used.
3. Grind the tendon with an equivalent volume of ice (*see* **Note 3**). Wash the tendon pieces three times in three volumes of ice water and store at –70°C until used.
4. Take 150 g of tendon pieces and add to 15 L of cold (2–8°C), sterile PBS:water in a Belco (Model 7764-00110) top stirring reaction vessel. Mix the solution (set motor to setting 10) for 2 h at 2–8°C (*see* **Note 4**).
5. Aspirate off the buffer and repeat the washing step three more times (*see* **Note 5**).
6. Add 15 L of cold (2–8°C), sterile 0.5 *M* acetic acid to the Belco vessel and mix gently (motor setting 5) for 72 h at 2–8°C.
7. Decant the acetic acid into 250-mL centrifuge bottles and spin at 25,000*g* for 30 min at 4°C in a Beckman J2-21 centrifuge, or equivalent.
8. Very carefully decant just the uppermost layers and combine them in another 15 L Belco vessel (*see* **Note 6**).
9. Pass the solution through an open mesh 5–6 µm filter in order to remove any tendon particles.
10. Make the solution 0.9 *M* with respect to NaCl by *slowly* adding one-third the volume of sterile, room temperature 3.6 *M* NaCl with continuous, gentle mixing (motor speed 5). After the NaCl has been added, continue mixing for 30 min.
11. Decant the solution into 500-mL centrifuge bottles and collect the precipitate by centrifugation at 10,000*g* for 30 min at 4°C.
12. Discard clear supernatants (*see* **Note 7**) and combine the precipitates in a 15 L Belco flask.
13. Add 2 L of cold, sterile 0.5 *M* acetic acid to the Belco flask and mix at 2–8°C until the precipitate has completely dissolved.
14. Repeat the precipitation step once more by making the solution 0.9 *M* NaCl. Redissolve this precipitate completely in 1 *M* acetic acid by vigorous mixing.
15. Transfer the collagen solution to 1 meter lengths of sterilized dialysis tubing.
16. Place the dialysis tube sections in 20 vol of 8.75 m*M* acetic acid at 2–8°C.
17. Saturate the solution with chloroform (approx 0.5 mL/L) and mix the acetic acid solution gently at 2–8°C for 24 h (*see* **Note 8**).
18. Discard the acetic acid/chloroform solution to waste and replace with 20 vol of fresh, cold 8.75 m*M* acetic acid. Mix for 24 h.
19. Repeat **step 17** twice.
20. Using sterile technique, spray one end of the dialysis tubing liberally with 70% ethanol solution. Use a sterile wipe to remove excess, cut open the dialysis tube and carefully poor into a suitable, sterile, screwcap container (*see* **Note 9**).

3.2. Production of Fibroblast Culture

A number of methods exist for the culture of fibroblasts. One convenient method for the production of these cells from discarded foreskin tissue for use in organotypic culture is described below *(3)*.

1. Wash the foreskin several times with the antimicrobial wash solution.
2. While holding the foreskin with sterile forceps, wash the foreskin for 1 min (no longer) in 95% ethanol with constant agitation.
3. Immediately remove the ethanol by washing in the antimicrobial solution.
4. Dissect away any subcutaneous tissue and wash the remaining tissue again in antimicrobial solution.
5. Transfer the foreskin to a 100-mm Petri dish containing antimicrobial solution. Mince the tissue into 1–3 mm^2 pieces (*see* **Note 10**).
6. Place a sterile microstir bar into a round bottom tube and add 5 mL of the enzyme solution.
7. Transfer the minced tissue and cap the tube. Incubate the tube at 37°C with gentle stirring. After 20 min, vigorously shake the tube.
8. After 30 min, allow the material to settle. Remove as much of the enzyme solution and discard. Add 4 mL of fresh enzyme solution and continue to incubate as above, shaking the tube periodically.
9. After a further 30 min incubation, allow the material to settle again and transfer the supernatant solution to a sterile 15-mL conical tube.
10. Add 5 mL of DMEM-10% NBCS to the 15-mL conical tube containing the supernatant and centrifuge for 5 min at 600*g*. Aspirate the supernatant, resuspend the pellet in 2 mL DMEM- 10% NBCS and place the tube on ice.
11. Add an additional 4 mL of enzyme solution to the tissue remnants from **step 9**. Repeat **steps 9** and **10**.
12. Repeat **step 11** until all the tissue is digested and only the stratum corneum of the epidermis remains.
13. Pool the cell fractions from step 10. Count the cells and determine viability. Add 1X 10^6 cells to each of several T-75 flasks. Add 10 mL of DMEM-10%NBCS to each flask. Incubate at 37°C, 10% CO2. Replace the media every 2–3 d.
14. When the cells are confluent, aspirate off the media and add 5 mL of trypsin/versene. Incubate the flask at 37°C, 10% CO_2. Strike flask sides sharply to dislodge the cells. Examine the cells to ensure they have rounded and left the dish. Add 5 mL of DMEM-10% NBCS to each flask.
15. Transfer the solution containing the cells to a sterile 15-mL conical tube and centrifuge 5 min at 600*g*.
16. Discard the medium and replace with 10 mL of DMEM-10%NBCS.
17. Suspend the cells and recentrifuge at 5 min at 600*g*.
18. Count the cells and transfer 1X 10^6 cells to each of several T-150 flasks. Add 20 mL of DMEM-10%NBCS. Incubate at 37°C, 10% CO_2.
19. Cells can continue to be passaged at confluence as described above (*see* **Note 11**).

3.3. Production of Cell-Populated Collagen Matrix

The following volumes are sufficient to produce six cell-populated collagen lattices using 2.5-cm diameter lattice inserts (*see* **Note 12**).

1. Measure out 18.5 mL of the collagen solution into a 50-mL conical tube and place on ice.
2. Prepare a suspension of dermal fibroblasts at a concentration of 2.5×10^5 cells/mL in DMEM with 10% NBCS.
3. Add 5.6 mL of cold (2–8°C) premix solution to the tube containing the collagen. Mix well by swirling (*see* **Note 13**).
4. Immediately place 1 mL of the neutralized collagen solution into each of six 2.5-cm diameter, 3-µm pore size culture inserts (Corning Costar, Cambridge, MA) (*see* **Note 14**) and ensure that the filter surface is completely covered. Allow this acellular layer to gel at room temperature (*see* **Note 15**).
5. Add 2 mL of the fibroblast suspension to the remainder of the neutralized collagen solution, swirl to mix (*see* **Note 13**) and immediately add 3-mL aliquots of the cellular collagen solution on top of the collagen gels in each of the culture inserts. Allow the collagen to gel without any disturbance or vibration (*see* **Note 15**).
6. Add sufficient DMEM with 10% NBCS to cover the top of the collagen gels and incubate at 37°C, 10% CO_2 until the lattices have contracted away from the sides of the culture insert (*see* **Note 16**).
7. Replace the medium with fresh DMEM with 10% NBCS at 2–3 d intervals.

4. Notes

1. Different collagen preparations (such as acid extracted rat tail yendon collagen) may be used, but these need to be carefully screened for gelation properties as well as cellular interactions and possible toxicity.
2. Rat tail tendon may also be used in place of the bovine digital extensor tendon.
3. As an alternative to grinding, either tendon can be sliced into pieces approx 1–3-mm thick. It is easier to achieve this if the tendon is partially frozen and a number-22 scalpel blade is used.
4. As an alternative to the Belco vessel, a conical flask and large magnetic stir bar may be used for volumes around 2 L. If the mixing is too vigorous then small insoluble particles can be produced which are difficult to remove at later stages.
5. The final PBS:water wash should have negligible UV absorbance at 280 nm. If this is not the case, repeat the wash step until the UV absorbance is negligible.
6. The material should separate into three layers, an upper layer, a middle layer that contains gelatinous particles and a lower layer containing the insoluble tendon pieces.
7. If air bubbles have been introduced at the precipitation step, this can cause some of the precipitated collagen to float at the end of the centrifugation step. This material may be removed with a sterile instrument and added to the rest of the precipitate.

8. The acetic acid may be mixed by either using a large stir bar or else recirculating the solution through a peristaltic pump.
9. Great care needs to be taken at this stage, and it is worth practicing with water-filled tubing in order to develop a good technique. It is easier if two people work together so as one supports the tubing to control the flow and the other opens the tubing and pours the collagen.
10. The tissue may be minced by using a sterile pair of forceps and either a scalpel or a pair of scissors.
11. Frozen vials of primary cells should be established and working cell stocks developed from this primary source, which can then be frozen at passage 5 or 6 and stored before use. Fibroblasts may be frozen using routine methods.
12. Various systems may be used to support the lattice. Tissue-culture treated dishes are not suitable as the collagen adheres to the base and walls and the lattice is only able to contract slightly. Petri dishes enable the lattice to contract without restraint (sometimes the collagen needs to be carefully released from the walls using a blunt instrument such as a fine glass rod or spatula). The preferred method is to use Transwell inserts produced by Corning Costar.
13. Care needs to be taken at this stage to not introduce air bubbles, which may become trapped as the collagen gels.
14. If smaller pore sizes are used, the collagen may release from the base, larger pores may allow the collagen to leak into the lower chamber before it gels.
15. Care needs to be taken here to keep the system free from vibrations, which may interfere with, or disrupt, the collagen gels as they are forming.
16. It will be necessary to release the collagen from the sides of the transwell by using a blunt, sterile instrument such as fine glass rod or spatula.

Acknowledgments

The author would like to thank all the Research Staff of Organogenesis, Inc. for their help.

References

1. Bell, E., Parenteau, N., Gay, R., Nolte, P., Kemp, P., Bilbo, P., et al. (1991) The living skin equivalent: its manufacture, its organotypic properties and its responses to irritants. *Toxic. In vitro* **5,** 591–596.
2. Bell, E., Rosenberg, M., Kemp, P., Gay, R., Green, G. D., Muthukumaran, N., and Nolte, C., (1991) Recipes for reconstituting skin. *J. Biomech. Eng.* **113,** 113–119.
3. Parenteau N. L. (1994) Skin equivalents, in *Keratinocyte Methods* (Leigh, I. and Watt, F., eds.), Cambridge University Press, Cambridge, pp. 45–55.
4. Eaglestein, W. H., Iriondo, M., and Laszlo, K. (1995) A composite skin substitute (Graftskin) for surgical wounds: a clinical experience **21,** 839–843.
5. Sabolinski, M. L., Alvarez, O., Auletta, M., Mulder, G., and Parenteau, N. L. (1996) Cultured skin as a 'Smart material' for healing wounds: experience in venous ulcers. *Biomaterials* **17,** 311–320.

6. Mochitate, K., Pawelek, P., and Grinnel, F. (1991) Stress relaxation of contracted collagen gells: disruption of actin filament bundles, release of cell surface fibronectin, and down-regulation of DNA and protein synthesis. *Exp. Cell Res.* **193,** 198–207.
7. Kemp, P. D., Bell, E., Falco, L., and Regan, K. (1995) Collagen compositions and methods of preparation thereof. European Patent 0 419 111 B1.

25

Solid Phase Assays for Studying ECM Protein-Protein Interactions

A. Paul Mould

1. Introduction

Solid-phase assays provide a simple, rapid and robust method for the analysis of protein–protein interactions; i.e., does protein A interact with protein B? In this assay, protein A (here termed "receptor") is adsorbed to the wells of an enzyme-linked immunosorbent assay (ELISA) plate (solid phase). The plate is then blocked using bovine serum albumin (BSA), and biotin-labeled protein B (here termed 'ligand') is added. After washing the wells to remove unbound ligand, bound ligand is detected by addition of an avidin-peroxidase conjugate followed by a colorimetric detection step.

This type of assay is particularly well suited for studying the interaction of ECM proteins with integrins. The first solid-phase integrin–ligand binding assay was described by Charo et al. *(1)* for studying fibrinogen binding to αIIbβ3. In our laboratory we have developed an extremely sensitive and highly versatile assay for fibronectin binding to α5β1 *(2)*. For example, we have described how the assay can be used to investigate the effects of divalent cations, activating and inhibitory MAbs, peptide inhibitors and mutations on ligand binding *(2–5)*. The pharmacological screening of inhibitors of integrin–ligand interactions is an important area in which this particular type of assay is finding use.

Our preferred method for labeling of ligands is biotinylation because of its safety and simplicity. One potential drawback is that if one or more lysyl residues in the ligand are crucial for receptor binding, their modification may render the ligand inactive. In this case, a possible solution may be to reduce the amount of biotinylation reagent so that some of the lysyl residues remain unmodified. Other labeling methods such as radioiodination can also be used.

From: *Methods in Molecular Biology, vol. 139: Extracellular Matrix Protocols*
Edited by: C. Streuli and M. Grant

Alternatively, if the ligand is a recombinant protein, a "tag" such as an epitope sequence or the Fc region of IgG can be incorporated for use in the detection of bound ligand. It should also be noted that solid-phase assays can only give a qualitative measure of the affinity of receptor–ligand binding; other techniques, such as surface plasmon resonance, must be used to calculate actual binding parameters.

It is essential that the specificity of the assay is tested carefully. The most important test for specificity is the ability of unlabeled ligand to compete with labeled ligand for binding to the receptor. Hence, in the presence of a large excess of unlabeled ligand, very little binding of labeled ligand should be observed. Some receptor–ligand interactions are divalent-cation dependent. Here, omitting divalent cations from the binding buffer should reduce binding to similar levels as to wells coated with BSA alone. Further tests for specificity can also be carried out, e.g., attempting to block the interaction using monoclonal antibodies to either receptor or ligand. If a specific interaction is observed, the sequences in the receptor and ligand involved in this recognition can be identified, using, e.g., proteolytic fragmentation, synthetic peptides, and site-directed mutagenesis.

2. Materials

2.1. Biotinylation of Ligand

1. Coupling buffer: 0.5 *M* NaCl, 0.1 *M* $NaHCO_3$. Dissolve 29.2 g of NaCl and 8.4 g of $NaHCO_3$ in 1 L of water; the pH should be approx 8.0 without further adjustment.
2. Sulfo-NHS biotin (Pierce).
3. Tris-buffered saline (TBS). Dissolve 8.77 g of NaCl and 3.03 g of Tris in ~900 mL H_2O, adjust pH to 7.4 using conc. HCl, and adjust final volume to 1 L by further addition of H_2O.
4. TBS-azide: Add 2.5 mL of 20% (w/v) sodium azide stock solution to 1 L of TBS.

2.2. Solid-Phase Assay

1. Dulbecco's PBS (Gibco-BRL, Gaithersburg, MD).
2. Blocking solution: TBS, 5% (w/v) BSA, 0.05% (w/v) sodium azide. This is conveniently made from TBS to which sodium azide is added from a 20% stock solution. BSA (Sigma, Madison, WI, fraction V) is added and dissolved by vigorous stirring. The solution is then centrifuged in 50-mL tubes (Becton-Dickinson, Rutherford, NJ) for 5 min at 3,500*g*, and filtered through a 20-mL disposable column (Bio-Rad, Richmond, CA). Store for <3 mo at 4°C.
3. Binding buffer: TBS containing 0.1% (w/v) BSA. Divalent cations 1 m*M* $MgCl_2$, 1 m*M* $CaCl_2$, or 1 m*M* $MnCl_2$ may be added. This buffer is conveniently prepared from TBS and stock solutions of 1 *M* $MgCl_2$ (20.3 g $MgCl_2 \cdot 6H_2O$/100 mL), 1 *M* $CaCl_2$ (14.7 g $CaCl_2 \cdot 2H_2O$/100 mL), or 1 *M* $MnCl_2$ (19.8 g $MnCl_2 \cdot 4H_2O$/100 mL; $MnCl_2$ is used in assays involving integrins). BSA (Sigma 99% pure grade) is then added and dissolved by stirring. Prepare fresh.

4. Extravidin peroxidase reagent (Sigma).
5. ABTS reagent: Prepare ABTS buffer: 0.05 *M* Na_2HPO_4 (0.69 g $Na_2HPO_4 \cdot 2H_2O$/100 mL), 0.1 *M* sodium acetate (1.36 g $CH_3COONa \cdot 3H_2O$/100 mL). Adjust pH to 5.0 using conc. HCl. Store at room temperature for <3 mo. ABTS solution: Dissolve 11 mg ABTS (Sigma) in 0.5 mL water. Prepare fresh. H_2O_2 solution: 67 μL 30% (w/w) H_2O_2 (Sigma) mixed with 7 mL water. Prepare fresh. Add 0.5 mL of ABTS solution to 10 mL of ABTS buffer and 100 μL of H_2O_2 solution. Mix thoroughly. This amount of reagent is sufficient for one full 96-well plate assay.
6. 2% (w/v) SDS. Dissolve 2 g sodium dodecyl sulfate in 100 mL water.

3. Methods

3.1. Biotinylation

1. Dialyze ligand into coupling buffer. About 0.5 mL of ligand at a concentration of approx 0.5 mg/mL gives sufficient material for a large number of assays.
2. Add an equal mass of sulfo-NHS biotin: protein (approx 0.25 mg) to the dialysate in a 1.5-mL Eppendorf tube and mix on a rotating platform for half hour at room temperature.
3. To remove unincorporated biotin, dialyze the solution against two changes of 1 L of TBS, and once against 1 L of TBS-azide.
4. Centrifuge the dialysate in a 1.5-mL Eppendorf tube for 15 min at maximum speed in a microcentrifuge. This removes any large aggregates or precipitate from the solution.
5. Store the supernatant at 4°C for <6 mo. Alternatively, many biotinylated proteins can be stored in aliquots at –70°C.
6. Measure the concentration of biotinylated protein using, for example, the BCA assay (Pierce).

3.2. Solid-Phase Assay

1. Dilute purified receptor to 1–10 μg/mL with Dulbecco's PBS (*see* **Note 1**).
2. Add the diluted receptor to the wells of an ELISA plate (100 μL /well) (*see* **Note 2**). Leave a set of wells empty for measuring binding of the ligand to BSA. We normally perform the assay using 4–6 replicates for each set of binding conditions.
3. Cover the plate with plastic film and store at room temperature overnight. Alternatively, the plate can be stored at 4°C for up to 1 wk.
4. Add 25 μL of blocking solution to each receptor-containing well using a multichannel pipet. Then remove the solution by aspiration, or by inverting the plate over a sink and flicking out the liquid (*see* **Note 3**).
5. Add 200 μL of blocking solution to each well (including those used for testing binding to BSA alone) using a multichannel pipet. Leave at room temperature for 1–3 h (*see* **Note 4**).
6. Aspirate or flick out the blocking solution, and wash the wells three times with 200 μL of binding buffer.

7. Aspirate or flick out the final wash. Remove residual liquid by inverting the plate and striking it hard several times onto adsorbent paper toweling.
8. Dilute the biotin-labeled ligand in binding buffer. Add 100 μL of this solution to each experimental well (*see* **Note 5**).
9. Cover the plate with plastic film and incubate at 37°C for 3 h. A cell-culture incubator is suitable for this purpose.
10. Aspirate the wells to remove unbound ligand.
11. Wash the wells three times with 200 μL of binding buffer. Remove residual buffer as in **step 7**.
12. Dilute Extravidin-peroxidase reagent 1:500 in binding buffer. Add 100 mL to each well using a repeating pipet. Incubate the plate for 10–15 min at room temperature. During this time prepare the ABTS reagent.
13. Aspirate the wells to remove unbound Extravidin-peroxidase.
14. Wash the wells twice with 200 μL of binding buffer and then twice with 400 μL of binding buffer. Remove residual buffer as in **step 7**.
15. Add 100 μL of ABTS reagent to each well using a repeating pipet. Allow the reaction to proceed until a strong (but not dark) green color is obtained (typically 10–30 min).
16. Stop the reaction by adding 100 μL of 2% SDS solution to each well using a repeating pipet.
17. Read the plate using an automatic plate reader at 405 nm.
18. Calculate mean and standard deviations of ligand binding to experimental wells. Subtract the level of binding to wells coated with BSA only.

4. Notes

1. In initial experiments, the concentrations of both receptor and ligand should be varied so that the conditions for optimal signal to background binding can be determined. Detailed protocols for the purification of integrin receptors have previously been described ***(2,6)***. For receptors containing detergents (e.g., Triton X-100), it is necessary to dilute the solution so that the detergent concentration is < 0.002% (w/v), otherwise the detergent interferes with adsorption of the receptor to the plate. Plates can be coated with receptor several days in advance.
2. Immulon 1B or 4HBX ELISA plates (Dynatech, Chantilly, VA) are suitable. More recently, we have found that half area EIA/RIA plates (Costar, Cambridge, MA) also work well, and have the advantage that similar results can be obtained with half as much receptor (i.e., 50 μL/well).
3. A small amount of blocking solution is added to the wells before aspirating the receptor solution because we have found that this renders the wells hydrophilic and prevents them drying out when they are aspirated. Drying out of the wells may destroy the activity of some of the receptor. For the aspiration we use a 21-gage hypodermic needle attached by tubing to a Buchner flask, which is connected to a water pump or vacuum line.
4. Longer blocking times (e.g., overnight) or alternative blocking reagents may be used if the background level of binding is high.

5. A concentration of ligand in the range 0.1–10 μg/mL should give a good response if the interaction is of high to moderate affinity. Other reagents (e.g., MAbs, peptides, or synthetic compounds) can be added simultaneously with the ligand at this stage to test for their effects on binding.

References

1. Charo, I. F., Nannizzi, L., Phillips, D. R., Hsu, M. A., and Scarborough, R. M. (1991) Inhibition of fibrinogen binding to GP IIb-IIIa by a GP IIIa peptide. *J. Biol. Chem.* **266,** 1415–1421.
2. Mould, A. P., Akiyama, S. K., and Humphries, M. J. (1995) Regulation of integrin α5β1-fibronectin interactions by divalent cations. Evidence for distinct classes of binding sites for Mn^{2+}, Mg^{2+}, and Ca^{2+}. *J. Biol. Chem.* **270,** 26,270–26,277.
3. Mould, A. P., Garratt, A. N., Askari, J. A., Akiyama, S. K., and Humphries, M. J. (1995) Identification of a novel monoclonal antibody that recognises a ligand-induced binding site epitope on the β1 subunit. FEBS Lett. **363,** 118–122.
4. Mould, A. P., Akiyama, S. K., and Humphries, M. J. (1996) The inhibitory anti-β1 integrin monoclonal antibody 13 recognises an epitope that is attenuated by ligand occupancy. Evidence for allosteric inhibition of integrin function. *J. Biol. Chem.* **271,** 20,365–20,374.
5. Mould, A. P., Askari, J. A., Aota, S., Yamada, K., Irie, A., Takada, Y., et al. (1997) Defining the topology of integrin α5β1-fibronectin interactions using inhibitory anti-α5 and anti-β1 monoclonal antibodies. Evidence that the synergy sequence of fibronectin is recognised by the amino-terminal repeats of the α5 subunit. *J. Biol. Chem.* **272,** 17,283–17,292.
6. Mould, A. P. (1998) Analyzing integrin-dependent adhesion unit 9-4, in *Current Protocols in Cell Biology* (Yamada, K. M. and Lipincott-Schwartz, J., eds.), Wiley, New York.

26

Tissue Engineering of Cartilage

Ronda E. Schreiber and Anthony Ratcliffe

1. Introduction

Cartilage is a dense connective tissue that functions to withstand and distribute load *(1)*. Articular cartilage lines the ends of long bones and distributes loads across the joints. It consists of a dense collagenous matrix (primarily collagen type II, with smaller amounts of other collagens, including types I, V, VI, IX, and XI), embedded in a high concentration of aggregating proteoglycan, aggrecan. The collagen provides tensile properties, and the proteoglycans confer compressive properties and resiliency. There is a sparse population of a single cell type, the chondrocyte, distributed throughout the tissue. These cells synthesize and maintain the cartilaginous matrix in a regulated fashion that involves breakdown of matrix components, release of proteolytic products from the tissue, and synthesis and incorporation of new components into the matrix.

Articular cartilage has a limited capacity to repair. Degeneration of this tissue in diarthrodial joints is a major component of osteoarthritis and related diseases. The options for treatment are limited, and many patients may be consigned to pain and joint dysfunction until joint replacement is a reasonable treatment option. A major motivation for tissue engineering cartilage is to provide a biological repair for cartilage degeneration.

One particularly promising method for cartilage tissue engineering is the growth of tissue by in vitro culture of chondrocytes after seeding them onto bioresorbable scaffolds *(2,3)*. After seeding, there is a period of cell proliferation, followed by matrix deposition. This growth phase has been achieved using a variety of methods, including implantation of the seeded scaffold subcutaneously in nude mice *(2)*, extended tissue culture *(4)*, and culture in perfusion chambers *(5–7)*. A variety of biodegradable scaffolds have been used, including poly (glycolic acid) and poly (L-lactic acid) polymers fabricated as non-

From: *Methods in Molecular Biology, vol. 139: Extracellular Matrix Protocols*
Edited by: C. Streuli and M. Grant © Humana Press Inc., Totowa, NJ

woven, fiber networks (felt). The selection of biomaterials can influence the growth of the construct ***(8,9)***. The extent and rate of tissue that is formed is dependent on the culture time and method, as well as the quality and quantity of cells applied to the scaffold. The methods described here will result in the formation of a uniformly distributed tissue containing proteoglycan, collagen, and water comprising up to 5%, 1% and 85–95%, respectively, of the total construct weight. These levels are equivalent to the lower levels of the corresponding matrix components of normal articular cartilage. The constructs grown in this manner consist of a smooth, hyaline-like cartilage that surrounds chondrocytes in lacunae, with a fibrous capsule localized at the periphery of the construct. While this tissue does not have the compressive and tensile properties of normal articular cartilage, tissue-engineered constructs have been successful in repairing cartilage defects in vivo ***(10)***.

The method below describes how a cartilage construct can be generated in vitro, using simple tissue-culture techniques. The tissue generated using this method will have type II collagen and the proteoglycan aggrecan as major components of its extracellular matrix. Metabolically active chondrocytes will superficially be distributed relatively homogeneously throughout the tissue. The tissue will also have a fibrocartilaginous layer surrounding the cartilage. These constructs can be used for cartilage transplantation studies and in vitro experimental studies.

2. Materials

2.1. Tissue Harvest and Chondrocyte Isolation (see Note 1)

All of the following materials must be cell culture grade and sterile.

1. Femurs (skinned): rabbits (</= 1 yr), sheep (</= 3 wk), or cows (</= 3 wk).
2. Plastic culture dishes: (10 cm diameter) and tubes (50 mL): Obtain tissue-culture grade vessels.
3. Isopropanol (70%, diluted in DI - H_2O).
4. Phosphate-buffered saline (PBS + Ca^{2+} and Mg^{2+}).
5. Culture medium A (gentamicin solution): Dilute gentamicin in PBS to 25 μg/mL.
6. Culture medium B (cell-culture medium): To Dulbecco's Modified Eagle's Medium (DMEM), add fetal bovine serum; FBS (10%), L-glutamine (2 m*M*), nonessential amino acids (0.1 m*M*), L-proline (50 μg/mL), sodium pyruvate (1 m*M*), and gentamicin (25 μg/mL).
7. Culture medium C (tissue digestion solution): Dissolve bacterial collagenase type II to a final concentration of 506.0 U/mL in culture medium B.
8. Orbital shaker: Obtain a unit that can be placed on an incubator shelf.
9. Cell filter (70-μm filters that fit atop 50-mL conical cell culture tubes).
10. Pipeter/pipet tips (with accuracy for pipetting 5 μL, 200 μL, and 1000 μL).
11. Trypan blue (0.04%).

12. Scalpel handels and blades (sizes 10, 15, and 20).

2.2. Cartilage Construct Growth

2.2.1. Preparation and Storage of Construct Scaffolds

1. Scaffold preparation. Synthetic polymer scaffolds can either be a) purchased or b) fabricated in the laboratory. The following two scaffold types support chondrocyte attachment, proliferation and cartilage matrix synthesis and deposition. They are fabricated as a) sheets or b) disks. Scaffold disks ranging from 2–3 mm in thickness and 8–10 mm in diameter have been used routinely to generate hyaline-like cartilage constructs.
   a. Entangled, nonwoven polyglycolic acid (PGA) felt scaffolds (Albany International Research Company, Mansfield, MA) (*see* **Note 2**).
      Felt, purchased in the form of 20 × 30 cm sheets needled to a final felt porosity of 97% with the following characteristics, has been successfully used in many laboratories *(4,5)*. This PGA scaffold degrades rapidly between 1 and 4 wk of culture *(4)*.
      - 2-mm thick
      - Nonheat plated
      - Extruded polymer fibers
      - 14 μm in diameter
      - Nonwoven
   b. Solvent-cast poly (lactic-co-glycolic acid) [PLGA] and polylactic acid [PLA] foam scaffolds (*see* **Note 3**). Solvent-cast, salt-leached polymer foams with a defined pore size range and porosity can be fabricated in the laboratory, based on published techniques *(11)*. Obtain the following to fabricate this scaffold type:
      - Glass container (20 mL capacity)
      - Glass pipets
      - Glass Petri dish (50 mm in diameter)
      - Methylene chloride (dimethyl chloride)
      - PLGA; Purchase this polymer with the following characteristics: molar ratio, between 50:50 and 85:15; inherent viscosity, 0.55-0.75
      - PLA; Purchase this polymer with the following characteristics: inherent viscosity, L-PLA 0.90–1.2, DD-PLA, 0.55–0.75
      - Teflon Petri dish (50 mm in diameter)
      - NaCl (common, laboratory grade)
      - Sieves (to select NaCl crystals of various sizes)
      - Deionized water (DI H_2O)
      - Oven (if scaffold of low crystallinity is desired)
      - Orbital shaker
2. Scaffold storage.
   a. Foil bags (4 mL capacity).
   b. Vacuum line or dry nitrogen tank.
   c. Dessicant.

2.2.2. Seeding and Culture of Cartilage Constructs

1. Mechanical punch: Obtain a unit that will punch polymer disks of less than or equal to 10 mm in diameter.
2. Petri dishes.
3. Culture tubes (50 mL).
4. Ethanol (100%).
5. Culture medium D (construct culture medium): Prepare culture medium B and add the following cell-culture grade components at the time of construct feeding: insulin (50 µg/mL) and ascorbate (50 µg/mL). Culture medium without insulin and ascorbate can be stored at 4°C for up to 2 wk. Insulin and ascorbate are added at the time medium is added to cultures. Medium is replenished every 3–4 d.
6. Culture dish:
   Construct seeding step: 24-well (15-mm-diameter wells) nontissue culture-treated dishes.
   Construct culture step: 6-well (35-mm-diameter wells) tissue culture-treated dishes.
7. Serological pipets (1 and 5 mL).
8. Chondrocyte suspension ($4–15 \times 10^6$ cells/mL culture medium D).
9. Orbital, vibrating shaker: 300 rpm.

3. Methods

3.1. Tissue Harvest and Chondrocyte Isolation

Note: Between $5–8 \times 10^7$ cells per gram of tissue are typically obtained from skeletally immature animals.

1. Skin femur, taking care not to puncture the joint capsule.
2. Swab the joint with 70% isopropanol.
3. Place joint and tissue harvesting materials in tissue-culture hood and procede with the remaining steps in the hood.
4. Make a medial parapateller incision in the distal femur to expose the articular surface of the condyle and pateller groove. Collect tissue within the patellar groove, patella, and condylar surfaces, being careful not to collect fibrous tissue, which has a distinct texture and is usually more yellow in color than that of the smooth, white, opaque cartilage. Dissect cartilage about 1 cm in length and 2–4 mm in depth (depending upon the donor species and age), being careful not to collect any calcified cartilage or into the subchondral bone.
5. Place tissue chips in 50-mL tubes containing sterile culture medium A.
6. After tissue dissection is completed, rinse once with culture medium A, then transfer the tissue into freshly prepared culture medium C (10 mL per gram of tissue).
7. Place closed tube(s), secured with parafilm, on their sides on an orbital shaker set at 200–300 rpm and incubate overnight at 37°C.
8. Examine suspension to determine degree of digestion. Solution should be slightly opaque with uniformly suspended cells.
9. Filter the cells through a cell filter (70 µm), discard trapped cells, and save cell filtrate.

10. Centrifuge cell suspension (filtrate from **step 9**) at 200–300*g* for 10 min between 4°–15°C (4°C is preferred).
11. Collect supernatant and set aside. Resuspend pellet in culture medium B. Obtain a single cell suspension of an estimated 5×10^6 cells per mL.
12. Recentrifuge supernatant as described above. If new cell pellet is formed, collect supernatant and recentrifuge. Resuspend new pellet with original cell suspension. Repeat as necessary until no additional pellet is obtained.
13. Remove an aliquot of the final cell suspension and suspend in trypan blue (1:1 dilution). Dispense in hemacytometer and count cells.
14. Centrifuge cell suspension as above, discard supernatant, and resuspend cell pellet in culture medium B to obtain a single-cell suspension of $2.5–9.5 \times 10^7$ cells/mL, depending on desired scaffold seeding density (*see* **Subheading 3.2.2., step 4**).
15. Place the remaining cell suspension on ice or at 4°C until scaffold seeding is to occur.

3.2. Cartilage Construct Growth

3.2.1. Preparation and Storage of Construct Scaffolds

3.2.1.1. Scaffold Preparation (*see* **Subheading 2.2.1.**)

Entangled, nonwoven PGA felt scaffolds. Obtain scaffolds from Albany International Research Company (Mansfield, MA) and store as described below.

Solvent-cast PLGA and PLA foam scaffolds (90% porous) (*see* **Note 4**). All work with methylene chloride should be performed in a fume hood.

1. In glass container (scintillation vials work well), mix polymer and methylene chloride (dimethylchloride) to a concentration of 1 g of polymer per 10 mL of solvent. Dissolve polymer in solvent, by incubating between 30 min to 1 h at room temperature, vortexing frequently.
2. Select NaCl crystals of a desired size range by passage through a series of sieves to dictate the pore size range of the scaffold. For example, to select particles between 106–150 microns in diameter, place a sieve that excludes particles greater than 150 microns above a sieve that excludes particles less than 106 microns in diameter. Pass NaCl through the top and bottoms sieves. Collect the NaCl that remains between the two sieves.
3. Weigh 4.5 g of sieved NaCl into Teflon dish.
4. In a fume hood, dispense 5 mL of the polymer solution with a glass pipet over the NaCl in a Teflon dish to obtain a 2-mm-thick foam. Mix well with a glass rod and manually shake dish to evenly distribute polymer and NaCl mixture.
5. Cover dish immediately with glass Petri dish cover to allow for slow solvent evaporation and place in a fume hood overnight.
6. To remove residual solvent from the polymer, place under high vacuum (up to 50 mm Hg) overnight.

7. Leach the NaCl to form pores in the polymer foams:
   - Place Petri dish containing polymer and NaCl in a large volume of DI-H_2O on orbital shaker and gently shake for 4 h, changing DI-H_2O every h.
   - Gently remove porous polymers from dishes.

3.2.1.2. Scaffold Storage

1. Place scaffold felts or foams in foil pouches and vacuum seal or sparge the pouches with dry nitrogen and seal.
2. Store the sealed pouches at room temperature for short-term storage (8 wk or less), or in a nonthaw noncycling freezer at –20°C to –70°C for longer term storage.

*3.2.2. Seeding and Culture of Cartilage Constructs (see **Note 5**)*

1. Prewet scaffolds by submerging in 100% ethanol for 15 min at room temperature. Rinse scaffold three times with large volumes of PBS to remove residual ethanol, then submerge scaffolds in culture wells containing approximately 5 mL of culture medium B per scaffold (10 mm × 2 mm) and incubate at 37°C and 5% CO_2 for 5–16 h. To hasten wetting, draw culture medium bidirectionally through the scaffold several times with a 1- or 5-mL pipet.
2. Prior to cell seeding, withdraw excess medium from scaffolds with a pipet and transfer each scaffold disc into an individual well of a 24 well dish.
3. Suspend chondrocytes in culture medium B at 4–15 × 10^6 cells/mL.
4. Dispense 2.5–9.4 × 10^7 cells/mL scaffold. Pipet cell suspension slowly onto scaffold, using a 1000-μL pipeter with a pipet tip to draw the suspension bidirectionally several times through the scaffold.
5. Place 24-well culture dish on orbital shaker, set at 300 rpm, and incubate overnight at 37°C and 5% CO_2.
6. Transfer seeded scaffolds to 6-well dishes containing porous well inserts and 10 mL of fresh culture medium D and incubate statically at 37°C and 5% CO_2.
7. Culture cartilage constructs for 2 wk or more, changing culture medium every 3–4 d, and culture plates once a week to prevent an overgrowth of chondrocytes in monolayer.

3.3. Cartilage Construct Analyses

Note: To evaluate the composition of cartilage constructs, routine histologic, biochemical, and molecular analyses are performed. Because these conventional techniques are well described elsewhere, this section includes general methods and references only (*see* **Note 6**).

3.3.1. Histological Analyses

1. Tissue processing of tissue-engineered cartilage constructs:
   a. Fixation: To fix tissue, immerse in 10% buffered formalin, incubate at room temperature for 24 h, and dehydrate through graded concentrations of ethanol (0%, 70%, 90%, 100%).

   b. Embedding and sectioning: Embed the fixed tissue in paraffin after dehydration with ethanol, as described above. Cut sections with a rotary microtome to 4 or 5 μm. and place onto microscope slide. Deparaffinize sections with xylene, then rehydrate by 5 min sequential incubations in 100%, 90%, and 70% ethanol. Incubate in water for 5 min. If preservation of newly seeded cartilage constructs with negligible levels of extracellular matrix is desired, embed tissue in methacrylate).
2. Histochemical staining for extracellular matrix components
   a. Conventional staining: To examine cell and tissue morphology, total collagen, and sulfated-glycosaminoglycan (S-GAG), stain with Hematoxylin and Eosin, Trichrome, ***(12)*** and Safranin-O ***(13)*** respectively, as described.
   b. Immunostaining (collagens and S-GAGs):
      i. To detect collagen type I and II, obtain antibodies specific for either collagen type I or II and appropriate enzyme-conjugated secondary antibodies and substrate. If no signal is detected, enzymatically pretreat the sections with chondroitinase A,B,C ***(14)***, with or without subsequent treatment with pepsin, trypsin, or a combination of these enzymes, as recommended by a histochemical vendor.
      ii. To detect chondroitin-4-sulfate and chondroitin–6-sulfate, use monoclonal antibodies 2B6 and 3B3, respectively, as described ***(14,15)***.

3.3.2. Biochemical Analyses

1. Construct sample preparation: Record the total construct weight (wet weight), as well as the dry weight of the constructs after drying with a SpeedVac (Savant Instruments, Holbrook, NY) or by lyophilization. Enzymatic digestion of dried cartilage constructs with papain is recommended, because it renders the samples amenable to the evaluations described below. Follow the procedure described by Kim et al. ***(16)***.
2. Cell number: The total cell number in cartilage constructs can be calculated after performing a DNA dye binding assay on papain digested cartilage constructs ***(16)***.
3. Total collagen: The total collagen in cartilage constructs can be calculated after determination of the hydroxyproline content from acid hydrolyzed papain digested cartilage constructs ***(17)***.
4. Total S-GAG: The total S-GAG content in cartilage constructs can be determined after performing a dye-binding assay on papain-digested constructs ***(18)***.

4. Notes

1. It is recommended to obtain articular cartilage from very juvenile, skeletally immature donors. Cells isolated from tissue obtained from adult donors or juvenile donors close to skeletal maturity of various species may produce a more fibrocartilaginous matrix, as compared to the hyaline-like cartilage produced by cells derived from tissue of very juvenile donors
2. Entangled, nonwoven polyglycolic acid (PGA) felt:
   Scaffolds composed of 100% PGA degrade very rapidly [4]. Relatively slower

degrading scaffolds fabricated similarly that are composed of PLA or PLA/PGA co-polymers are also available from Albany International Research Company.

3. Solvent-cast polylactic-glycolic acid (PLGA) and polylactic acid (PLA) foam:
   The method described in this chapter is for the fabrication of polymers with 90% porosity. To create polymers with different porosities, alter the ratio of NaCl:polymer (wt:wt). For example, to obtain a 95% porous scaffold, combine 9.5 parts NaCl and 0.5 parts polymer at a ratio of 9.5:0. Alternatively, maintain a constant polymer concentration and alter the volume of polymer solution and NaCl amount to reach the desired porosity.
4. Quenching foam scaffolds (to replace a crystalline structure with an amorphous one):
   If a crystalline polymer form is undesirable, (some polymers are brittle when fabricated in a crystalline form) a polymer can be made more amorphous by quenching. To quench polymer, proceed as follows.
   - Heat oven to above the polymer melting temperature (T_m).
   - Incubate polymer foams containing salt in oven for 1 h.
   - Quickly remove polymers from the oven and plunge into liquid nitrogen for at least 60 s.
   - Remove from liquid nitrogen and allow polymers to equilibrate to room temperature. Proceed to leaching step.
5. Construct seeding: Using the method described in this chapter, the efficiency of cell seeding is between 80–90%.
6. Construct culture: The described methods will routinely result in a relatively homogeneous tissue containing up to 5% S-GAG, 1% total collagen, and 85%–95% water of the total construct weight. These are levels that approach the lower levels of normal articular cartilage.

References

1. Mow, V. C., Ratcliffe, A., and Poole, A. R. (1992) Cartilage and diarthodial joints as paradigms for hierarchical materials and structures. *Biomaterials* **13,** 67–97.
2. Cima, L. G., Vacanti, J. P., Vacanti, C., Ingber, D., Mooney, D., and Langer, R. (1991) Tissue engineering by cell transplantation using degradable polymer substrates. *J. Biomech. Eng.* **113,** 143–151.
3. Puelacher, W. C., Vacanti, J. P., Ferraro, N. F., Schloo, B., and Vacanti, C. A. (1996) Femoral shaft reconstruction using tissue-engineered growth of bone. *J. Oral Maxillofac. Surg.* **3,** 223–228.
4. Freed, L. E., Vunjak-Novokovic, G., Biron, R. J., Egles, D. B., Lesnoy, D. C., Barlow, S. K., and Langer, R. (1994) Biodegradable polymer scaffolds for tissue engineering. *Biotechnology* **12,** 689–693.
5. Dunkelman, N. S., Zimber, M. P., LeBaron, R. G., Pavelec, R., Kwan, M., and Purchio, A. F. (1995) Cartilage production by rabbit articular chondrocytes on polyglycolic acid scaffolds in a closed bioreactor system. *Biotech. Bioeng.* **46,** 299–305.
6. Sittinger, M., Bujia, J., Minuth, W. W., Hammer, C., Burmester, G. R. (1994) Engineering of cartilage tissue using bioresorbable polymer carriers in perfusion culture. *Biomaterials* **15,** 451–456.

7. Sittinger, M., Schultz, P. Keyszer, G., Minuth, W. W., and Burmester, G. R. (1997) Artificial tissues in perfusion culture. *Int. J. Artific. Organs* **20,** 57–62.
8. Grande, D. A., Halberstadt, C., Naughton, G., Schwartz, R., and Manji, R. (1997) Evaluation of matrix scaffolds for tissue engineering of articular cartilage grafts. *J. Biomed. Mater. Res.* **34,** 211–220.
9. Sittinger, M., Reitzel, D., Dauner, M., Hierlemann, H., Hammer, C., Kastenbauer, E., et al. (1996) Resorbable polyesters in cartilage engineering: affinity and biocompatibility of polymer fiber structures to chondrocytes. *J. Biomed. Mater. Res.* **33,** 57–63.
10. Freed, L. E., Marquis, J. C., Nohria, A., Emmanual, A., Mikos, A. G., and Langer, R. (1993) Neocartilage formation in vitro and in vivo using cells cultured on synthetic biodegradable scaffolds. *J. Biomed. Mater. Res.* **27,** 11–23.
11. Thomson, R. C., Yaszemski, M. J., and Mikos, A. G. (1996) Polymer scaffold processing, in *principles of tissue engineering* (Lanza, R. P., Langer, R., Chick, W. L., eds.) pp. 263–272.
12. Sheehan, D. and Hrapchak, B. (1980) *Theory and Practice of Histotechnology,* 2nd ed. (Sheehan, D. and Hrapchak, B., eds.), Batelle, Columbus, OH.
13. Rosenburg, L. C. (1971) Chemical basis for the histological use of safranin O in the study of articular cartilage. *J. Bone Joint Surg.* **53,** 69–82.
14. Couchman, J. R., Caterson, B., Christner, J. E., and Baker, J. R. (1984) Mapping by monoclonal antibody detection of glycosaminoglycans in connective tissues. *Nature* **307,** 650–652.
15. Sorrell, J. M., et al. (1988) Epitope-specific changes in chondroitin sulfate/dermatan sulfate proteoglycans as makers in the lymphopoietic and granulopoietic compartments of developing bursae of frabricius *J. Imunol.* **140,** 4263–4270.
16. Kim, Y.-H., Sah, R. L. Y., Doong, J. Y., and Grodzinsky, A. J. (1988) Fluorometric assay of DNA in cartilage explants using Hoescht 33258. *Anal. Biochem.* **174,** 168–176.
17. Woessner, J. F. (1961) The determination of hydroxyproline in tissue and protein samples containing small proportions of this amino acid. *Arch. Biochem. Biophys.* **93,** 440–447.
18. Farndale, R. W., Buttle, D. J., and Barrett, A. J. (1986) Improved quantitation and discrimination of sulphated glycosaminoglycans by use of dimethylmethylene blue. *Biochem. Biophys. Acta* **883,** 173–177.

27

Tissue Recombinants to Study Extracellular Matrix Targeting to Basement Membranes

Patricia Simon-Assmann and Michèle Kedinger

1. Introduction

The basement membrane that separates an epithelium or other parenchymal tissues from the connective tissue has been postulated for a long time to be of epithelial origin. Because of difficulties in interpreting results obtained mostly by autoradiographic labeling at the light microscopy level, new analytical tools and models were important to develop. In particular, Lipton *(1)* using an in vitro coculture system was probably the first to provide evidence for a dual origin of the basement membrane: indeed a distinct basement membrane structure developed in myoblast cultures only after the addition of muscle fibroblasts.

Various experimental techniques are currently used to study the expression of basement membrane molecules. They include immunohistochemistry, biochemical approaches, or detection of transcripts by *in situ* hybridization on isolated tissue compartments or cell lines. Apart from the minor limitations of each model (such as cellular contamination in the case of isolated epithelial or mesenchymal cell preparations, abnormal cell behavior or loss of differentiation of cultured cells, threshold sensitivity), they all share a major drawback. Indeed, the fact that a tissue compartment expresses a given basement membrane molecule does not necessarily imply that this molecule is deposited at the basement membrane region. Autoradiographic studies (incorporation of radioactive precursors) that circumvent this problem, unfortunately do not allow discrimination between individual components. The strategy designed by Sariola et al. *(2,3)* that is grafting of avascular murine embryonic kidneys onto quail chorioallantoic membrane deserves special attention. The major advantage of this model is that it allows the authors to follow the

From: *Methods in Molecular Biology, vol. 139: Extracellular Matrix Protocols*
Edited by: C. Streuli and M. Grant © Humana Press Inc., Totowa, NJ

capillary ingrowth in these interspecies chimeric kidneys using species-specific antibodies.

To distinguish the deposition of a single molecule at the subepithelial basement membrane, we designed a new experimental model using recombinants between chick and mouse embryonic intestines. Isolation of pure intestinal endodermal and mesenchymal compartments was performed by enzymatic and mechanical treatments. Following constructions of interspecies endodermal/mesenchymal associations, developmental growth was achieved by in vivo transplantation. Immunocytochemistry using species-specific antibodies recognizing either chick or mouse basement membrane molecules was then performed on cryosections made through the developed hybrid intestines.

The use of this experimental design permits determination of the precise chronological expression/deposition at the intestinal basement membrane region of the individual constituents: some of them are strictly of epithelial or mesenchymal origin and others of dual origin (***4,5*** and for a review, *see* ***6***). This specific technique can be completed by the use of cocultures in vitro, in which one of the tissue compartments can be modified (overexpression or inhibition of basement membrane components or of regulatory molecules) to analyze the consequences on the basement membrane composition and on the resulting extracellular matrix–cell signaling ***(7,8)***. Interspecies or heterotopic (from different levels of the gastrointestinal tract) recombinants have also been used successfully to approach the regulation of the expression of functional markers, i.e., digestive enzymes ***(9–11)***.

2. Materials

2.1. Tissue Dissection

2.1.1. Dissection of the Embryos

1. Paraffin support: Mix melted paraffin (Histomed standard, Labo Moderne, Paris) with activated charcoal (Merck, Darmstadt, Mannheim, Germany). Pour the mixture into small glass dishes. After cooling, cover them with aluminium foil. Sterilize at 110°C for 1 h. Let cool at room temperature. These supports provide accentuated contrast for the dissection of transparent embryos.
2. 9% NaCl solution: 9 g NaCl made up to 1 L. Sterilize by autoclaving.

2.1.2. Preparation of Gelified Medium

75% Ham's F-10 medium (Gibco, Life Technologies, Gaithersburg, MD), 25% agar solution at 1 g/100 mL, and 2 mg/mL gentamicin (Septigen 40, Schering Plough, Kenilworth, NJ).

1. Hank's solution: 137 m*M* NaCl, 5 m*M* KCl, 2 m*M* $CaCl_2$, 1 m*M* $MgCl_2$, 0.8 m*M* $Na_2H\ PO_4$, 0.22 m*M* KH_2PO_4, 0.28 m*M* $MgSO_4$, 5 m*M* glucose.

2. Dissolve 1 g agar (Bacto-Agar, Difco Laboratories, E. Molesley, Surrey, UK) per 100 mL Hank's solution in a boiling water bath; sterilize solution by autoclaving. Once cooled, the solution can be kept at 4°C until use.
3. The day before the experiment, prepare the number of dishes required containing the gelified medium. Melt the agar solution in a boiling water bath. Then rapidly add to the Ham's F10 solution (+ antibiotic) the adequate amount of melted agar. Mix thoroughly.
4. Dispense approximately 2 mL Ham's F10/agar/antibiotic mixture into 3-cm diameter culture dishes.
5. Allow to gel overnight at 4°C.

2.2. Pregrafting Culture: Preparation of Enriched Gelified Medium

1. Prepare chick embryo extracts by mechanical homogenization of 9-d-old embryos. Clarify by centrifugation. Store by aliquots at –20°C.
2. Mix 9 mL of Dulbecco's modified Eagle's medium (DMEM; Gibco, Life Technologies), 1 mL of 9-d-chick embryo extract, 2 mg/mL gentamicin to 2.5 mL of agar solution (as prepared in **Subheading 2.1.2.**). To be prepared the day before the experiment. Keep at 4°C.

2.3. Dissociation of Embryonic Intestines

1. Collagenase solution *make fresh each time*: dissolve 2.5 mg collagenase A (0.5 U/mg) from Clostridium histolyticum (Boehringer Mannheim, Germany) in 10 mL CMRL 1066 medium (Gibco, Life Technologies) (*see* **Note 1**).
2. Blocking solution: mix Ham's F10 medium to newborn bovine serum (1:1) (Gibco, Life Technologies).

2.4. Immunocytochemistry

1. 0.1 *M* phosphate buffered saline solution (PBS): dissolve 13.6 g KH_2PO_4 and 14.2 g Na_2HPO_4, in 1 L H_2O. Adjust pH to 7.2. Store at 4°C.
2. Antibody dilution solution: add 0.01% NaN_3 as a preservative agent to the 0.1 *M* PBS solution.
3. Mounting and antifading solution:
   - solution A: dissolve 0.136 g KH_2PO_4 and 0.876 g NaCl in 100 mL water. Adjust pH to 7.4. Store at 4°C.
   - solution B: dissolve 1.59 g Na_2CO_3 and 2.93 g $NaHCO_3$ in 100 mL water. Adjust pH to 9.0. Store at 4°C.
   - mix 10 mL of solution A with 100 mg p-phenylenediamine (Sigma, Madison, WI): **caution**-toxic and carcinogenic. Adjust pH to 8.0 with solution B. Then add 90 mL glycerol. Store in aliquots at –20°C.
   - the solution has to be thrown away when it turns violet.

Note: All solutions and materials must be sterile; manipulations must be done under sterile conditions.

3. Methods

3.1. Creation of Intestinal Interspecies Associations (Fig. 1)

1. Careful planning is required to get embryos at the correct stages on the day of the experiment. In particular, incubate fertilized eggs from white Leghorn chicken at 38°C in a humidified incubator 5 days before the experiment
2. Remove the 5-d-old chick embryos from the eggs (day 0 being the first day of incubation). At the same day, 12/13 day foetal mice are removed from the placenta after laparotomy of the anaesthetized pregnant mothers (day 0 being the day when a vaginal plug is observed). Care must be taken while embryos are removed out of the eggs or placenta because the intestine is still externalized at these stages.
3. Fasten the embryos with pins on the paraffin support and add sterile 9% NaCl solution. Dissect out the intestinal chick or mouse rudiments with forceps and Pascheff-Wolff iris scissors (Moria Dugast SA, Paris, France) under the dissecting microscope. Place the intestinal anlagen onto gelified medium with the help of a curet (3-mm diameter) and a needle-mounted probe. Remove vessels and adherent tissues with the iris scissor.
4. Incubate the embryonic intestines in collagenase solution (about 6–10 rudiments/ per 2 mL dish) at 37°C for 1 h in a humidified incubator (5% CO_2, 95% air) to disrupt the basement membrane that separates the endoderm from the mesenchyme *(12)*.
5. Then transfer the intestines into a dish containing the blocking solution for at least 30 min at room temperature to stop the action of collagenase.
6. Place the intestines on gelified medium and open the mesenchymal tubes lengthwise with a microscalpel under a dissecting microscope; the endoderm can then be pushed out of the mesenchymal gutter with forceps.
7. Cut the mesenchymes into small fragments ($\approx$1–2 mm length) and transfer them with the help of the curet and of the needle-mounted probe onto a dish containing the enriched gelified medium.
8. Place in each individual mouse mesenchymal segment an equally long endoderm derived from the chick intestine. Cover with a second mouse mesenchymal segment. In parallel, the inverse associations are performed in between chick mesenchymes and mouse endoderms. Control chick or mouse intestinal segments are prepared by reassociating endoderm of each species with its own mesenchyme (*see* **Note 2**).
9. Label the interspecies mesenchymal/endodermal recombinants and control segments with a few carbon particles to identify grafted tissue; incubate overnight in a humidified incubator (5% CO_2, 95% air) at 37°C.

3.2. Grafting Procedure: Intracoelomic Grafting (Fig. 2)

1. 3-d-old chick embryos are used for grafting experiments. Put a mark on the upper part of the shell to check that the egg is maintained in the correct position throughout the experiment. Incubate the fertilized eggs at 38°C in the humidified incubator in an horizontal position. The day before grafting, make a hole at the sharp

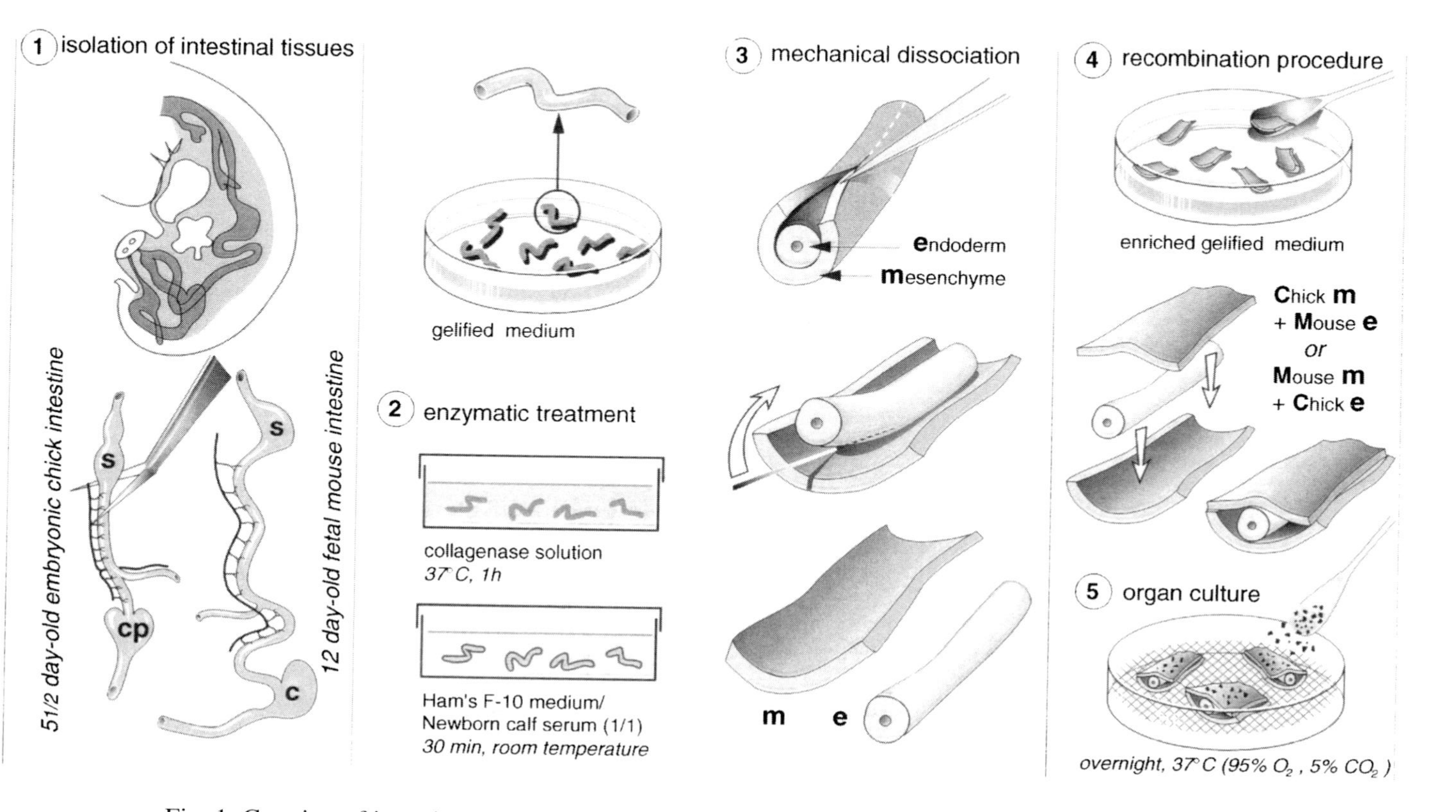

Fig. 1. Creation of intestinal interspecies associations. s, stomach; cp, caecal primordia; c, caecum

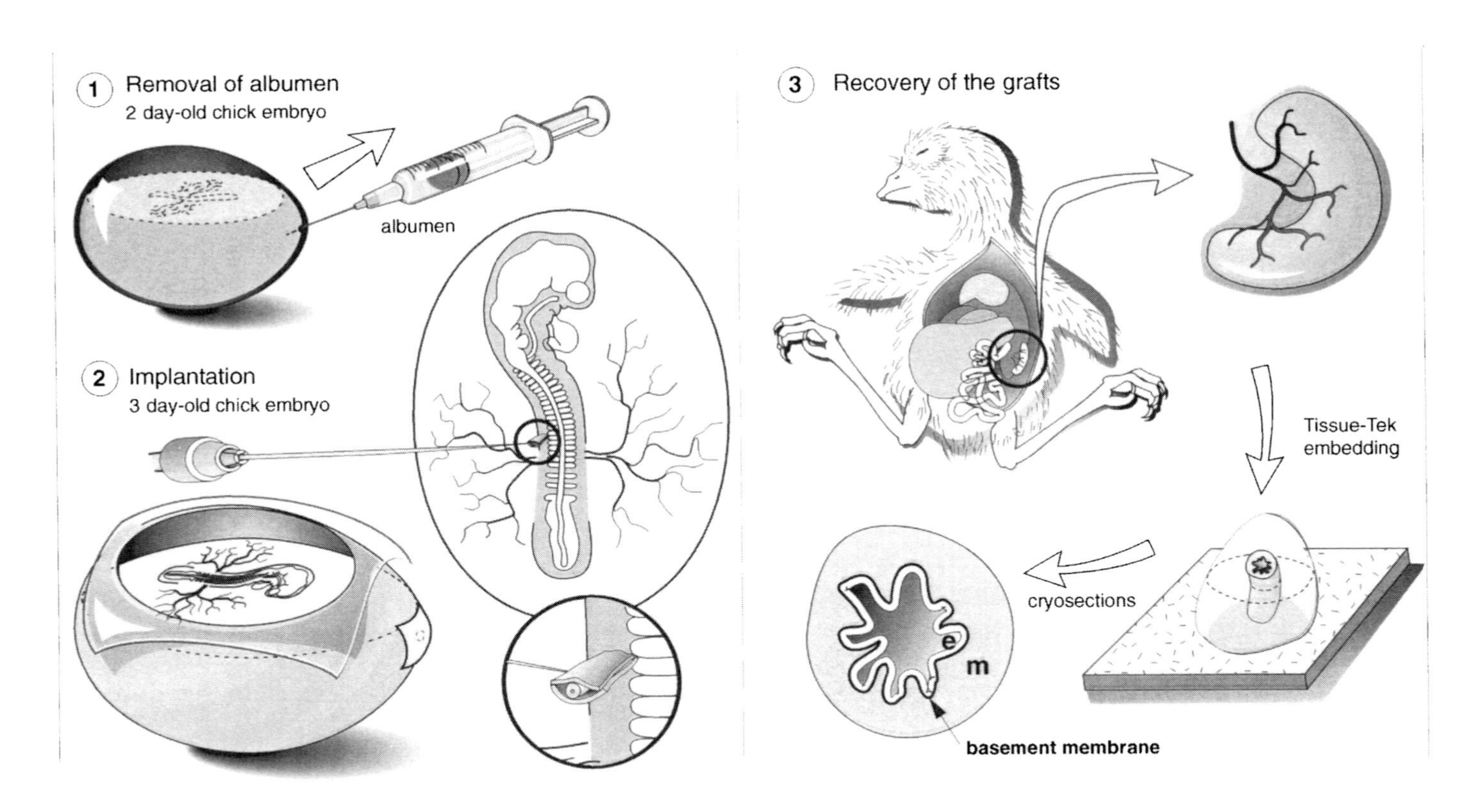

Fig. 2. Grafting procedure and recovery of the grafts. e, epithelium derived from the endoderm; m, mesenchyme-derived tissues.

end of the egg and then drop the chorioallantoic membrane by aspirating about 5 mL of albumen with a sterile syringe. Cover the opening with sterile scotch tape and incubate the eggs at 38°C for an additional 24 h in the same position (*see* **Notes 3** and **4**).

2. Grafting procedure:
   - Cut an opening of approx 1-cm diameter in the shell above the chorioallantoic membrane using forceps and curved scissors.
   - To gain access to the embryo, open tactfully the vitellin and the amniotic membranes with the needle-mounted probe and with the help of forceps.
   - Make a sharp incision in the coelom of the chick embryos near the intersection of the major blood vessels and then carefully and gently implant the graft.
   - Seal the opening in the shell with sterile scotch tape and reincubate the eggs at 38°C in the humidified incubator up to the wished developmental stage of the explants.
3. The grafts are recovered at various times after implantation.
   - Decapitate the chick embryo after its removal out of the egg.
   - Fasten the embryo laterally with pins on both sides; make an incision longitudinally along the middle of the embryo.
   - Locate the grafts with the aid of the carbon particles; the implants which have developed their own vascularization, can be found anywhere within the lateral coelom, the visceral loops or underneath the lungs. Recover the grafts and handle for subsequent analysis.

3.3. Immunodetection of Basement Membrane Molecules on Interspecies Intestines

1. Preparation of the sections (**Fig. 2**)
   - Prepare a cork support of about 1 cm^2; deposit on it a drop of Tissue-Tek compound (O.C.T. Compound, Miles Inc., USA)
   - Embed the hybrid intestine vertically in the Tissue-Tek compound and immediately freeze in isopentane (Prolabo) prealably cooled in liquid nitrogen. These inclusions can be kept for several years at –40°C.
   - Cut transverse sections of about 5–6-µm thick at –25°C using a cryostat and place the sections on SuperFrost/Plus Microscope slides (Menzel-Gläser, Germany). Store at –20°C until use.
2. Immunofluorescence
   - Prepare the required amount of antibody dilutions in NaN_3-containing PBS. These aliquots can be kept up to 1–2 mo at 4°C
   - Add onto each section, an aliquot of first antibodies and incubate the sections for 1 to 2 h at room temperature in a humidified chamber.
   - Wash the sections several times with PBS.
   - Add the corresponding secondary trimethyl-rhodamine or fluorescein isothiocyanate (TRITC or FITC)- labeled antibodies at the optimal dilution indicated by the manufacturer. Place the humidified chamber in dark or cover it with an aluminium foil.

- Wash the sections thoroughly several times with PBS
- Add a drop of the antifading solution and mount under a coverslip.
- The slides can be kept at –20°C until observation under a photomicroscope equipped with an epi-illumination fluorescent system.

Note: Prior to immunocytochemical analysis of the hybrid intestines, check the antibody specificity on control chick and mouse intestinal fragments.

4. Notes

1. The quality and activity of the collagenase batch used for the dissociation of the endoderm from the mesenchyme are very important. Indeed, after enzymatic treatment, the endoderm must be easily removed mechanically using microchirurgical instruments to avoid mesenchymal cell contamination. The use of quail rudiments makes it possible to confirm that there is no contamination during the dissociation step, as quail cells can be recognized by nuclei marker ***(13)***.
2. The techniques of dissociation/reassociation of rudiments and of grafting need some dexterity and experience.
3. Because of a severe lethality of the chick embryos as a consequence of the grafting procedure, plan to incubate twice as many host eggs as needed.
4. There are multiple variations of this method:
   - The endoderm or the mesenchyme can be replaced by established cell lines or by primary cell cultures ***(14,15)***.
   - Associations can be grafted under the kidney capsule ***(5)*** or under the skin ***(11)*** of adult nude mice (nu/nu Swiss mice); in these conditions, exogenous regulation of basement membrane formation and of epithelial cell differentiation is submitted to the hormonal supply provided by the adult host.
   - The associations can be deposited directly on the chorioallantoic membrane of two successive 9-d host embryos allowing a longer developmental growth ***(16)***.
   - Such heterospecific associations can also be performed using rat or human tissues: in these cases, the intestines have to be taken at 14 d and 8 wk of gestation, respectively, in order to get the optimal conditions for dissociation.

Acknowledgments

The authors would like to thank Dr. Katy Haffen for the major contribution in the development of this technique. They also thank Cathy Leberquier and Christiane Arnold for their continuous and dextrous help in performing these experiments. A special thank to Bernard Lafleuriel (Curri Visualisation, Université Louis Pasteur, Strasbourg, France) for generating the figures. Isabelle Gillot is gratefully acknowledged for secretarial help.

References

1. Lipton, B. H. (1977) Collagen synthesis by normal and bromodeoxyuridine-modulated cells in myogenic culture. *Dev. Biol.* **61,** 153–165.

2. Sariola, H., Timpl, R., Von Der Mark, K., Mayne, R., Fitch, J. M., Linsenmayer, T. F., and Ekblom, P. (1984a) Dual origin of glomerular basement membrane. *Develop. Biol.* **101,** 86–96.
3. Sariola, H., Peault, B., Le Douarin, N., Buck, C., Dieterlin-Lievre, F., and Saxen, L. (1984b) Extracellular matrix and capillary ingrowth in interspecies chimeric kidneys. *Cell Differ.* **15,** 43–51.
4. Simon-Assmann, P., Bouziges, F., Arnold, C., Haffen, K., and Kedinger, M. (1988) Epithelial-mesenchymal interactions in the production of basement membrane components in the gut. *Development* **102,** 339–347.
5. Simon-Assmann, P., Bouziges, F., Vigny, M., and Kedinger, M. (1989) Origin and deposition of basement membrane heparan sulfate proteoglycan in the developing intestine. *J. Cell Biol.* **109,** 1837–1848.
6. Simon-Assmann, P., Lefebvre, O., Bellissent-Waydelich, A., Olsen, J., Orian-Rousseau, V., and De Arcongelis, A. (1998) The laminins: role in intestinal morphogenesis and differentiation. *Ann. NY Acad. Sci.* **859,** 46–64.
7. De Arcangelis, A., Neuville, P., Boukamel, R., Lefebvre, O., Kedinger, M., and Simon-Assmann, P. (1996) Inhibition of laminin d_1-chain expression leads to alteration of basement membrane assembly and cell differentiation. *J. Cell Biol.* **113,** 417–430.
8. Lorentz, O., Duluc, I., De Arcangelis, A., Simon-Assmann, P., Kedinger, M., and Freund, J. N. (1997) Key role of the Cdx2 homeobox gene in extracellular matrix-mediated intestinal cell differentiation. *J. Cell Biol.* **139,** 1553–1565.
9. Kedinger, M., Simon, P. M., Grenier, J. F., and Haffen, K. (1981) Role of epithelial-mesenchymal interactions in the ontogenesis of intestinal brush-border enzymes. *Dev. Biol.* **86,** 339–347.
10. Yasugi, S. (1993) Role of epithelial-mesenchymal interactions in differentiation of epithelium of vertebrate digestive organs. *Develop. Growth Differ.* **35,** 1–9.
11. Duluc, I., Freund, J. N., Leberquier, C., and Kedinger, M. (1994) Fetal endoderm primarily holds the temporal and positional information required for mammalian intestinal development. *J. Cell Biol.* **126,** 211–221.
12. Gumpel-Pinot, M., Yasugi, S., and Mizuno, T. (1978) Différenciation d'épithéliums endodermiques associés à du mésoderme splanchnique. *C. R. Acad. Sci. Paris* **286,** 117–120.
13. Le Douarin, N. (1973) A Feulgen-positive nucleolus. *Exp. Cell Res.* **77,** 459–468.
14. Haffen, K., Lacroix, B., Kedinger, M., and Simon-Assmann, P. M. (1983) Inductive properties of fibroblastic cell cultures derived from rat intestinal mucosa on epithelial differentiation. *Differentiation* **23,** 226–233.
15. Kedinger, M., Simon-Assmann, P., Lacroix, B., Marxer, A, Hauri, H. P., and Haffen K. (1986) Fetal gut mesenchyme induces differentiation of cultured intestinal endoderm and crypt cells. *Develop. Biol.* **113,** 474–483.
16. Kramer, B., Andrew, A., Rawdon, B. B., and Becker, P. (1987) The effect of pancreatic mesenchyme on the differentiation of endocrine cells from gastric endoderm. *Development* **100,** 661–671.

28

Fluorescence Assays to Study Cell Adhesion and Migration In Vitro

Paola Spessotto, Emiliana Giacomello, and Roberto Perris

1. Introduction

1.1. Cell Adhesion as the Basis for the More Complex Process of Cell Migration

The basis for cell movement and the maintenance of organized epithelial structures formed by "stationary" cells is a phenomenon known as cell adhesion, i.e., the establishment of the firm anchorage of a cell to an underlying substratum. Both cell adhesion and the subsequent process of cell locomotion occur via the interaction of cell-adhesion molecules, integrins, cell-surface proteoglycans, nonintegrin receptors, and selections with each other or with extracellular ligands such as various ECM constituents. The number of extracellular matrix (ECM) components discovered to be implicated in cell adhesion and migration phenomena is rapidly growing. This fact, in conjunction with the progressively unveiled diversity in the mode by which these ECM molecules affect cell behavior, has raised the demands on the in vitro assays aimed at the analysis of cell adhesion. Probably, the requirement for a higher accuracy and versatility; for an increased capability to most closely reproduce the in vivo situations; and for the uncompromised possibility to apply the assays to the analysis of ex vivo cells; represent the primary demands that have forced investigators to ameliorate the currently available cell-adhesion protocols, as well as device novel alternative ones.

In order to move, a cell needs to adhere to the underlying substratum by a reversible and adequately strong binding *(1,2)*. Thus, in some instances the adhesive interactions could be strong, whereas in others it could be relatively weak, as for instance in the case of rapidly locomoting cells and circulating lymphocytes. Accordingly, for a comprehensive understanding of the mecha-

From: *Methods in Molecular Biology, vol. 139: Extracellular Matrix Protocols*
Edited by: C. Streuli and M. Grant © Humana Press Inc., Totowa, NJ

nisms regulating cell migration, the binding force of a cell to the substratum must be considered to be a critical parameter to establish. A number of recent studies have centered upon this problem and have provided data on the relative force of cell adhesion exhibited by diverse cell types onto cellular and ECM substrates, both under static *(3–9)* and dynamic conditions (i.e., under shear stress; **ref.** ***10***).

1.2. Cell Adhesion Assays Under "Static" Conditions

When it comes to cell adhesion under static conditions, a number of procedures to quantify this process have been described in the literature and all of them have capitalized on the use of mechanical forces to remove the nonbound or weakly bound cells *(11)*. The methods by which these forces have been applied to the bound cells include simple fluid flushing, the application of buoyance *(12)* or rotating motions *(7,13)*, and centrifugation *(4–8,14,15)*. It was early on proposed that, under static conditions, the application of a strictly perpendicular removal force could yield a precise way to exert a definable and measurable detachment force onto a population of cells *(4–8)*. Thus, although considering all of the involved biophysical constrains and variables, in our opinion this method still emerges as the preferred one. However, the centrifugation assay procedure devised originally with the intent of quantitating weak cell bindings, and the forces with which cells could bind to a given substratum *(16)*, had a number of drawbacks and limitations. We describe here a novel cell adhesion assay which we have denoted Centrifugal Assay for Fluorescence-based Cell Adhesion (CAFCA) (TECAN AG, Ropperswil, Switzerland) (***17***; **Fig. 1**), and which we regard to be the cell adhesion assay of choice for the following reasons. The assay exploits the use of differential centrifugal forces to achieve maximal accuracy and reproducibility, while maintaining the possibility to precisely estimate the relative cell adhesion strengths. What we arbitrarily denote adhesion force (A_{fd}) can be calculated as dyn/cell, following the generation of a suitable "force-dependent curve" to retrieve the force required to detach 50% of the bound cells. The formula to adopt is then: $A_{fd} = (D_c - D_m) \times V_c \times F_c$ (where D_c is the specific cell density, intermediate value = 1.07 mg/cm^3; D_m is the specific density of the medium, 1 mg/cm^3; V_c is the volume of the cell; and F_c is the centrifugal force yielding 50% cell detachment).

CAFCA is based on two centrifugation steps: the first one to allow for a synchronized cell-substratum contact; and the second one (in the reverse direction) to allow for removal of the unbound/weakly bound cells under controlled conditions (**Fig. 1**). The assay is unique in that it combines the possibility of accurately estimating the cell-binding avidities while allowing a precise assessment of the ratio of bound vs nonbound cells within a given cell population. CAFCA is rapid (total assay time may be <1 h) and is applicable to a

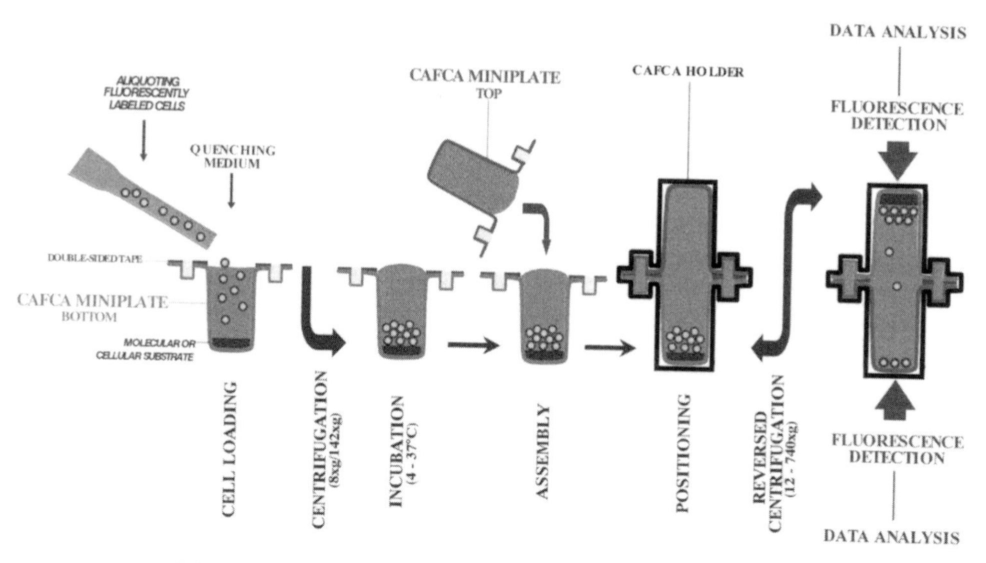

Fig. 1. Schematic representation of the CAFCA procedure.

small number of cells, without limiting the total number of samples/conditions that can be examined. It is based on the use of a commercially available microplate fluorometer and a few associated accessories (**Fig. 2**), including, for maximal convenience, a dedicated software. Finally, as previously suggested for cell–cell adhesion assays *(18–20)*, the employment of vital fluorochromes to label the cells adds to the above advantages that characterize CAFCA. This is because it renders this procedure an ideal assay for the analysis of freshly isolated normal and diseased human cells.

1.3. Fundamentals of Cell Migration Assays

As hinted above, cell movement, either in the form of a chemotactic/chemokinetic response to soluble factors, or as a phenomenon governed by a haptotactic response to ECM-associated components, is one of the primary cellular processes during embryonic development, the maintenance of a healthy adult organism and the progression of a number of pathological events such as inflammation and tumor metastasis. Although cell migration has been studied in numerous in vivo model systems, it has been proven difficult to investigate the molecular events responsible for the movement of cells and the governing control mechanisms involved resorting to cell-culture techniques. A variety of methods have also been described for the studying of cell movement in vitro and several of these protocols employ different types of 3-dimensional ECMs *(21–25)*. Cells capable of invading these ECMs can be monitored at

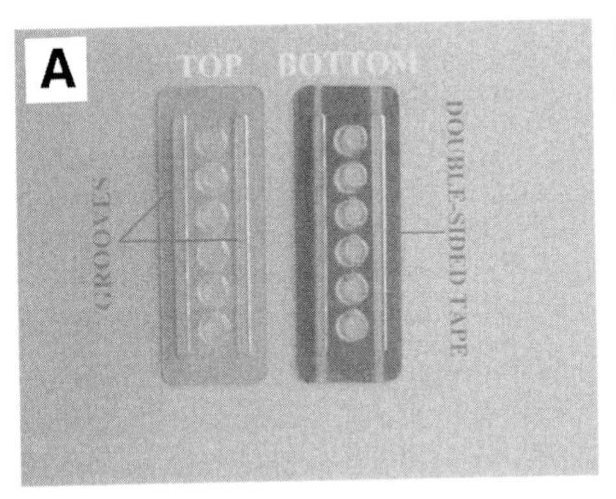

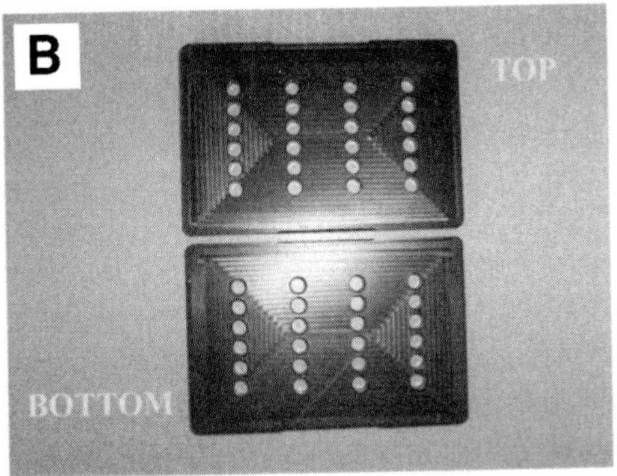

Fig. 2. Photographs showing the CAFCA accessories. **(A)** Top and bottom 6-well CAFCA miniplates having two parallel grooves (1 mm large and 2 mm deep) on each side of the wells. Note that the bottom CAFCA miniplate is provided with an adhesive surface. **(B)** CAFCA holders in black hard plastic to be used as centrifugation and fluorescence detection supports.

chosen time-intervals, or continuously in real-time by utilizing video time-lapse microscopy in conjunction with light-sensitive CCD cameras *(7,26,27)* or, preferably, by computer-assisted confocal laser microscopy systems (*28,29*; multiple photon confocal laser microscopes are currently available that cause minimal fluorescence fading during continuous cell tracking). These latter microscopic systems have the additional potential to allow for true 4-dimensional analyses *(30)*. The common final goal of both types of analyses is to be able to score the depth at which the invading cells localize within the 3-dimensional matrix, and the relative speed with which they accomplish this movement. Establishment of these two quantitative parameters provides a measure for the invasive capability of the cells.

It is implicit that both qualitatively and quantitatively the mechanisms of cell locomotion are most accurately analyzed utilizing the aforementioned methods. However, these cannot efficaciously be applied to the rapid and easily performed screening tests, because they are laborious, time-consuming, and invariably involve the use of sophisticated instruments. Thus, the classical Boyden chambers, and their more recent derivatives, the so-called Transwells, provide alternative means of assessing cell motility in vitro. Originally, these systems were intended for the quantification of the chemotactic motility response of leukocytes *(31–35)*, but later were demonstrated also to be suitable for the assessment of similar phenomena in mesenchymal and epithelial cells *(36–38)*. Assays involving the use of Transwells are all based on the passage of cells through a porous, inert micromembrane of polycarbonate or polypropylene, which may or may not be coated with single ECM molecules, mixtures of molecules, complex matrices such as Matrigel and Humatrix *(39)*, cell monolayers, or reconstituted tissues and tissue slices.

1.4. Assessing Transfilter Cell Migration

Scoring chemotaxis, random and directed cell movement, and invasion through the porous membranes of Transwells has conventionally relied upon visual counting of stained cells *(37,38)*. Obviously, visual counting of cells dispersed over a membrane is subjective and time-consuming. Furthermore, it is virtually impossible to determine the true percentage of cell migration and/or invasion because of the difficulties in scoring the total cell number contained in the system by simple visual counting. In order to be able to accomplish a more precise cell counting, colorimetric procedures that are based on the use of dyes, such as toluidine blue *(38)*, or cell viability markers, such as 3-(4,5-dimethylthiazol–2-yl)–2,5-diphenol tetrazolium bromide (MTT; **ref. *40***) have been adopted. After dye incorporation, cells are solubilized and the detected absorbance is directly correlated with the number of cells attached to the underside of the membrane (i.e., transmigrated cells). Major drawbacks with these colorimetric methods are the potential of a variable background staining associated with protein-coated membranes; the variable toluidine blue staining of different cell types; and the extra incubation times needed to accomplish the procedure. Furthermore, if the migratory capacity of leukocytes is studied, additional steps may be needed to accomplish the procedure. For instance, it may be necessary to centrifuge the plate to allow the transmigrated leukocytes to more firmly adhere to the bottom of the well, such as to permit the removal of the culture medium and the addition of the suitable dye solution.

Isotope-based assays provide an alluring alternative to the rather unprecise colorimetric procedures because they are sensitive and afford an efficient means of quantitating the number of transmigrated cells *(37,38)*. However, apart from the inconvenient and hazardous nature of isotopes, as well as the time-consuming step of prelabeling cells metabolically, an additional problem is that associated with the fact that migration assays are long-term assays that may be protracted for days. Thus, the spontaneous release of radioactivity by the moving cells may be the cause of overestimations of the actual number of migratory/invasive cells. It has more recently been realized that this problem can be circumvented by employing fluorescent cell tagging, via calcein-AM *(34,35,41)* or BCECF-AM *(35,41)*. A suitable protocol to follow in this case is to utilize the fluorochrome calcein AM to label by spontaneous uptake those cells that have transversed the porous membrane of the Transwell. However, this labeling step has to be preceded by removal of the nonmigrated/invading cells from the upperside of the membrane (**Fig. 3**). Although the sensitivity of this method is high, it does not resolve a fundamental problem of the procedure, namely to be able to precisely quantify the relative ratio transmigrated vs nontransmigrated cells within the same population, and to determine the kinetics with which the transmigration occurs. It should be also noted that there is an intrinsic caveat in using fluorescent dyes with relative low intracellular retention time when performing long-term migration assays, in that the sensitivity of the assay invariably declines with time.

We describe here an improved fluorimetric assay that has the potential to circumvent the above problems and has been denominated fluorescence-assisted transmigration invasion and motility assay (FATIMA, TECAN AG). The system may utilize conventional single unit Transwells, Unicell 24 plates (**Fig. 3**) or a newly devised variant of these latter plates denoted FATIMA plates (Whatman/Polyfitronics, Boston, MA) (**Fig. 4**). The assay may be performed according to two different protocols depending on which of these units is employed (**Figs. 3** and **4**). The unique trait of the FATIMA plates is the incorporation of a specific porous membrane, which shares most of the properties of the conventional polycarbonate membranes, but is devised such as to shield fluorescent light in the wavelength range 450–550 nm. FATIMA is based on the indiscriminate cell labeling with lipophilic carbocyanines dyes (e.g., DiI, DiO, DiA, DiR and derivatives, Molecular Probes, Inc.), or the more specific molecular tagging through GFP-based vectors.

Lipophilic fluorescent dyes are particularly suitable for long-term labeling and tracking of cells in vivo and in vitro ***(43–45)*** and are therefore the preferred fluorochromes when cell migration assays are run under extended periods of time. Additional advantages of these fluorochromes include their minimal spontaneous release from cells and the limited increase in the global fluorescence intensity obtained upon cell division (when using isotope-tagged cells this parameter has to be taken into account). When working with

Fig. 3. *(opposite page)* Schematic representation of the FATIMA procedure for anchorage-dependent *(left)* and suspension-growing *(right)* cells, based on the use of conventional Transwells or plates that carry transparent membranes. Fluorescence detections *(blue boxes)* are performed at three different times: initially to determine the total number of cells aliquoted into the Transwells; after removal of the nontransmigrated cells; and at the end of the experiment after removal of all cells from the Transwells. These latter fluorescence measurements serve to determine the background fluorescence, i.e., including that deriving from any possible dye release from the cells during migration and the fluorescence that may remain associated with the substrate and/or polycarbonate membrane. When working with anchorage-dependent cells, complete removal of "nontransmigrated cells" is an absolute requirement for obtaining reliable

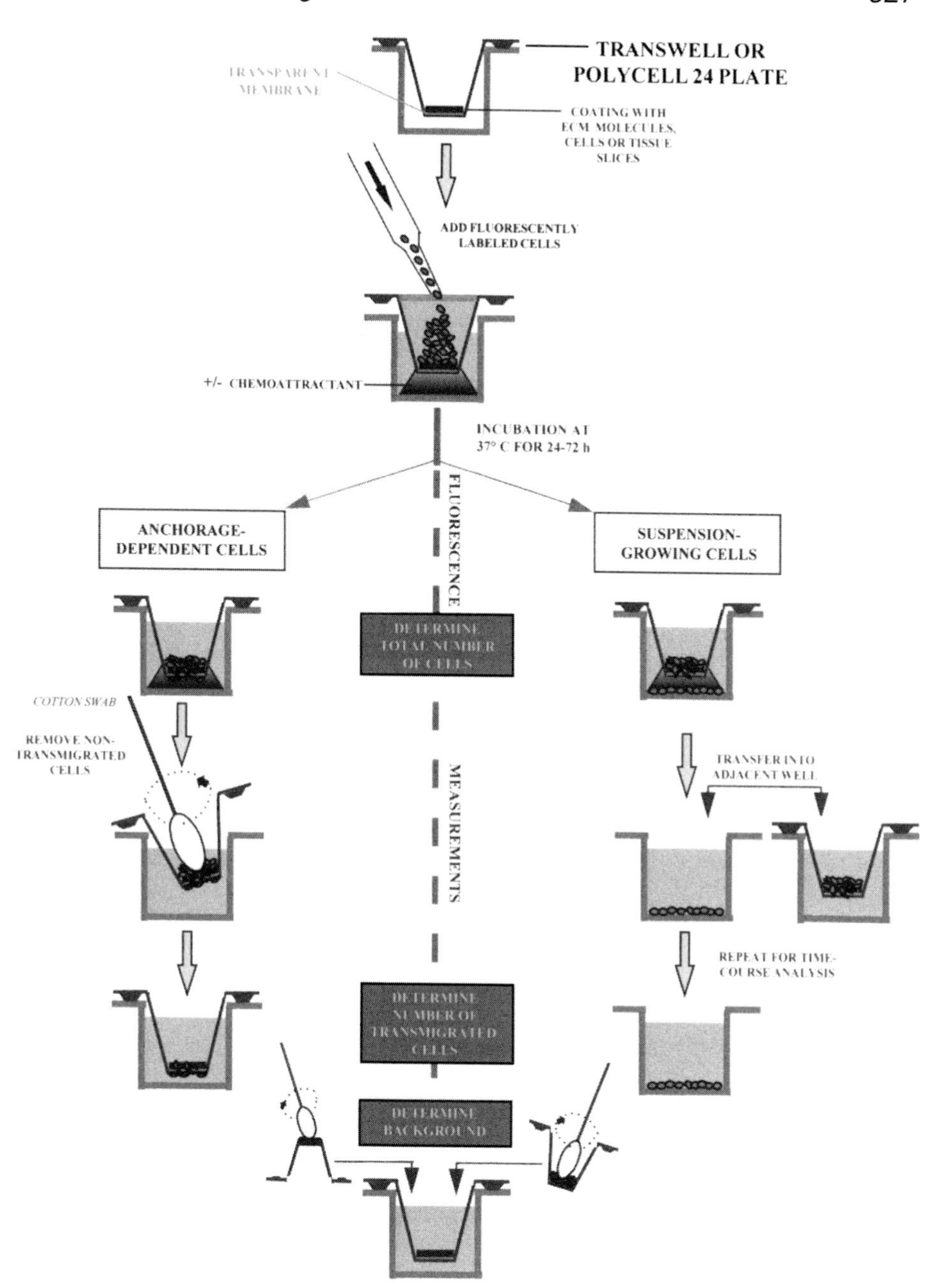

results. This because, in this case, the majority of the transmigrated cells remains bound to the underside of the membrane. Thus, fluorescence measurement from either the top or bottom side of the membrane would unavoidably cause detection of the fluorescence signals emitted from both cell populations.

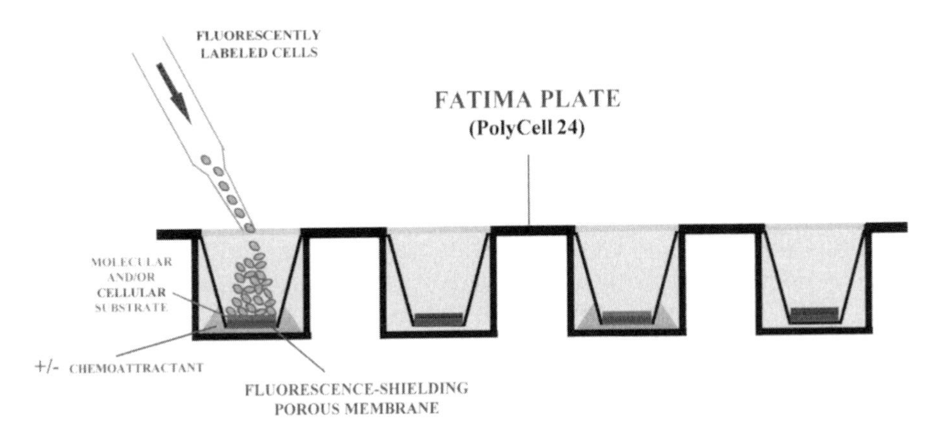

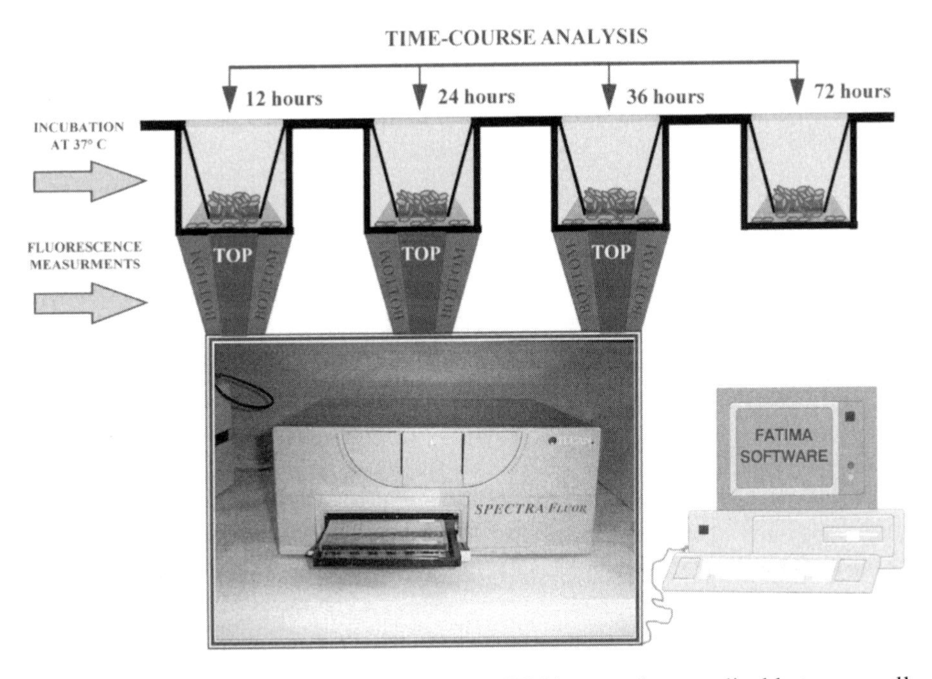

Fig. 4. Schematic representation of the FATIMA procedure applicable to any cell type and based on the use of Unicell 24 plates carrying a unique fluorescence shielding membrane (Whatman/Polyfitronics) i.e., FATIMA plates. In this case, there is no need to mechanically (or enzymatically) manipulate the wells (compare with **Fig. 3**), once the cells have being aliquoted into them. Thus, independent fluorescence detection from the top (*nontransmigrated cells*) and bottom (*transmigrated cells*) side of the plate can efficiently be used to determine the exact percentage transmigrated cells out of the total amount of cells introduced into the system. Moreover, kinetic studies may readily be performed for both anchorage-dependent and suspension-growing cells within the same well. This system is well-suited for large-scale analyses of cell migration.

FATIMA, a suitable computer-interfaced microplate fluorometer with independent top- and bottom-side measurement capabilities, such the SPECTRAFluor Plus (TECAN AG), should be employed to determine the levels of corresponding "top/bottom" fluorescent signals of the plate. FATIMA further has the advantage that it takes into account also "spatial" parameters. Especially when analyzing anchorage-dependent cells, we have noticed that transmigrated cells tendentially accumulate at the perimeter of the membrane. Thus, by utilizing a microplate fluorometer equipped with the capability of an automated multiple point-measurement mode to detect fluorescence in different areas of the well surface (e.g., the SPECTRAFluor Plus is capable of measuring 25 different points in a well of a 24-well plate) there is the possibility of rapidly quantifying the spatial distribution of the migrated cells. This possibility can further be exploited when migratory substrates with a spatially heterogenous configuration are investigated. Apart from the possibility of accomplishing kinetic studies of cell movement, there is an unparalleled capability of FATIMA; this is the possibility of determining within few minutes and simultaneously the precise ratio transmigrated cells in up to 24 inserts.

2. Materials

2.1. CAFCA

1. Cells to be assayed for their binding capability to molecular substrates and cells to be assayed in a monolayer arrangement for their capability to bind to other cells.
2. 0.05 *M* bicarbonate buffer, pH 9.6: prepare a $NaHCO_3$ solution of 4.2 g/L and one solution of Na_2CO_3 of 5.3 g/L and mix the two components to reach the desired pH; this takes less Na_2CO_3 than $NaHCO_3$.
3. ECM component(s) to test as substratum.
4. Bovine serum albumin (BSA, fraction V).
5. PBS: 8 g NaCl, 0.2 g KCl, 1.44 g $Na_2HPO_4 \cdot H_2O$, 0.2 g KH_2PO_4 in 1 L apyrogen H_2O.
6. PBS containing 5 m*M* EDTA (0.84 g/L Na_2-EDTA in PBS; to be used only for anchorage-dependent cells).
7. Culture medium, RPMI (for suspension-growing cells) and DMEM (for anchorage-dependent cell) both supplemented with 15% FCS.
8. Vital fluorochrome calcein AM (dissolved in DMSO at 2 mg/mL as stock solution; Molecular Probes, Inc.).
9. Serum-free culture medium containing 0.1–0.5% polyvinylpyrrolidone (M_r 360,000; PVP).
10. Serum-free medium containing 0.1–0.5% PVP and with 2% v/v India ink.
11. CAFCA miniplates (TECAN Group, Austria; **Figs. 1** and **2**).
12. Semihydraulic vice (TECAN Group; **Fig. 2**).
13. CAFCA holders (TECAN Group; **Figs. 1** and **2**).
14. Assembly devices (TECAN Group; **Fig. 2**).

15. Microplate fluorometer capable of reading independently from top and bottom of the plate (SPECTRAFluor Plus; TECAN Group).

2.2. FATIMA

1. Cells to be assayed for their capability to transmigrate/invade a 2- and 3-dimensional molecular substrate (*see* **Note 14**).
2. Bicarbonate buffer as used for CAFCA
3. Fluorescent cell labeling reagents: A) *Labelling dye solution*: 0.25 *M* sucrose (0.85 g per 10 mL apirogen H_2O), store at 4°C. The solution should be sterilized using a 0.2-μm filter. B) *Stock dye solution*: Fast DiI™ and Fast DiO™ (Molecular Probes, Inc.) are prepared in ethanol at 1–5 mg/mL and stored at –20°C. Centrifugation of the concentrated solution before addition to the cells is recommended to remove the undissolved dye crystal (*see* **Note 15**).
4. Transwells (Costar, Cambridge, MA), Unicell 24 plates, FATIMA plates (Whatman Polyfiltronics) carrying membranes with different pore sizes (*see* **Note 16**).
5. Purified extracellular matrix molecules. Matrigel should be handled as indicated in the instructions provided by the supplier (*see* **Note 17**).
6. PBS and EDTA as used for CAFCA.
7. Serum-free DMEM or RPMI as used for CAFCA.
8. 2% bovine serum albumin (BSA; stock solution). 2.0 g of BSA in 100 mL DMEM/RPMI. Sterilize using a 0.2-μm filters and store in aliquots at –20°C.
9. *Working assay solution:* RPMI or DMEM containing 0.1–0.5% BSA.
10. Conditioned medium from fibroblastic cells to use as generic chemoattracting agent. We normally use the conditioned medium from the NIH 3T3 fibroblasts grown in DMEM containing 10% FCS. When cells are at 70–80% confluency, remove the supernatant gently, wash with PBS and replace medium with 10 mL serum-free DMEM. After 24 h, collect the supernatant, centrifuge, sterilize using a 0.2-μm filter, add BSA (0.1–0.5 % final concentration) and store at –20°C until use.
11. Cotton swabs to remove nontransmigrated cells from the upper side of the porous membrane of the transmigration unit (not necessary when using FATIMA plates).
12. Microplate fluorometer, such as the SPECTRAFluor Plus, capable of reading independently from top and bottom of the plate.

3. Methods

3.1. CAFCA

3.1.1. Procedure for Cell-Substratum Adhesion

1. Prepare the "coating solution" composed of the ECM molecule(s) of interest dissolved at 0.01–100 μg/mL (total protein concentration) in the bicarbonate buffer and aliquot 50 μL in each well of the *bottom* CAFCA miniplate (*see* **Fig. 1** and *see* **Note 1**).
2. Incubate the *bottom* CAFCA miniplates at 4°C for 8–16 h (*see* Note 2).
3. Dissolve the BSA at 1% (w/v) in the bicarbonate buffer and heat up the solution to 56°C for 15 min to denatured the protein (*see* **Note 3**).

4. Remove the coating solution from the wells, wash them at least 2–3 times with bicarbonate buffer, and fill them with 200 µL of the BSA blocking solution (or analogous solution). Incubate at room temperature for at least 2 h (*see* **Note 4**).
5. Fluorescent cell labeling: *Suspension-growing cells:* Rinse the cells once and resuspend them in 300 µL of RPMI with 15% FCS. Add calcein AM (at final concentration of 1–10 µ*M*; the recommended calcein AM-cell ratio is 2 µ*M*/10^6 cells) and incubate the cells at 37°C for 10–20 minutes to allow the calcein AM to be metabolized. The incubation time with calcein AM may in some cases need to be extended, especially when working with ex vivo cells. The optimal labeling is achieved when the cell pellet attains a yellowish color.

   Anchorage-dependent cells: Remove the culturing medium and extensively rinse the plates with PBS, followed by incubation with 2 mL PBS (for a 35-mm plate) containing 5 m*M* EDTA. Incubate the cells in the presence of the EDTA for up to 15 min (most cells should detach from the culture dish within 5 min, but we have found that some cells may require somewhat longer incubation times). Collect the cell suspension, wash the cells by centrifugation to remove all EDTA, and resuspend them in 300 µL of DMEM (or another preferred medium). Cells can then be labeled with calcein AM as indicated for suspension-growing cells (*see* **Note 5**).
6. Remove the blocking agent from the wells and wash them at least twice with 200 µL of the cell-adhesion medium containing PVP. Fill then the wells with 200 µL of the cell adhesion medium in which 2% (v/v) India ink has been added (**Fig. 1**) (*see* **Note 6**).
7. Collect the fluorescently labeled cells by centrifugation, rinse them twice with cell-adhesion medium containing PVP and resuspend them in the chosen cell-adhesion medium at the appropriate concentration. Aliquot 50 µL of the cell suspension in each well (*see* **Note 7**).
8. Place the *bottom* CAFCA miniplates in the apposited bottom CAFCA black holder (**Fig. 2**) and centrifuge them at 142*g* for 5 min, followed by incubation 37°C for 20 min (**Fig. 1**) (*see* **Note 8**).
9. Fill the wells of the *top* CAFCA miniplate with the same PVP- and India ink-containing medium as used for the bottom CAFCA miniplates, such as to assure to form a bulging meniscus (**Fig. 1**). Place a *bottom* CAFCA miniplate into the apposited light-metal assembly device with the handle-rod (*see* **Fig. 2**). Fill also the wells of the *bottom* CAFCA miniplates with a similar excess of liquid (*see* **Note 9**).
10. Take a *top* CAFCA miniplate and reverse it upside down in the air (**Fig. 1**) by holding it from the ends (not from the middle as it tends to bend). Avoid touching the double-sided tape, especially when wearing gloves (e.g., such as those that may be worn when handling infectious cells and/or solutions). Because of the high liquid surface tension in such a narrow well, the liquid will not fall out from the upside-down oriented well. Bring the *top* CAFCA miniplate in this reversed orientation in proximity of the corresponding *bottom* CAFCA miniplate, making sure to align perfectly the "*top*" and "*bottom*" wells. First, gently bring in con-

tact the two liquid meniscuses and then press the *top* CAFCA miniplate onto the *bottom* one allowing the former to attach firmly to the double-sided tape of the *bottom* plate. At this point an "air-bubble-free" communicating chamber should have formed (**Fig. 1**). Place the second light-metal device (**Fig. 2**) over the top CAFCA miniplate and place the entire unit under the vise (**Fig. 2**) to tighten the affixed plates.

11. Remove the light-metal devices and place the assembled CAFCA miniplate unit into the assembled black CAFCA hard-plastic holders (**Fig. 2**) and tighten the holders. These units can now be reversed centrifuged for 5 min at the desired centrifugal force (standard force = 46*g*; **Fig. 1**) (*see* **Note 10**).
12. Measure the fluorescence signal emitted by cells in wells of the top (*nonbound cells*) and bottom (*substrate-bound cells*) sides of the CAFCA miniplates independently (**Fig. 1**), ideally using a microplate fluorometer such as the SPECTRAFluor Plus capable of detecting the fluorescence emanated from both the top and bottom side of the microplate. The percentage bound cells, out of the total amount of cells introduced into the system, can be calculated as: bottom fluorescence value/bottom fluorescence + top fluorescence values. If the above indicated computer-interfaced instrument is used, a dedicated CAFCA software is provided by the supplier that automatically runs the calculations (*see* **Note 11**).

3.1.2. Procedure for Cell–Cell Adhesion Assay

1. Seed the cells to be used as the underlying (substrate) cell monolayer into the CAFCA miniplate wells (make to use the presterilized type) at a concentration earlier estimated to yield confluency by the following day, or whenever the experiment is intended to be performed (*see* **Note 12**).
2. Once it is time to run the experiment, the procedure is the same as described for cell-substratum adhesion, i.e., **steps 5–12** (*see* **Note 13**).

3.2. FATIMA

3.2.1. Preparation of the Membranes

1. *Cell motility:* Coat the membrane of the FATIMA plate, Unicell 24, or Transwell (Costar) as indicated for CAFCA, considering that polycarbonate may adsorb protein 10–50-fold less efficiently than the PVC plastic. The amount of coating solution should be sufficient to cover the entire membrane surface. We recommend to use a 100-µL volume for the individual units of FATIMA plates, Unicell 24 plates, and Transwells with a 6.5-mm insert diameter (0.33 cm^2 area; corresponding to those fitting a conventional 24-well plate), and a 250-µL volume for Transwells with a 12-mm insert diameter (1.0 cm^2 area). These correspond to the only commercially available units carrying membranes with 12-µm pore size and fitting a 12-well plate. After coating, remove excess liquid and wash the inserts twice with serum-free DMEM or RPMI.
2. *Cell invasion:* Dilute the thawed Matrigel to the desired final concentration using cooled serum-free medium. Make a 50-µg/mL (I) and 62.5-µg/mL (II) solution for 6.5-mm and 12-mm diameter inserts, respectively. Aliquot 100 µL (I) or

250 μL (II) of solution into the upper surface of the membrane of the inserts. Make sure to keep all undiluted solutions on ice to avoid undesired gelification. Leave the plates in a cell-culture hood to allow the Matrigel solution to air-dry. It takes about 5 h to dehydrate the indicated amount of Matrigel under a continuous laminar flow. For convenience, it is possible to leave the inserts to air-dry overnight in the absence of laminar flow. Reconstitute Matrigel with 100–200 μL of serum-free DMEM or RPMI at room temperature for 90 min under constant rotation. Remove the excess medium from the membranes before adding the cells.

3.2.2. Cell Labeling

1. When assaying anchorage-dependent cells, rinse the adherent cells with PBS, add 5 m*M* EDTA in PBS and incubate at 37°C in 5% CO_2 for 2–5 min (*see* **Note 18**).
2. Collect the detached cells and wash them twice in serum-free DMEM. When assaying suspension-growing cells, collect the cells directly from the flask and wash them twice with serum-free RPMI.
3. Prepare the working dye solution by diluting the concentrated stock solution in 0.25 *M* sucrose to reach the final concentration of 1–10 μ*M* (*see* **Note 18**).
4. Resuspend the cell pellet ($\leq 5 \times 10^6$ cells) in 200 μL of dye solution.
5. Incubate the cell suspension at 37°C in 5 % CO_2 for 30–45 min (*see* **Note 19**).
6. Wash the cells twice with the working assay solution.

3.2.3. Procedure for FATIMA Using Unicell 24 Plates (Fig. 3)

1. Resuspend the cells in working solution at 2×10^6 cells/mL (leukocytes), or at 1×10^6 cells/mL (anchorage-dependent) and fill the upper portion of the membrane with 100 μL (for 6.5-mm diameter inserts) or 500 μL (for 12-mm diameter inserts) of the cell suspension.
2. Fill the lower part of the unit, i.e., the 24-well plate using as a tray for the Unicell 24 plate through the openings in the insert wall, with 600 μL (6.5 mm) or 1.5 mL (12 mm) of control medium (working assay solution); conditioned medium (as a positive control); and/or other potential stimulating agents diluted in working assay solution.
3. Incubate the plates at 37°C in 5% CO_2 for the desired time (*see* **Note 20**).
4. At the end of the assay, or at the desired time-intervals, and without manipulating the inserts, determine the fluorescence levels using the microplate reader. The detected fluorescence value corresponds to the total number of cells (**Fig. 3**).
5. Remove the nonmigratory/noninvading cells from the upper surface of the membrane with a cotton swab (**Fig. 3**). Perform this operation for anchorage-dependent cells assuring to dislodge all cells residing near the border of the insert, without removing the inserts from their original well. For suspension-growing cells, removal of nonmigratory cells from the upper side of the membrane is not necessary since the 24-insert Unicell 24 unit, can temporary be transferred to an adjacent empty well or to another plate (**Fig. 3**).
6. Measure the fluorescence levels (in the microplate reader), which now corresponds to the migrated cells (**Fig. 3**).

7. Assign for each set of assays a "blank." For this purpose, utilize one of the inserts in which both cells on top of the membrane and cells at its underside are wiped off with a cotton swab and the fluorescence is read in the absence of cells.
8. Calculate the percentage migrated/invaded cells by subtracting the "blank" from the fluorescence values corresponding to the migrated cells.

3.2.4. Procedure for FATIMA Using the Specifically Devised FATIMA Plates (Fig. 4)

1. **Steps 1–3** are identical to the procedure described in **Subheading 3.2.3.**
2. Measure the fluorescence from the top (corresponding to nonmigrated cells) and bottom (corresponding to transmigrated cells) side of the plate (*see* **Note 21**).
3. Repeat the fluorescence measurement at different time-intervals (kinetic analysis). TECAN AG, the supplier of the FATIMA system, provides a dedicated software that automatically performs the calculations of the ratios transmigrated cells/time unit (*see* **Note 22**).

4. Notes

1. The ideal coating concentration to use for each individual ECM molecule may vary. This depends upon the cell type and the intrinsic capacity of the ECM molecule to become adsorbed onto plastic. We find that a coating concentration range of 0.01–10 µg/mL is suitable for a wide range of ECM proteins when carrying out dose-dependency tests of cell-substratum attachment. In our experience, a coating concentration of 0.01 µg/mL can be adopted as a starting concentration when designing a coating concentration curve. A lower concentration usually yields in sufficient amounts of the ECM molecule (we have tested >30 different ones) becoming bound to plastic. If there is a specific interest in preparing a cell-adhesion substratum containing a mixture of ECM molecules, the total protein concentration in the coating solution may obviously be raised to assure that a sufficient amount of the less represented molecules in the mixture is obtained. However, caution should be taken when coating with single or mixtures of ECM components that have an intrinsic propensity to self-assemble spontaneously (or form heterogenous assemblies), even in the absence of divalent cations, physiological pH and temperature, or other assembly-promoting factors. Examples of ECM components with this tendency are fibronectin, vitronectin, von Willebrand factor, laminins, and collagens. The precise amount of protein bound to the specific PVC-based CAFCA miniplates has been determined for a number of ECM components (*5,6,17*).
2. We find that protein coating at 4°C overnight in the indicate bicarbonate buffer yields the optimal adsorbance of the molecules onto plastic both in terms of amount and "active" configuration of the immobilized molecule. It also largely prevents unwanted multimeric complex formation of the molecules.
3. A 1% solution of denatured BSA is a suitable blocking agent, i.e., its acts well in saturating areas of the plastic left uncoated by the ECM molecule (this is especially important when carrying out accurate dose-dependency tests), and it is pre-

ferred over native BSA for most cell types. Our experience, however, and that reported by others, indicate that this may not always be the case. Alternative blocking agents to consider may then be human serum albumin, α-casein and ovalbumin, in their native or denatured form. Avian neural crest cells, for instance, bind to some extent to BSA ***(5,6)***, but not to ovalbumin, whereas human and murine lymphocytic cells may bind to α-casein and ovalbumin, but fail to bind to denatured BSA. On the other hand, fibroblastic cells such as the human rhabdomyosarcoma RD-KD and the human embryonic kidney 293 cells bind to both native BSA and ovalbumin, but do not interact with α-casein and denatured BSA. Thus, it may be necessary to identify empirically the suitable blocking agent for each given cell type. In this context the most difficult situation that we have faced has been that of B lymphocytes freshly isolated from patients affected by chronic lymphocytic B cell leukemia or adult myeloid leukemia. In these cases, we have found that 1% human serum albumin plus 0.5% Tween-20, was the sole saturating agent that consistently yielded a low background binding. Finally, we find that when the adhesion assays are run in the presence of the stimulating Mn^{2+} divalent cation, or activating anti-integrin antibodies, the nonspecific interaction of the cells with these blocking proteins has a tendency to be exaggeratedly enhanced. Caution should be taken when carrying out cell-adhesion assays under these conditions.
4. Use wells filled only with blocking agent as negative control. We normally adopt a "background" cell binding of <10% for these control wells as threshold value for judging that the experiment was successful.
5. Calcein AM is a colorless polyanionic fluorescein derivative that, upon digestion by cytoplasmic esterases becomes fluorescent (λ_{ex} 485, λ_{em} 535). The cleavage of the AM group by acetylases causes the calcein molecule to become negatively charged and prevents it from rapidly diffusing out of the cells. In fact, in comparison with the thiol-reactive fluorescent dyes CellTrackers (Molecular Probes, Inc.), which can alternatively be used as vital cell tracers, we find that calcein AM exhibits higher retention time. On the other hand CellTrackers generally produce a somewhat more intense cell labeling and can effectively be used for multiple cell labeling. Calcein AM is inoquous to the cells and is not known to influence their adhesive behavior.
6. The function of the PVP is to provide a higher viscosity of the medium, such as to match the relative density of the cells, and is an inert substitute to the previously utilized BSA ***(4–8)***. Thus, the exact concentration of PVP may vary depending upon the cell type analyzed. The range of 0.1–0.5% is the one that, in most cases, would satisfy the viscosity equilibrium requirement. For lymphocytes we normally use 0.1%, whereas for larger cells as fibroblasts, epithelial cells, endothelial cells, and various tumor cells, we use 0.5%. When analyses of cation-dependency of cell binding are to be carried out, we recommend using as cell adhesion medium: 0.25 m*M* Tris-HCl, pH 7.4, with 0.15 *M* NaCl to which the different cations can be added at the desired concentrations. In this condition, there is no detectable precipitation of cations, even when applied at rather high

concentrations. For optimal fluorescent measurements purposes (**Fig. 1**), India ink (at the optimal concentration of 2%) turns out to be the most effective, inert fluorescence quencher. We have ascertained for a number of cell types that incubation in medium containing this concentration of India ink does not affect the proliferation rate or cell adhesion behavior of cells.

7. The number of cells seeded per well may vary depending upon the size of the specific cell type. We find that a concentration of 30,000 cells/well is ideal for suspension-growing cells, whereas 1000–5000 cells/well are optimal when examining larger anchorage-dependent cells. The minimal amounts of cells per well that can be used in CAFCA are ≤1,000/well and ≤100/well for lymphocytes and fibroblastic cells, respectively.
8. This is one of the first critical steps of the procedure. First, we find that a force of 142*g* is an ideal centrifugation force to bring all cells (both lymphocytic and fibroblastic) contained by each well in simultaneous contact with the substratum. Moreover, we find that the centrifuges sold by Juan (France) are the most convenient ones since they are precise and can accommodate 4 × 4 CAFCA miniplates (i.e., 4 × 96-well plates) per centrifugation. The length of the subsequent incubation at 37°C may be varied as desired, although 15–20 min is the time-period that we find necessary to allow for a stable cell adhesion to ensue.

   If there is a specific interest in analyzing the receptor-ligand interaction without involvement of the cytoskeleton or signal transduction phenomena, we find that the entire procedure of CAFCA may effectively be accomplished at 4°C, as shown for its predecessing adhesion assay ***(4,5,8)***. In such a case, all the solutions should be precooled to 4°C and the centrifugations run at 4°C in a cooled centrifuge. Here, there is no need to perform an incubation of the cells after the centrifugation step, but if there is a specific reason to do so, it obviously should be carried out at 4°C.
9. This is an extremely important step that may determine the final outcome of the experiment. Care has to be taken to adequately fill the wells of the bottom and top CAFCA miniplates such as to avoid air-bubble formation during the subsequent assembly of the plates. This means that wells of both plates have to contain an excess of liquid (forming a bulging meniscous), such as to assure that no air is trapped between the upper surfaces of the wells during face-to-face assembly (**Fig. 1**).
10. Differential centrifugation forces to detach the "weakly bound" or nonbound cells are used to determine the relative binding avidity of the cells to the substratum. It should be noted that firmly bound cells, i.e., those binding with high avidity may not be removed with forces below those ascertained to retain viable cells (we have determined the threshold to be a force of ≤750*g*). Thus, when a "force-dependent" centrifugation curve is generated, a suitable force range to adopt is ~10–~750*g*. We have observed that forces <40*g* may not be sufficient to displace the nonbound fibroblastic cells to the top CAFCA miniplate wells: this is an absolute requirement for being able to detect physically separated fluorescence signals (i.e., those emitted by substrate-bound cells in the wells of the *bottom*

CAFCA miniplates and those emitted by cells in the wells of the *top* CAFCA miniplate), and thereby accurately determine the ratio bound vs nonbound cells (*see below*). On the other hand, when working with small cells, such as lymphocytes and neutrophils, weak binding interactions are possible to detect by lowering the centrifugal force down to ~10*g*.

11. If there is a specific interest in knowing the exact number of cells bound to the substratum and the number of cells that have failed to do so, a calibration curve based on a serial dilution of equivalently labeled cells can be run in parallel to extrapolate the corresponding cell numbers on the basis of the corresponding relative fluorescence levels.
12. For the achievement of optimal results, care should be taken to produce a cell monolayer as homogeneous as possible, i.e., avoiding in as much as possible to leave uncovered areas of the plastic. This would minimize the possibility for nonspecific binding to areas of the well not covered by cells. It has to be taken in consideration that in most cases, serum components that passively adsorb onto uncoated plastic may promote cell adhesion (even when using fibronectin- and/or vitronectin-depleted serum). Thus, it may generally be advantageous using minimal concentrations of serum during the pregrowing of the "substratum" cell monolayer. Moreover, if underlying cell monolayer can be produced by growing the cells in serum-free medium, there is a possibility to carry out a "substrate saturation" with a suitable blocking agent, similar to that described for cell adhesion to ECM molecules (**step 4**). However, if this is not possible, it is advisable to select a substrate molecule allowing the attachment of the cells selected to form the underlying monolayer, but not adhesion of the cells to assayed for their capability to bind to this monolayer, even after a "conditioning" by serum-contained factors (i.e., binding of these factors to the selected substrate molecule). For instance, when we run the assay for examining the lymphocyte-endothelium interaction, we find that we can pregrow most types of endothelial cells on a von Willebrand factor substrate, whereas most lymphocytes fail to significantly interact with this ECM molecule, independently of whether or not serum components in the endothelial growth medium have bound to it or not. Alternatively, it may be possible to pregrow the underlying monolayered cells in a "panning-like fashion" onto wells precoated with a suitable antibody direct against a cell surface-component specific for these cells. This provided that antibody ligation of the targeted cell surface component does not cause significant changes in the cell–cell adhesion behavior of the cells. It is also important to ascertain that cells of the underlying cell monolayer do not detach from the substrate during the reverse centrifugations. This is efficiently controlled for each single experimental situation by pretagging cells of the underlying cell monolayer with a red- or blue-fluorescent CellTracker dye (Molecular Probes, Inc.) and then determining the respective fluorescent signals in each well. Most microplate fluorometers, including the SPECTRAFluor Plus, are equipped with several filter sets to allow for multiple fluorescence detection.
13. The procedure for cell–cell adhesion requires a more delicate parameter setting because of the marked variability when two or more cell types are involved in

the assay. Thus, it is preferable to run the first centrifugation step at a lower force than that used for cell-substratum adhesion, and additionally, this has to be empirically set according to the cell type. In the case of lymphocytes binding to the endothelium, for instance, we have observed that a force of 8–10*g* is an adequate force to allow synchronized contact with the endothelial cell monolayer, while minimizing the number of lymphocytes that are becoming constrained between the single endothelial cells. The reversed centrifugal force may similarly need to be differently adjusted when compared to that applied to cell-molecule adhesion. Normally, a higher force is required to discriminate between true and nonspecific cell binding. To ascertain that the parameter settings are optimal for each individual experimental condition, it is advisable to observe the bound cells under an inverted microscope, with or without the use of fluorescence.

14. FATIMA (*Fluorescence-Assisted Transmigration Invasion and Motility Assay*) is a versatile in vitro "cell migration assay" based on the transversing of tagged cells through an inert porous micromembrane. The lather serves the sole purpose to function as a physical barrier for differentiating "motile" versus "nonmotile cells"; meaning cells that have exhibited the capability to actively locomote under the influence, or in the absence, of a chemoattracting agent. According to our definitions *Transmigration* (T) in the acronym refers to the process whereby a cell is penetrating a single-cell monolayer (e.g., an endothelium or epithelium) or a thin tissue section (<5–10 µm in thickness). *Invasion* (I) refers to the process whereby a cell penetrates and move through a multicellular structure, i.e., a cellular multilayer, an explanted/ in vitro reconstituted tissue; a thicker tissue section vibratome (<20 µm in thickness), or a 3-dimensional ECM. *Motility* (M) refers to the movement of a cell on a bidimensional ECM substrate.
15. Presently, there is a vast assortment of fluorochromes for intracellular labeling (*see* the Molecular Probes' product catalog). We have recently carried out an extended comparison between the presently available main categories of fluorescent dyes having the capability to become taken up spontaneously by nonneuronal cells (i.e., neurons may also be tagged by retrograde transport of single fluorochromes or fluorescent microspheres through their axons/projections). These include vital dyes based upon intracellular esterase activity; thiol-reactive fluorescent compounds; and numerous lipophilic dyes. When carrying out long-term migration assays, the otherwise very convenient vital dyes based on esterase activity and thiol-reactivity are improper as they are completely released by the migrating cells within less than 24 h. On the other hand, the lipophilic dyes, which remain within cells for weeks, diffuse passively into the cells and may be taken up to a significant extent also by dying cells. On the basis of this unavoidable trade-off concerning the fluorochrome choice, we find that lipophilic dyes are the optimal cell labeling agents to use for long-term assays. The wide variety of lipophilic tracers currently available also provides a great flexibility as well as the possibility to accomplish experiments involving multiple labeling of two or more cell populations.
16. Both Unicell 24 and FATIMA plates are provided in a 24-well format and carry either a transparent, polycarbonate-based porous membrane (Unicell 24), or a

specifically devised fluorescence-shielding membrane of similar material (FATIMA plate). Alternatively it may be possible to use the conventional single-well Transwell units which can be combined in the number desired by being inserted into the wells of a conventional 24-well plate. Standard Unicell 24 and conventional Transwells may be suited for migration analyses in which there is a specific interest in monitoring the process by combined fluorescence and phase-contrast microscopy, whereas FATIMA plates are specifically devised for accurate and high-throughout put fluorescence-based assays. In all cases, the choice of pore size of the membrane of the insert depends largely upon the cell type and its relative size. For example, if neutrophils or smaller lymphocytes are studied, inserts carrying 3–5-μm pore size membranes are recommended. For anchorage-dependent cells it is preferable to use 8–12-μm pore size membranes. It is advisable to run some pilot assays to determine the type of membrane that allows for the minimal "passive" transmigration rate, i.e., in the absence of membrane-coating and/or chemoattracting agent.

17. Matrigel is currently the most commonly used 3-dimensional ECM substrate. However, it should be emphasized that it is a poorly characterized murine sarcoma-derived basement membrane ECM. If there is a specific need to work with human material, the isolation of a human homolog to Matrigel, *Humatrix* ***(39)***, has recently been described from smooth muscle cells. Interstitial-like ECMs of desired compositions may readily be prepared by incorporating selected ECM molecules during in vitro fibrillogenesis of interstitial collagens (mainly collagen type I, III, and V; normally used at 0.5–1.5 mg/mL; **refs.** ***23*** and ***46***). Other possibilities to produce interstial ECM-like structures are through the generation of fibrin clots from fibrinogen or the use of artificial biopolymeric matrices ***(47)***. Finally, if there is a specific interest in assaying native ECMs, several protocols are described in the literature for the production of such matrices in vitro derived from cultured cells or tissues.
18. Some experimental protocols suggest that labeling of anchorage-dependent cells while attached to plastic is ideal. This should result in an improved viability compared to labeling in suspension after detachment of the cells from their growth substrate. We have noticed, however, that for several anchorage-dependent cell lines labeling in mobilized phase is less efficient than when done in suspension. Therefore, we recommend to carry out cell labeling after detachment and by diluting the fluorochrome in sucrose as indicated.
19. In some cases it may be advantageous to label the cells on ice to allow the dye to incorporate into the plasma membrane under reduced rate of endocytosis, thus reducing the potential of dye accumulation in cytoplasmatic vesicles. Cell tagging at this lower temperature requires a longer labeling time and may lead to a compromising tagging. Leukocytes normally prefer a physiological labeling temperature and a higher dye concentration. They also incorporate more efficiently DiI-derivatives than other lipophilic dye variants, but, in some cases, may also be refractory to these dyes. Labeling at 4°C may, however, be preferable when working with freshly isolated human leukocytes in which the tendency to self-aggregate and become metabolically activated has to be prevented.

20. For highly migratory cells incubation times can be rather short, i.e., 2–4 h, whereas poorly migratory cells may require up to 72 h to accomplish significant transmigrations. It is advisable to perform some pilot time-course studies to empirically determine the optimal assay time for a given cell and/or experimental condition. Kinetics of the transmigration/invasion process is possible when using FATIMA plates (**Fig. 4**), which allow for independent top and bottom fluorescence measurements. This time-related information can additionally be used as a valuable parameter for establishing the migratory capacity of the cells.
21. The currently available membrane of the FATIMA plate shields the fluorescence in the wavelength range of 450 to 550 nm. This implies that this plate can optimally be employed when working with cells tagged with a fluorescent label having emission spectra within that range, and hence, there is presently a certain technical limitation when it comes to perform FATIMA experiments with multiple labeled cells.
22. The FATIMA software is, similarly to the CAFCA one, a commercially available (TECAN AG), Excel-based software that integrates the *XFluor* software needed for the handling of the SPECTRAFluor Plus microplate fluorometer (TECAN AG) and a specifically designed software for data elaboration.

Acknowledgments

The work was supported by TECAN Austria and grants from Consiglio Nazionale delle Ricerche (CNR; p.f. ACRO), Associazione Italiana della Ricerca sul Cancro (AIRC) and Fondo Sanitario Nazionale (Ricerca Finalization 1995–1996). We are greatful to Johanna Neumayer and John Lipsky for her numerous contributions to the development of the assays and Alfonso Colombatti for his helpful suggestions.

References

1. DiMilla, P. A., Stone, J. A., Quinn, J. A., Albelda, S. M., and Lauffenburger, D. A. (1993) Maximal migration of human smooth muscle cells on fibronectin and type IV collagen occurs at an intermediate attachment strenght. *J. Cell. Biol.* **122,** 729–737.
2. Palecek, S. P., Loftus, J. C., Ginsberg, M. H., Lauffenburger, D. G., and Horwitz, A. F. (1997) Integrin-ligand binding properties govern cell migration speed through cell-substratum adhesiveness. *Nature* **385,** 537–540.
3. Roy, P., Petroll, W. M., Cavanagh, H. D., Chuong, C. J., and Jester, J. V. (1997) An in vitro force measurement assay to study the early mechanical interaction between corneal fibroblasts and collagen matrix. *Exp. Cell Res.* **232,** 106–117.
4. Lotz, M. M., Burdsal, C. A., Erickson, H. P., and McClay, D. R. (1989) Cell adhesion to fibronectin and tenascin: quantitative measurements of initial binding and subsequent strengthening response. *J. Cell. Biol.* **109,** 1795–1805.
5. Lallier, T. and Bronner-Fraser, M. (1991) Avian neural crest cell attachment to laminin: Involvement of divalent cation dependent and independent integrins. *Development* **113,** 1069–1084.

6. Perris, R., Kuo, H. J., Glanville, R. W., and Bronner-Fraser, M. (1993) Collagen type VI in neural crest development: distribution in situ and interaction with cells in vitro. *Dev. Dyn.* **198,** 135–149.
7. Lallier, T., Deutzmann, R., Perris, R., and Bronner-Fraser, M. (1994) Neural crest cell interactions with laminin: structural requirements and localization of the binding site for $\alpha1\beta1$ integrin. *Dev. Biol.* **162,** 451–464.
8. Yin, Z., Gabriele, E., Leprini, A., Perris, R., and Colombatti, A. (1997) Differential cation regulation of the $\alpha5\beta1$ integrin-mediated adhesion of leukemic cells to the central cell-binding domain of fibronectin. *Cell. Growth Differ.* **8,** 1339–1347.
9. Garcìa, A. J., Huber, F., and Boettiger, D. (1998) Force required to break $\alpha5\beta1$ integrin-fibronectin bonds in intact cells is sensitive to integrin activation state. *J. Cell. Biol.* **273,** 10,988–10,993.
10. Puri, K. D., Chen, S., and Springer, T. A. (1998) Modifying the mechanical property and shear threshold of L-selectin adhesion independently of equilibrium properties. *Nature* **329,** 930–933.
11. Bongrand, P., Claesson, P. M., and Curtis, A. S. G., eds. (1994) *Studying Cell Adhesion.* Springer-Verlag, Berlin Heidelberg.
12. Goodwin, A. E. and Pauli, B. U. (1995) A new adhesion assay using buoyancy to remove nonadherent cells. *J. Immunol. Methods* **187,** 213–219.
13. St. John, J. J., Schroen, D. J., and Cheung, H. T. (1996) An adhesion assay using minimal shear force to remove nonadherent cells. *J. Immunol. Methods* **170,** 159–166.
14. Segat, D., Pucillo, C., Marotta, G., Perris, R., and Colombatti, A. (1994) Differential attachment of human neoplastic B cells to purified extracellular matrix molecules. *Blood* **83,** 1586–1594.
15. Perris, R., Perissinotto, D., Pettway, Z., Bronner-Fraser, M., Mörgelin, M., and Kimata, K. (1996) Inhibitory effects of PG-H/aggrecan and PG-M/versican on avian neural crest cell migration. *FASEB J.* **10,** 293–301.
16. McClay, D. R., Wessel, G. M., and Marchase, R. B. (1981) Intercellular recognition: quantitation of initial binding events. *Proc. Natl. Acad. Sci. USA* **78,** 4975–4979.
17. Giacomello, E., Neumayer, J., Colombatti, A., and Perris, R. (1999) CAFCA—A centrifugal assay for fluorescence-based cell adhesion adapted to the analysis of *ex vivo* cells and capable of determining relative binding avidities. *Biotechniques* **26,** 758–762.
18. Maeda, Y., Tanaka, K., Koga, K., Zhang, X. Y., Sasaki, M., Kimura, M., and Nomoto, K. (1993) A simple quantitative in vitro assay for thymocyte adhesion to thymic epithelial cells using a fluorescein diacetate. *J. Immunol. Methods* **157,** 117–123. .
19. Vaporciyan, A. A., Jones, M. L., and Ward, P. A. (1993) Rapid analysis of leukocyte-endothelial adhesion. *J. Immunol. Methods* **159,** 93–100.
20. DeClerck, L. S., Bridts, C. H., Mertens, A. M., Moens, M. M., and Stevens, W. J. (1994) Use of fluorescent dyes in the determination of adherence of human leukocytes to endothelial cells and the effect of fluorochromes on cellular function. *J. Immunol. Methods* **172,** 115–124.

21. Albini, A., Iwamoto, Y., Kleinman. H. K., Martin, G. R., Aaronson, S. A., Kozlowski, J. M., and McEwan, R. N. (1987) A rapid in vitro assay for quantitating the invasive potential of tumor cells. *Cancer Res.* **47,** 3239–3245.
22. Terranova, V. P., Hujanen, E. S., Loeb, D. M., Martin, G. R., Thornburg, L., and Glushko, V. (1986) Use of a reconstituted basement membrane to measure cell invasiveness and select for highly invasive tumor cells. *Proc. Natl. Acad. Sci. USA* **83,** 465–469.
23. Perris, R., Krotoski, D., and Bronner-Fraser, M. (1991) Collagens in avian neural crest development: Distribution in vivo and migration-promoting ability in vivo. *Development* **113,** 969–984.
24. Tuan, T. L., Song, A., Chang, S., Younai, S., and Nimni, M. E. (1996) In vitro fibroplasia: matrix contraction, cell growth, and collagen production of fibroblasts cultured in fibrin gels. *Exp. Cell Res.* **223,**127–134.
25. Sundquist, K. G., Hauzenberger, D., Hultenby, K., and Berström, S. E. (1993) T lymphocyte infiltration of two- and three-dimensional collagen substrata by an adhesive mechanism. *Exp. Cell Res.* **206,** 100–110.
26. Grunwald, J. (1987) Time-lapse video microscopic analysis of cell proliferation, motility and morphology: application for cytopathology and pharmacology. *Biotechniques* **5,** 680–687.
27. Krull, C. E., Collazo, A., Fraser, S. E., and Bronner-Fraser, M. (1995) Segmental migration of trunk neural crest: time-lapse analysis reveals a role for PNA-binding molecules. *Development* **121,** 3733–3743.
28. Schönermark, M. P., Bock, O., Büchner, A., Steinmeier, R., Benbow, U., and Lenarz, T. (1997) Quantification of tumor cell invasion using confocal laser scan microscopy. *Nature Med.* **3,** 1167–1171.
29. Lamb, R. F., Hennigan, R. F., Turnbull, K., Katsanakis, K. D., McKenzie, E. D., Birnie, G. D., and Ozanne, B. W. (1997) AP–1-mediated invasion requires increased expression of the hyaluronan receptor CD44. *Mol. Cell. Biol.* **17,** 963–976.
30. Thomas, C. F. and White, J. G. (1998) Four-dimensional imaging: the exploration of space and time. *Trends Biotech.* **16,** 175–182.
31. Harvath, L., Falk, W., and Leonard, E. J. (1980) Rapid quantitation of neutrophil chemotaxis: use of a polyvinylpyrrolidone-free polycarbonate membrane in a multiwell assembly. *J. Immunol. Methods* **37,** 39–45.
32. Bignold, L. P. (1987) A novel polycarbonate (Nucleopore) membrane demonstrates chemotaxis, unaffected by chemokinesis, of polymorphonuclear leukocytes in the Boyden chamber. *J. Immunol. Methods* **105,** 275–280.
33. Pilaro, A. M., Sayers, T. J., McCormick, K. L., Reynolds, C. W., and Wiltrout, R. H. (1990) An improved in vitro assay to quantitate chemotaxis of rat peripheral blood large granular lymphocytes (LGL). *J. Immunol. Methods* **135,** 213–224.
34. Sunder-Plassmann, G., Hofbauer, R., Sengolge, G., and Hore, W. H. (1996) Quantitation of leukocyte migration: improvement of a method. *Immunol. Invest.* **25,** 49–63.

35. Schratzberger, P., Kahler, C. M., and Wiedermann, C. J. (1996) Use of fluorochromes in the determination of chemotaxis and haptotaxis of granulocytes by micropore filter assays. *Ann. Hematol.* **72,** 23–27.
36. Gehlsen, K. R. and Hendrix, M. J. (1986) In vitro assay demonstrates similar invasion profiles for B16F1 and B16F10 murine melanoma cells. *Cancer Lett.* **30,** 207–212.
37. Repesh, L. A. (1989) A new in vitro assay for quantitating tumor cell invasion. *Invasion Metast.* **9,** 192–208.
38. Muir, D., Sukhu, L., Johnson, J., Lahorra, M. A., and Maria, B. L. (1993) Quantitative methods for scoring cell migration and invasion in filter-based assays. *Anal. Biochem.* **215,** 104–109.
39. Kedeshian, P., Sternlicht, M. D., Nguyen, M., Shao, Z. M., and Barsky, S. H. (1998) Humatrix, a novel myoepithelial matrical gel with unique biochemical and biological properties. *Cancer Lett.* **123,** 215–226.
40. Imamura, H., Takao, S., and Aikou, T. (1994) A modified invasion–3-(4,5-dimethylthiazole–2-yl)–2,5-diphenyltetrazolium bromide assay for quantitating tumor cell invasion. *Cancer Res.* **54,** 3620–3624.
41. Garrido, T., Riese, H. H., Quesada, A. R., Mar Barbacid, M., and Aracil, M. (1996) Quantitative assay for cell invasion using the fluorogenic substrate 2',7'-bis(2-carboxyethyl)–5(and–6)-carboxyfluorescein acetoxymethylester. *Anal. Biochem.* **235,** 234–236.
42. Marchetti, D., Menter, D., Jin, L., Nakajima, M., and Nicolson, G. L. (1993) Nerve growth factor effects on human and mouse melanoma cell invasion and heparanase production. *Int. J. Cancer* **55,** 692–699.
43. Godement, P., Vanselow, J., Thanos, S., and Bonhoeffer, F. (1987) A study in the developing visual system with a new method of staining neurons and their processes in fixed tissue. *Development* **101,** 697–713.
44. Serbedzija, G. N., Fraser, S. E., and Bronner-Fraser, M. (1990) Pathways of trunk neural crest migration in the mouse embryo as revealed by a vital dye labelling. *Development* **108,** 605–612.
45. Ragnarson, B., Bengtsson, L., and Haegerstrand, A. (1992) Labeling with fluorescent carbocyanine dyes of cultured endothelial and smooth muscle cells by growth in dye-containing medium. *Histochemistry* **97,** 329–333.
46. Turley, E. A., Erickson, C. A., and Tucker, R. P. (1985) The retention and ultrastructural appearances of various extracellular matrix molecules incorporated into three-dimensional hydrated collagen lattices. *Dev. Biol.* **109,** 347–369.
47. Kim, B. S. and Mooney, D. J. (1998) Development of biocompatible synthetic extracellular matrices for tissue engineering. *Trends Biotech.* **16,** 224–230.

29

Analyzing Cell-ECM Interactions in Adult Mammary Gland by Transplantation of Embryonic Mammary Tissue from Knockout Mice

Teresa C. M. Klinowska and Charles H. Streuli

1. Introduction

1.1. Background

Understanding the function of ECM and cell–matrix interactions in mammalian development has reached new levels of sophistication with the introduction of gene knockout technology. Indeed, two of the chapters in this volume provide detailed methods for producing mice with deletions in specific ECM genes (*see* Chapters 13 and 14). However, in some knockout mice, animals die during late embryogenesis or shortly after birth. In such cases, it is possible to analyze embryonic developmental phenotypes, but it is less easy to determine the in vivo role of cell–matrix interactions in adult tissues.

Although this problem has been partially solved by the development of tissue-specific knockouts (*see* Chapter 13), the approach relies on appropriate tissue-specific promoters. In many cases, genes that uniquely characterize specific cell types within complex tissues have not been identified. Thus, knockout technology can be restrictive when analyzing cell–matrix interactions in specific cases of tissue development and/or homeostasis.

A significant proportion of mammary gland development occurs postnatally. Epithelial cells within this tissue are organized as two-layered ductal structures consisting of a central layer of luminal epithelial cells and a basal layer of myoepithelial cells contacting basement membrane. These ductal networks are embedded within mammary stroma. The formation of ducts and development of lactational alveoli are highly dynamic events that occur during various developmental stages of the mammary gland. The mechanism of tissue mor-

From: *Methods in Molecular Biology, vol. 139: Extracellular Matrix Protocols*
Edited by: C. Streuli and M. Grant © Humana Press Inc., Totowa, NJ

phogenesis, the biochemical signal transduction pathways that regulate transcription of mammary specific genes, and the survival of mammary cells (i.e., suppression of apoptosis) are all dependant on cell–matrix interactions within the tissue ***(1–6)***. Thus, the tissue has great potential for deciphering the roles of specific ECM proteins and their cellular receptors, e.g., integrins, in the control of many aspects of phenotype.

One significant advantage of studying the mammary gland is that the mammary epithelium from one mouse can be transplanted into the stroma of a syngeneic host *(7)*. Transplanted epithelium forms a ductal network within the mammary stroma, and, if the recipient mice are mated the epithelium develops lactational alveoli. The host mammary epithelium is poorly developed until puberty, thus if it is surgically removed prior to transplantation, all the new transplanted epithelium which populates the stroma will have the genotype of the donor.

This transplantation strategy is particularly powerful for examining the role of ECM proteins and their receptors in the mammary glands of transgenic or knockout animals which would otherwise suffer embryonic mortality. We have used the technique to analyze the function of α6 integrin using mammary epithelium from knockout mice ***(8,9)***. These mice have a severe skin blistering defect and die at or shortly after parturition. However, by transplanting mammary epithelium from affected animals to syngeneic hosts, it has been possible to analyze the role of α6 integrin in mammary development. Such a technique is equally applicable to the analysis of ECM function. A further advantage of transplantation and retransplantation is that it allows considerable expansion of the epithelial cell population derived from just one mammary rudiment. This is particularly useful when large numbers of cells are needed for subsequent analysis (*see* **Note 1**).

1.2. Summary of the Technique

At birth, the murine mammary gland consists of a small epithelial rudiment located under the nipple. Development then proceeds very slowly until the onset of puberty (approximately 3 wk postpartum) when the epithelium grows rapidly into the subcutaneous fat pad and subsequently populates the entire available stroma. To enable transplantation, the endogenous mammary rudiment of the host gland is removed at 21 d after birth leaving the fat pad devoid of epithelium. This is known as a "cleared" fat pad. Mammary tissue from another syngeneic animal is then transplanted into the cleared fat pad where it will grow and form a functional glandular epithelium (*see* **Note 2**).

2. Materials

2.1. Isolation of Embryonic Mammary Tissue

1. Eared (blunt tipped) scissors
2. Thick forceps
3. Fine forceps (#5)
4. Watchmaker's springbow scissors
5. Dissecting microscope (Leica)
6. Adjustable fibre optic lights (Euromex, Arnhem, Holland)
7. Phosphate buffered saline (PBS): 10 m*M* Na_2HPO_4, 1.76 m*M* KH_2PO_4, 137 m*M* NaCl, 1.33 m*M* KCl pH 7.0
8. L15 medium (Sigma, Madison, WI, #L4386)
9. Petri dishes
10. Cryovials (Nalgene)
11. Freezing mix (3 parts medium, 1 part serum, 1 part DMSO)
12. Cryo 1°C freezing container "Mr Frosty" (Nalgene)
13. Glass microscope slides
14. Telly's fix: 70% ethanol, 5% formalin, 5% glacial acetic acid
15. Acetone
16. Ethanol
17. Meyer's hematoxylin stain for whole mounts: 0.25 g haematoxylin, 50 mg sodium iodate, 12.5 g aluminium potassium sulphate per liter distilled water.
18. Acidified 50% alcohol (50% ethanol acidified with 25 mL 1 *N* HCl per liter)
19. Methyl salicylate

2.2. Transplantation of Mammary Tissue into Recipient Mice

1. Gaseous anaesthetic: Anaesthetic box perfused with 0_2 (2 L/min), NO_2 (1 L/min), 2% halothane
2. Liquid anaesthetic: One part Hypnorm (Janssen, Beerse, Belgium) and one part Midazolam (Hypnovel; Roche, Basel, Switzerland) to six parts sterile water for injection (Fresenius Health Care Group, Basingstoke, UK). Use 10 μL/g body weight by intraperitoneal injection. Anaesthesia should last around 40 min, which is adequate for a transplantation experiment. If necessary, the period of anaesthesia can be extended by additional doses of 0.3 μL/g Hypnorm alone every 30–40 min and additional doses of Midazolam every 4 h.
3. Needles (27 gage 1/2 in) and 1-mL syringes
4. Shaver (Wahl)
5. Cotton wool
6. Chlorhexidine gluconate solution (Preston Pharmaceuticals, Preston, UK)
7. Cork board (Fisons)
8. Elastic bands
9. Pins

10. Cauterizer with fine tip (Rimmer Bros., London)
11. Sterile swabs
12. Glass slides, Telly's fix, acetone, ethanol, Meyer's haematoxylin, acidified 50% alcohol, and methyl salicylate for whole mounts (*see* **Subheading 2.1.**).
13. Trypan blue solution: 0.1% in saline
14. Needle holders
15. Sutures: 5/0 Vicryl 13 mm 3/8 curved needle (Ethicon, Sommerville, NJ)
16. Analgesia: Bupranorphine (Temgesic, Reckit and Colman) 1:100 in sterile water for injection. Use 10 μL/g body weight by subcutaneous injection. Bupranorphine also has the effect of reversing the action of the anaesthesia.
17. Heated pads (International Market Supplies, Congleton, UK).
18. Vetbed (Petsmart)

2.3. Analysis of Phenotype

1. Scissors, forceps, and cotton swabs (*see* **Subheadings 2.1.** and **2.2.**).
2. Glass slides, Telly's fix, acetone, ethanol, Meyer's haematoxylin, acidified 50% alcohol and methyl salicylate for whole mounts (*see* **Subheading 2.1.**).
3. Cryovials, freezing mix, cryo 1°C freezing container (*see* **Subheading 2.1.**).
4. 4% paraformaldehyde in PBS
5. 0.2 *M* glycine in PBS
6. Alcohol
7. Chloroform
8. Paraffin wax
9. Xylene
10. EM fix: 2.5% gluteraldehyde, 2% paraformaldehyde, 0.1 *M* sodium cacodylate pH 7.4
11. 0.1M sodium cacodylate buffer pH 7.4
12. 1% osmium tetroxide in cacodylate buffer
13. Desiccated alcohol
14. Propylene oxide
15. Agar100 resin (Agar Scientific, UK)
16. 2% uranyl acetate
17. 3% lead citrate
18. Aluminium foil
19. O.C.T mounting medium (TissueTec)
20. Small metal block
21. Polystyrene container (for liquid nitrogen)
22. Liquid nitrogen
23. BrdU labeling and detection kit e.g., Amersham RP202 (we use 0.2 mL of the labeling reagent for 20 g mouse).

3. Methods

3.1. Isolation of Embryonic Mammary Tissue

3.1.1. Isolation of Embryos

1. Embryos are obtained by caesarian section from mothers killed by cervical dislocation at the appropriate age of gestation (*see* **Note 3**).
2. The ventral flank of the female is opened to expose the bicornate uterus. This is then gently opened using scissors and fine forceps to reveal the embryos.
3. The embryos should be removed from their foetal membranes and killed by cervical dislocation before dissection. Any remaining attached umbilical cord or placenta should also be removed, with the aid of a dissecting microscope and fiberoptic illumination, if necessary.
4. To prevent the skin from drying out, the embryos should be kept moist with squirts of PBS.
5. For transportation, embryos with heads removed can be shipped in L15 medium on ice for 24–48 h (but see **Note 4**).
6. The embryonic tail should be removed for genotyping and each embryo given a unique identifying code to aid correlation with genotype.

3.1.2. Sexing

The sex of the embryos is determined by examining the anogenital distance which in male mice is larger than in females. Males also have a slight bump between the urogenital ridge and the anus, which is smaller in the female. In older embryos (>E15) the lack of obvious nipples in males can also be used to confirm anogenital sexing. Sexing embryos is quite difficult because the differences are small and therefore requires practice (*see* **Note 5**).

3.1.3. Removing the Mammary Rudiment

1. The relative location of the nipples in the embryo is the same as in the adult female (**Fig. 1A**). The mother can therefore be used as a reference aid. The five pairs of glands lie on either side of the midline in two approximately straight lines. The first pair are high on the neck (#1 glands; see **Note 6**), pairs two and three close to the forelimbs, pair four on the abdomen, and pair five in the inguinal region.
2. A dissecting microscope and adjustable fiberoptic illumination is required to view the nipples. The embryo should be placed in a Petri dish on its back. A small piece of tissue underneath the embryo may be useful to prevent it slipping during dissection. The nipples appear as small white circles on the surface of the skin (**Fig. 1B**). It may aid their location to adjust the angle of the light and move the embryo from side to side.

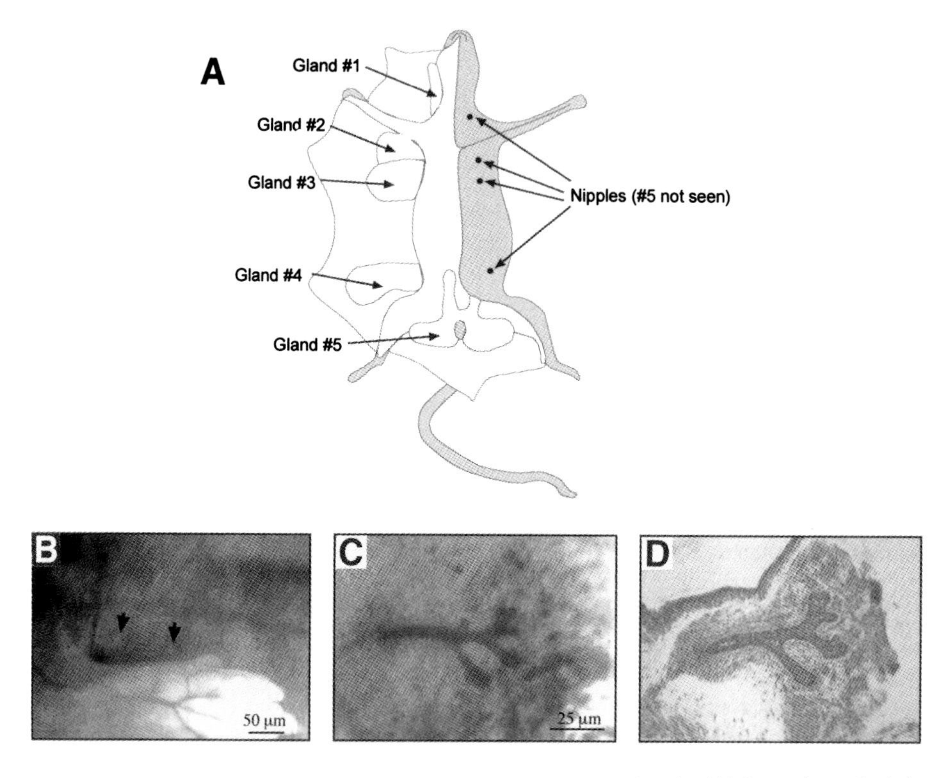

Fig. 1. Location and morphology of murine mammary glands. **(A)** Location of adult mammary glands and nipples; **(B)** Embryonic nipples of glands #2 and #3 *in situ* (E17.5). Arrows indicate the pale circular nipples on the skin; **(C)** Whole mount of embryonic rudiment (E17.5) stained with haematoxylin; **(D)** Wax section of rudiment in **(C)** stained with haematoxylin and eosin.

3. Once located, remembering that the mammary rudiment extends just a short distance into the skin from the nipple (**Fig. 1C,D**), the skin should be gently lifted using fine forceps and the nipple and gland cut away using watchmakers scissors (*see* **Note 7**).
4. If the gland is to be used for transplantation, immediately it should be kept in serum-free medium on ice. If not, it should be immediately frozen (*see* **Subheading 3.1.4.**).
5. On the first few attempts, to confirm that the mammary gland has been correctly isolated, whole mounts can be used to stain the mammary rudiment (**Fig. 1C**). To do this, the isolated gland should be gently spread, skin side down, on a glass slide, allowed to dry to the slide for 30 s, and placed in Telly's fixative. The subsequent procedure is exactly the same as for whole mount of adult mammary glands (*see* **Subheading 3.3.2.**), although because the tissue is much smaller, the time in each solution may be reduced.

3.1.4. Freezing Mammary Rudiment for Transport or Storage

1. As mentioned in **Note 4**, it is highly preferable to transport dissected mammary rudiment frozen. It may also be necessary to keep the gland frozen until a suitable recipient is available. To do this, the tissue should be placed in a cryovial containing 0.5 mL freezing mix and frozen slowly to –80°C (*see* **Note 8**).
2. Liquid nitrogen should be used for longer term storage.

3.1.5. Recovery of Frozen Tissue

1. Frozen tissue should be rapidly thawed and washed several times in serum-free medium (e.g., L15) to remove any traces of serum or DMSO before transplantation.
2. Thawed tissue should be kept in serum-free medium on ice until transplantation.

3.2. Transplantation of Mammary Tissue into Recipient Mice

3.2.1. Breeding of Recipient Mice

To avoid tissue rejection, transplantation should always be into syngeneic-recipient animals. Nude mice can be used as recipients, however, their mammary glands are small and postoperative infections may cause problems. Most transgenic animals are a cross between C57BL6 and 129 strains of mice and, therefore, F1 progeny of 129xC57BL6 are ideal recipients (*see* **Note 9**).

3.2.2. Clearing the Fat Pad and Transplantation

1. Recipient mice must have their endogenous mammary epithelium removed by 21 d after birth. As this is normally the time of weaning from the mother, it is usual to wait until 21 d before operating. The #4 or abdominal glands are the easiest to clear for transplantation. Clearing the fat pad can be done at the same time as transplanting tissue and this is preferable because the animal then only undergoes one operation. However, if this is not possible for logistical reasons, the fat pad can be cleared and the animal left until required for transplantation.
2. We routinely use 21-d-old F1 progeny C57BL6x129 mice which are quite skittish. To minimize the stress of an intraperitoneal (ip) injection, the mice are subdued in an anaesthetic box perfused with halothane, N_2O, and O_2 before ip injection of 10 μL/g body weight anaesthetic.
3. Once anaesthetized, the abdomen is shaved and swabbed with a small amount of chlorhexidine solution.
4. The mouse is then restrained on its back on a cork dissecting board using small elastic bands tied in a slip knot around each paw and secured at the other end with a small pin.
5. A small Y-shaped incision is then made in the ventral skin from just under the rib cage to slightly down each of the hind limbs using eared scissors. If any small blood vessels are accidentally nicked they are immediately cauterized to minimize blood loss.

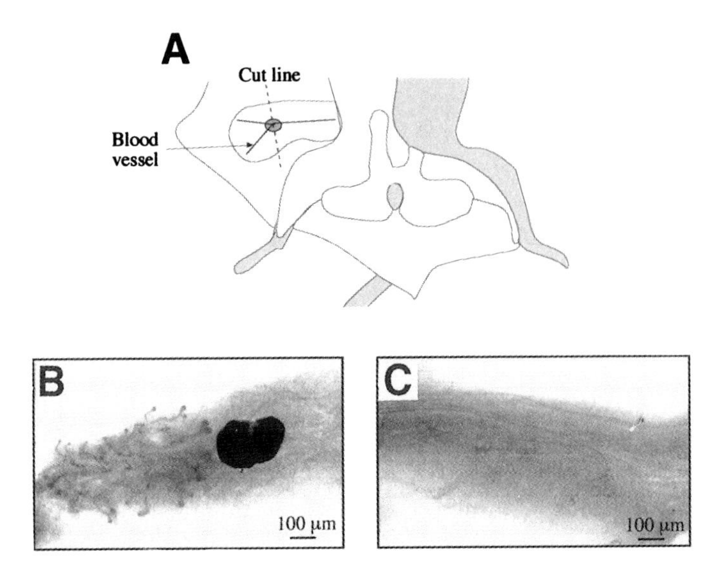

Fig. 2 Clearing the mammary gland of epithelium. **(A)** Diagram of the abdominal #4 mammary gland indicating the approximate line to cut to remove the distal portion of the gland containing epithelium. Circle indicated at junction of blood vessels is lymph node; **(B)** Whole mount of mammary gland from 21-d-old mouse stained with haematoxylin showing the extent of the ductal network and the darkly stained central lymph node; **(C)** Stained whole mount of cleared fat pad 6 wk after clearing.

6. The skin is gently retracted on one side using a sterile swab until the lymph node of the #4 gland is visible as a small oval in the center of the gland (**Fig. 2A**). To secure the skin out of the way, a small-gage needle may be used to pin it to the board. The major blood vessel which forks over the lymph node is cauterized and the fatty tissue distal to the lymph node is removed using the cauterizer, taking care not to damage the skin.
7. The removed tissue can be whole mounted for examination to check that all the epithelium has been cleared (*see* **Subheading 3.3.2.**). On the first few attempts, it is prudent to leave this cleared gland untransplanted and check after some weeks that it remains epithelium free (**Fig. 2B,C**). This gives confidence for the future that any epithelium seen after transplantation is likely to be the result of a successful transplant rather than a failed clearing. Once this has been established, it is preferable to transplant tissue at the same time as clearing the fat pad.
8. Mammary rudiment, either fresh or thawed (*see* **Subheading 3.1.5.**) is placed in a very dilute solution of trypan blue to aid identification of the tissue during transplantation. A small pocket is made in the recipient fat pad quite proximal to the abdomen using fine forceps and the transplanted tissue placed inside using another pair of fine forceps. The top of the pocket is held against the inserting

forceps as they are withdrawn to keep the transplanted tissue in place. The transplanted tissue should be clearly visible in the opaque mammary fat pad as a small blue lump. To increase the chances of success, several mammary rudiments may be transplanted into one recipient fat pad remembering that they should all come from the same donor embryo to clarify subsequent interpretation of results.

9. The contralateral #4 fat pad is then cleared and transplanted as above if required.
10. The skin is sutured closed.
11. The mouse is injected subcutaneously with analgesia (10 μL/g body weight) and left on a heated pad overnight in a box lined with Vetbed rather than sawdust to recover.
12. Transplanted mice should be permanently marked or housed individually to aid subsequent identification.

3.3. Analysis of Phenotype

3.3.1. Harvesting Transplanted Mammary Tissue

1. Transplanted tissue should be left *in situ* for a minimum of 5–6 wk to see significant growth. To stimulate maximum proliferation the recipient animal can be mated once it reaches 6 wk of age, and the mammary tissue harvested during late pregnancy.
2. To access the transplanted glands, the mouse is killed by cervical dislocation and the ventral skin opened and gently retracted using scissors, forceps and a cotton swab to reveal the abdominal #4 glands. The distal end of the gland can be separated gently from the skin using scissors and, held with forceps, lifted to aid separation of the rest of the gland from the skin.
3. Once removed, the entire gland can be spread on a glass microscope slide for whole mount analysis or dissected into smaller pieces for wax histology, electron microscopy, cryosectioning, and immunostaining or protein/RNA/DNA analysis.
4. Alternatively, if passaging the tissue for retransplantation is required (*see* **Note 1**), small pieces (approximately 1–2 mm^3) of gland-containing epithelium (*see* **Note 10**) should be cut and frozen (*see* **Subheading 3.1.4.**) or immediately retransplanted.

3.3.2. Whole Mount Analysis

1. This technique is used on whole gland or pieces of gland to reveal the epithelial architecture (*see* **Figs. 1C** and **2B**). The mammary tissue is gently spread on an uncoated glass microscope slide and allowed to dry for approximately 1 min. It is then placed in Telly's fix for at least 2 h.
2. The tissue is defatted in three changes of acetone (1 h each) and rehydrated through 100%, 95%, and 70% alcohol (at least 1 h each).
3. The nuclei are stained with Meyer's haematoxylin for approximately 30 min. The exact time depending on the age of the staining solution. The haematoxylin is "blued" in running tap water for approximately 20 min. To improve contrast, the tissue may require destaining with acidified 50% alcohol before dehydration through 70%, 95% 2X 100% alcohol.

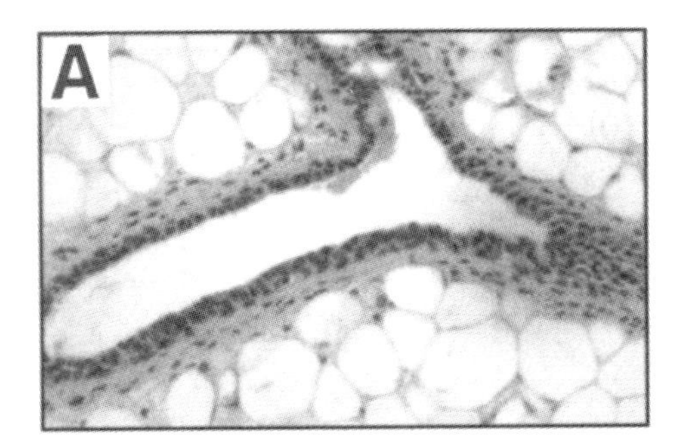

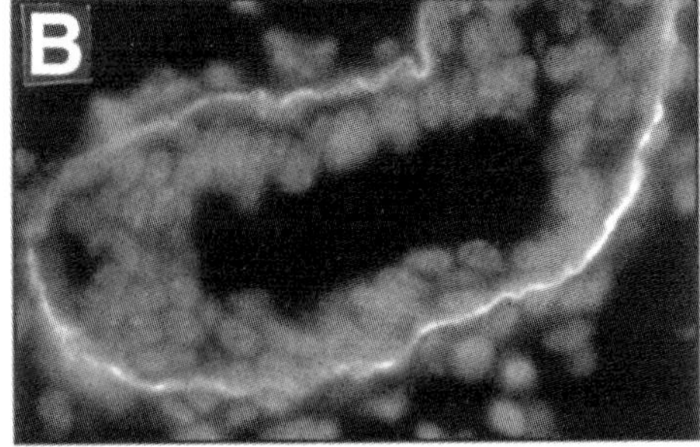

Fig. 3 Analysis of phenotype. **(A)** Haematoxylin and eosin stained paraffin section of virgin mammary gland; **(B)** Cryosection of virgin transplanted gland stained for laminin-1. Nuclei counterstained with Hoechst.

4. Finally, the tissue is cleared for examination with 50% methyl salicylate/50% alcohol overnight and stored in 100% methyl salicylate.

3.3.3. Wax Embedding

1. Wax-embedded material can either be obtained directly from fresh tissue or alternatively pieces of interest can be cut from stained whole mounts and subsequently embedded in wax.
2. If sections are required for *in situ* hybridization, care should be taken to ensure all solutions are RNase free.
3. Fresh tissue is fixed in 4% paraformaldehyde in PBS for 1 h at 4°C. Free-aldehyde groups are blocked by incubation in 0.2 *M* glycine for 2 h at 4°C.
4. The tissue is dehydrated through 70% alcohol for at least 2 h followed by 30 min incubations in two changes each of 90% and 100% alcohol. This is replaced by a 50:50 mix of alcohol:chloroform (30 min), followed by two 45-min incubations in 100% chloroform.
5. The tissue is then blotted and transferred to wax 1 at 62°C for 10–15 min, changed into wax 2 (62°C, partial vacuum, 30 min), and finally the vacuum is increased to maximum for the final 30 min or until no further bubbles emerge from the tissue. The tissue is oriented in molten wax in moulds and left to harden overnight.
6. It is subsequently sectioned on a rotary microtome at 5 μm and the sections mounted on glass slides. Standard haematoxylin and eosin staining protocols can be used to highlight the histological architecture (**Fig. 3A**).
7. Pieces of whole-mounted material require washing with two changes of xylene over 1 h to remove any methyl salicylate before placing in wax 1 as above.

3.3.4. Electron Microscopy

1. Small pieces of tissue (1 mm^3) are fixed overnight at 4°C in EM fix, washed four times in 0.1 *M* cacodylate buffer, and postfixed for 1 h at room temperature in 1% osmium tetroxide in cacodylate buffer.

2. The osmium is washed off well with buffer and the tissue dehydrated by 20 min incubations in 50%, 70%, 80%, 90%, 95%, two changes of 100%, and two changes of desiccated 100% alcohol.
3. The tissue is then placed in propylene oxide for 20 min and then left in 50% propylene oxide:50% Agar100 medium hardness resin overnight at 4°C.
4. After two changes of Agar100 resin over at least an hour each, the tissue is oriented in specimen vials and the blocks left to harden at 60°C for 20 h.
5. Ultrathin sections can be stained with 2% uranyl acetate (16 min) followed by 3% lead citrate (6 min) with thorough washing in water between and after staining.

3.3.5. Cryoembedding

1. Pieces of tissue or even the entire gland can be frozen for cryosectioning. The tissue is frozen in a foil cup of an appropriate size (just bigger than the tissue) made by moulding aluminium foil over a suitable object such as a marker pen lid or the cap of a small bottle. This cup is filled with O.C.T mounting medium and the tissue inside oriented appropriately for sectioning.
2. The cup is then placed on a metal block precooled in a bath of liquid nitrogen and left until the O.C.T has become hard and opaque.
3. Tissue is stored at –20°C until required.
4. Cryosections (7 μm) are used for immunostaining using standard fixation and staining protocols (**Fig 3B**).

3.3.6. Protein/RNA/DNA Analysis

Small pieces of tissue can be snap frozen in aluminium foil parcels in liquid nitrogen and kept under nitrogen until required. The tissue is then ground down and protein, RNA, or DNA extracted using standard protocols.

3.3.7. Cell Culture

Mammary epithelial cells can be isolated from transplanted mammary glands using established protocols ***(10)*** and their biology studied in tissue culture.

3.3.8. Proliferation Indices

If information on the proliferative index of the mammary gland is required, then the mouse can be injected ip with a solution of bromodeoxyuridine (BrdU) 2 h before death. This will incorporate into the DNA of any cells in S-phase during this period. The tissue is then harvested as normal and processed either for wax embedding and sectioning or cryosectioning and the incorporated BrdU detected by immunocytochemistry.

3. Notes

1. Transgenic or knockout mammary epithelium can be serially transplanted once it has been established that it can form ductal networks within mammary gland.

Tissue should be harvested around 8 wk after initial transplantation, cut into small pieces, and then used to repopulate more host mammary gland. Serial transplantation of mammary tissue is only successful for a limited number of generations because of cellular senescence. The proliferative potential of normal mammary cells declines with serial transplantation and is lost after 5–6 serial transplants ***(11)***.

2. It is possible to mate the recipient mice and thereby examine development of the transplanted tissue. However, it should be remembered that because the ductal network does not connect with the nipple, a fully lactational phenotype will not be achieved as accumulation of milk will induce immediate involution of the glandular epithelium.
3. Determining the appropriate embryonic age at which to isolate mammary tissue depends on several factors. First, the viability of the embryos; a phenotype which is lethal at, e.g., embryonic day 15 (E15) requires tissue to be harvested before that time. Second, the ease of sexing the embryos: from around E14 onward, viable mammary tissue can only be isolated from females (G. Cunha, personal communication). And third, the ease of finding the nipples on the developing skin. These last two factors must be traded against each other. The difference in anogenital distance, which is the main aid to determining the sex of the embryos, is more obvious in larger embryos. However, at later stages of embryogenesis, the developing hair follicles in the skin cause the formation of small bumps which can be hard to distinguish from nipples. I have found F1 embryos (C57BL6x129) of E16.5 to E17.5 to be the easiest to isolate mammary tissue from.
4. Shipping embryos markedly reduces the viability of the mammary tissue. It is therefore preferable to be able to isolate the mammary tissue from the embryo immediately and process it for cryopreservation if transportation is required. This is most easily done by travelling to the site of the transgenic mouse colony and performing the tissue isolation there.
5. Sexing mice by anogenital distance is a standard animal husbandry technique used on newborn mice. Although the distances are smaller in embryonic mice, it may be useful to have the differences between the sexes at birth pointed out by an experienced animal technician.
6. The #1 mammary glands lie just above the submandibular salivary glands and great care should be taken when isolating the mammary rudiments not to take any salivary tissue unintentionally as this will also transplant successfully. The salivary glands can easily be distinguished in whole mounts (and even unstained under the dissecting microscope) by their lung-like lobular appearance.
7. In embryos with a skin detachment phenotype, the skin should not be lifted away from the body of the animal. Instead the scissors should be used to cut into the skin around the nipple to remove the gland. This phenotype applies to α6 integrin null animals and may also be apparent in some mice with altered expression of basement membrane proteins.
8. We achieve slow freezing using a Nalgene "Mr Frosty" tub containing isopropanol which cools at 1°C per minute (*see* **Subheading 2.1.**).

9. Complications can arise if later progeny are backcrossed onto another strain, which is sometimes done to increase fecundity. If this is the situation, providing all strains are inbred, at least six generations of backcrossing are required before transplantation is possible.
10. It is usually quite difficult to see unstained mammary epithelium through the fatty stroma, even with the aid of a dissecting microscope. However, sometimes (especially in thinner areas of the gland) it is possible to check the success of transplantation and establish which areas of the gland have been populated by outgrowing epithelium.

Acknowledgments

We would like to thank Kathy van Horn, Phyllis Strickland, Gary Silberstein, Charles Daniel, and Paul Edwards for their invaluable technical advice. This work was supported by grants from the BBSRC and The Wellcome Trust. CHS is a Wellcome senior research fellow.

References

1. Barcellos-Hoff, M. H., Aggeler, J., Ram, T. G., and Bissell, M. J. (1989) Functional differentiation and alveolar morphogenesis of primary mammary cultures on reconstituted basement membrane. *Development* **105,** 223–235.
2. Steuli, C. H., Bailey, N., and Bissell, M. J. (1991) Control of mammary epithelial differentiation—basement membrane induces tissue-specific gene expression in the absence of cell cell interaction and morphological polarity. *J. Cell Biol.* **115,** 1383–1395.
3. Pullan, S., Wilson, J., Metcalfe, A., Edwards, G. M., Goberdhan, N., Tilly, J., et al. (1996) Requirement of basement membrane for the suppression of programmed cell death in mammary epithelium. *J. Cell Sci.* **109,** 631–642.
4. Edwards, G. M. and Streuli, C. H. (1999) Activation of integrin signalling pathways by cell interactions with extracellular matrix, in *Adhesive Interactions of Cells* (Garrod, D. R., North, A., and Chidgey, M. A. J., eds.), JAI, Stamford, CT; *Advances in Molecular and Cell Biology* **28,** 235–266.
5. Farrelly, N., Lee, Y.-J., Oliver, J., Dive, C., and Streuli, C. H. (1999) Extracellular matrix regulates apoptosis in mammary epithelium through a control on insulin signaling. *J. Cell Biol.* **144,** 1337–1348.
6. Klinowska, T. C. M., Soriano, J. V., Edwards, G. M., Oliver, J. M., Valentijn, A. J., Montesano, R., and Streuli, C. H. (1999) Laminin and beta 1 integrins are crucial for normal mammary gland development in the mouse. *Developmental Biol.* **215,** 13–32.
7. DeOme, K. B., Faulkin, L. J., Jr., Bern, H. A., and Blair, P. E. (1959) Development of mammary tumors from hyperplastic alveolar nodules transplanted into gland-free mammary fat pads of female C3H mice. *Cancer Res.* **19,** 515–520.
8. Georges Labouesse, E., Messaddeq, N., Yehia, G., Cadalbert, L., Dierich, A., and Le Meur, M. (1996) Absence of integrin alpha 6 leads to epidermolysis bullosa and neonatal death in mice. *Nat. Genet.* **13,** 370–373.

9. Klinowska, T. C. M., Alexander, C., Georges-Labouesse, E., Van der Neut, R., Sonnenberg, A., and Streuli, C. H. (1999) Alpha 6, alpha 3 and beta 4 integrin null mammary tissue undergoes normal development in the mouse in vivo. In preparation.
10. Pullan, S. and Streuli, C. H. (1996) The mammary gland epithelial cell, in Epithelial Cell Culture (Harris, A., ed.), Cambridge University Press, Cambridge, UK, pp. 97–121.
11. Daniel, C. W., De Ome, K. B., Young, J. T., Blair, P. B., and Faulkin, L. J., Jr. (1968) The in vivo life span of normal and preneoplastic mouse mammary glands: a serial transplantation study. *Proc. Natl. Acad. Sci. USA* **61,** 53–60.

Index

D

E

V